Atherosclerosis in Primates

Primates in Medicine

Vol. 9

Series Editors
E. I. GOLDSMITH, New York, N.Y.
J. MOOR-JANKOWSKI, New York, N.Y.

S. Karger · Basel · München · Paris · London · New York · Sydney

Atherosclerosis in Primates

Volume Editor
J. P. STRONG, New Orleans, La.

111 figures and 68 tables, 1976

S. Karger · Basel · München · Paris · London · New York · Sydney

<h1 style="text-align:center">Primates in Medicine</h1>

Vol. 4: REYNOLDS, H. H. (ed.): Chimpanzee: Central Nervous System
and Behavior A. Review.
VIII + 159 p., 11 fig., 19 plates, 1969.
ISBN 3–8055–0563–9
Vol. 5: HARRISSON, B.: Conservation of Nonhuman Primates in 1970.
VI + 98 p., 3 tab., 1971.
ISBN 3–8055–1243–0
Vol. 6: KRATOCHVIL, C. (ed.): Chimpanzee: Immunological Specificities of Blood.
VI + 146 p., 6 fig., 27 tab., 1972.
ISBN 3–8055–1389–5
Vol. 7: MURPHY, G. P. (ed.): Transplantation in Primates.
VII + 140 p., 36 fig., 27 tab., 1972.
ISBN 3–8055–1408–5
Vol. 8: KALTER, S. S.: The Baboon, Microbiology, Clinical Chemistry and Some
Hematological Aspects.
VIII + 171 p., 8 fig., 42 tab., 1973.
ISBN 3–8055–1442–5

Cataloging in Publication
Atherosclerosis in primates
Editor: J. P. STRONG. Basel, New York, Karger (c1976)
(Primates in medicine, v. 9)
1. Arteriosclerosis 2. Primates
I. STRONG, JACK PERRY, 1928- ed.
W1 PR522E v. 9/WG 550 A868
ISBN 3–8055–2195–2

All rights, including that of translation into other languages, reserved.
Photomechanic reproduction (photocopy, microcopy) of this book or parts
thereof without special permission of the publishers is prohibited.
©
Copyright 1976 by S. Karger AG, Basel (Switzerland), Arnold-Böcklin-Strasse 25
Printed in Switzerland by Effingerhof AG, Brugg
ISBN 3–8055–2195–2

Contents

Dedication

This book is dedicated to Nicholas T. Werthessen and to the late Russell L. Holman, two distinguished scientists who did much to stimulate early research on atherosclerosis in nonhuman primates and who, by organizing an expedition to Kenya in 1958 to survey the prevalence and extent of arterial lesions in the baboon, were primarily responsible for the introduction of this primate animal into atherosclerosis research.

Preface

A comprehensive up-to-date monograph on atherosclerosis investigations in primate animals seems to be needed by investigators of atherosclerosis in man as well as by primatologists. The present volume should satisfy the need by reviewing recent accomplishments in the field and by relating them to human medical research.

The authors who have participated in preparing this monograph include many of the pioneers who began investigating arterial disease in monkeys in the 1950s and 1960s. In an attempt to meet an early deadline and for ease of day-to-day communication, the editor chose contributors from North America who were willing and able to meet a relatively short deadline. Some authorities in the field were not able to participate because of other pressing commitments. Their work and work by investigators in United Kingdom, continental Europe, Africa, and Asia is, however, referred to in the monograph.

Dr. H. C. STARY convinced the editor to undertake the mission of organizing the monograph. Dr. THOMAS CLARKSON and Dr. HENRY MCGILL were most helpful with suggestions for the content. The authors were uniformly cooperative in giving priority to their writing assignments even though international symposia, national meetings, and the Christmas holidays were competing for their time.

The editor thanks all of those who have contributed to the completion of this monograph, especially RHEA DUPEIRE for her invaluable assistance. The editor takes responsibility for any errors of judgment concerning material which was included.

JACK P. STRONG

Prim. Med., vol. 9, pp. 1–15 (Karger, Basel 1976)

Atherosclerosis in Primates
Introduction and Overview[1]

JACK P. STRONG

Department of Pathology,
Louisiana State University Medical Center, New Orleans, La.

Contents

I. Introduction

The purpose of this monograph is to provide an up-to-date review of atherosclerosis research involving nonhuman primates. It is designed to serve the needs of primatologists as well as investigators in the field of atherosclerosis. Most of the chapters emphasize recent studies; however, reference is also made to pioneering experiments and surveys.

A. Nomenclature

The reader not familiar with the language of arterial disease needs an introduction to its nomenclature. Atherosclerosis is the underlying cause of coronary heart disease (coronary occlusion, coronary thrombosis, or myo-

[1] The author's investigations of atherosclerosis in primates are supported by the USPHS, NIH grant HL 08974.

cardial infarction) and of the most frequent type of stroke (cerebral thrombosis and cerebral infarction). Coronary heart disease is the leading cause of death in the USA and most other technically advanced countries. Stroke ranks third, after cancer, as a cause of death in these same countries. Knowledge of the etiology and pathogenesis of atherosclerosis is therefore important in developing means of retarding or preventing the process.

In order to avoid confusion due to differing terminology for arterial disease, I shall define and discuss selected terms as they are used in this monograph.

Arteriosclerosis is a generic term that includes practically any arterial disease which leads to thickening and hardening of arteries of any size.

Atherosclerosis is a specific form of arteriosclerosis. The most distinctive feature of atherosclerosis is the accumulation of *lipid* in the *intima* of *large elastic arteries* (aorta) and *medium-sized muscular arteries* (coronary, femoral, carotid, and others). In addition to lipid, cells, connective tissue fibers, and various blood products accumulate in the lesions. A number of complications, such as thrombosis, hemorrhage into a plaque, and ulceration, can also occur in the lesions. The hallmarks of atherosclerosis are its intimal location in the initial stage, involvement of large and medium-sized arteries, and the accumulation of fat. Atherosclerosis is the most frequent form of arteriosclerosis that causes clinically significant disease.

Mönckeberg's medial calcific sclerosis, characterized by calcification of the medial layer of muscular arteries, and *arteriolosclerosis*, characterized by thickening, fibrosis, hyalinization, and narrowing of arterioles, are other types of arteriosclerosis quite distinct from atherosclerosis. Thus, they are beyond the scope of this monograph. Medial and arteriolar lesions have sometimes caused confusion in interpreting experimental studies, principally those in which rabbits and rats have been used. Only the intimal lesions which contain lipid and connective tissue elements in large elastic and medium-sized muscular arteries are models of human atherosclerosis.

The term *atheroma* has been used in several different ways. Atheroma sometimes refers to the entire process of atherosclerosis and is sometimes used to describe a specific lesion. Some pathologists use the word 'atheroma' to mean a large atherosclerotic plaque with a pool of necrotic cells, lipid, and connective tissue in the base. Others would call this lesion an advanced fibrous plaque or a complicated lesion. Atheroma is also used to refer to any lesion of atherosclerosis, including fatty streaks, fibrous plaques, or complicated or calcified lesions. To avoid ambiguity, one should use the term only if its exact meaning is made clear.

The following working definitions are offered for different types of atherosclerotic lesions detectable grossly after staining vessels with Sudan IV or other fat stains.

Fatty streak. A fatty intimal lesion that is stained distinctly by Sudan IV and shows no other underlying change. Fatty streaks are flat or only slightly elevated and do not significantly narrow the lumina of blood vessels.

Fibrous plaque. A firm, elevated intimal lesion which in the fresh state is gray-white, glistening, and translucent. Human fibrous plaques characteristically contain fat. A thick fibrous connective tissue cap containing varying amounts of lipid covers a more concentrated 'core' of lipid. If a lesion also contains hemorrhage, thrombosis, ulceration, or calcification, that lesion is classified according to one of the next two categories.

Complicated lesion. An intimal plaque in which there is hemorrhage, ulceration, or thrombosis with or without calcium.

Calcified lesion. An intimal plaque in which insoluble mineral salts of calcium are visible or palpable without overlying hemorrhage, ulceration, or thrombosis.

The term *raised atherosclerotic lesion* is sometimes used to include fibrous plaques, complicated lesions, and calcified lesions. Raised lesions are contrasted with fatty streaks, which typically show little or no elevation above the surrounding intimal surface.

Although this classification implies a pathogenetic sequence, it can be used as a descriptive classification regardless of the ideas of pathogenetic interrelationships among the lesions.

Other lesions which are sometimes considered as atherosclerosis or as lesions predisposing to atherosclerosis include fibromuscular intimal thickening, gelatinous or edematous intimal lesions, and organizing mural thrombi on an otherwise normal intima. The pathogenetic relationship of these lesions to atherosclerosis and its clinical manifestations is less well-established, and these lesions should be referred to by their descriptive terms until their significance is clarified.

Experimental arterial lesions should be quantitated and reported according to the corresponding descriptive categories listed above. Data on prevalence, extent of intimal surface involved, or other quantitative measures should not be lumped together under the term atherosclerosis or arteriosclerosis. *Atherosclerosis* might well be reserved as a general term for the entire process in titles and headings in scientific reports, very much as the term *cancer* is used.

B. General Reference Sources

The history of our knowledge about atherosclerosis, from ancient times to the recent past, is told in an entertaining account by LONG [28]. The morphology and pathogenesis of human atherosclerotic lesions as described by DUFF and MCMILLAN [14] is still one of the best reviews of this topic available. The gross and microscopic features of typical coronary and aortic human lesions at various ages were illustrated by MCGILL *et al.* [34]. Data on the worldwide distribution of atherosclerotic lesions among different populations were published in 1968 [33]. STRONG *et al.* [61] reviewed the development of atherosclerotic lesions by age, sex, and race, the geographic variation in prevalence and extent of atherosclerosis, and the relationship of atherosclerotic lesions to risk factors for coronary heart disease. A monograph on arterial smooth muscle cells by GEER and HAUST [18] contains an extensive review of publications related to the nature of cells in atherosclerotic lesions, descriptions of the histologic and ultrastructural features of arterial lesions, and a generous selection of electron micrographs illustrating atherosclerotic lesions. The published proceedings of recent symposia on atherosclerosis [24, 51, 72] contain review articles and reports of current investigative work in atherosclerosis.

Readers are also referred to selected references concerning pertinent aspects of primatology which were not included in this monograph. The proceedings of a symposium on the feeding and nutrition of nonhuman primates [22] is useful for investigators who plan long-term studies with primates. A comprehensive illustrated handbook of primates by NAPIER and NAPIER [39] contains valuable information concerning the taxonomy and nomenclature of primates.

Since this volume emphasizes recent advances involving atherosclerosis in primates, readers are referred to previous reviews of this subject. A volume on comparative atherosclerosis edited by ROBERTS and STRAUSS [49] contains three reviews on experimental or naturally occurring atherosclerosis in primates.

These review articles by TAYLOR [66], CLARKSON [8], and STRONG [58] are illustrated with color photographs of primate lesions and cover work completed prior to 1964. Subsequent reviews have been devoted entirely or in large part to the subject of primate atherosclerosis [11, 16, 19, 27, 29, 56, 59, 60]. CLARKSON [7] emphasizes primate in his recent review of animal models of atherosclerosis.

II. Atherosclerosis in Free-Ranging Primates

Investigators experimentally produced lipid-containing arterial lesions in primates before systematically surveying free-ranging primate species for arterial lesions, but the occurrence of such lesions in control animals demonstrated the need for base line data to interpret experimental results. A number of species has now been surveyed for the prevalence and extent of sudanophilic lesions (fatty streaks) and more advanced atherosclerotic lesions.

PORTMAN and ANDRUS [45] surveyed the extent of atherosclerotic lesions in several species of New World monkeys, but only the Cebus and woolly monkeys were represented by more than ten animals per species. They found no arterial lesions in 36 Cebus monkeys and only minimal lesions in 15 woolly monkeys.

McGILL *et al.* [36] conducted the prototype survey of arterial lesions in free-ranging primates when they examined 163 baboons trapped from the wild in Kenya. They performed complete autopsies on these animals with special attention to the vascular system. They found sudanophilic aortic lesions comparable to human fatty streaks in about three fourths of the 67 adult animals, and in about one fifth of the adolescent animals. Very few of the baboons had fibrous plaques, complicated lesions or calcified lesions comparable to advanced atherosclerotic lesions in the adult human. The coronary arteries of some of the animals contained fibromuscular intimal thickening, but they did not have lipid-containing intimal lesions characteristic of human coronary atherosclerosis. Some idea of the health status of free-ranging animals can be gained from the high rate of parasitic infection encountered in this survey [64].

STRONG and TAPPEN [65] evaluated aortic lesions in two species of Central African monkeys, *Cercocebus albigena* and *Cercopithecus ascanius*, and found differences in the prevalence and extent of lesions of these monkeys which had widely differing habits. The *Cercocebus albigena*, which are exclusively arboreal monkeys, have more extensive aortic fatty streaks than the *Cercopithecus ascanius* which frequently come to the ground.

Workers from Bowman Gray School of Medicine and the Louisiana State University School of Medicine found lipid-containing aortic lesions in squirrel monkeys, *Saimiri sciureus*, in about one-half of 220 free-ranging animals [38]. Lipid-containing coronary artery lesions were found in some of the adult squirrel monkeys but in contrast to the human, they occurred only in the smaller intramyocardial branches. These investigators also surveyed the prevalence and extent of lesions in five additional species of New World monkeys [41]. The marmoset, *Saguinus nigricollis*, was essentially free of aortic and coronary artery lesions. *Cebus apella* and *Cebus albifrons* occasionally had aortic lesions but the lesions were less frequent than in squirrel monkeys or baboons. Adequate numbers of adult woolly monkeys (*Lagothrix* sp.) and spider monkeys (*Ateles* sp.) could not be obtained for thorough evaluation; however, results obtained in a limited number of spider and woolly monkeys indicate that these species may have an unusually high prevalence of sudanophilic aortic lesions.

The rhesus monkey, *Macaca mulatta*, is the primate species most widely used in experimental atherosclerosis. CHAWLA *et al.* [6] reported that 3% of 150 free-ranging rhesus monkeys had aortic fatty streaks, but the gross inspection for lesions was made without staining for fat, and it is possible that small or non-elevated fat-containing lesions were not detected. From study of control animals in many experiments, it seems likely that the rhesus monkey is relatively free of naturally occurring lesions. Examination of rhesus monkeys which have served as control animals in the author's laboratory has shown that some animals which have been in captivity for only a short time have small sudanophilic aortic intimal lesions. The extent of these lesions, however, is much less than experimentally induced lesions. This finding agrees with ARMSTRONG's [2] statement concerning lesions in the rhesus that the 'ordinary experience of the investigator will be the discovery of minimal atherosclerotic change in his controls'. I believe that a survey of wild-caught rhesus monkeys with gross sudan staining of aortas would reveal a higher prevalence of fatty streaks than that reported by CHAWLA *et al.* [6].

The cynomolgus monkey *(Macaca fascicularis)* has recently been systemically surveyed for naturally occurring arterial lesions. PRATHAP [47] found that sudanophilic fatty streaks involving up to 10% of the intimal surface were found in 90% of 73 freshly captured wild adult Malaysian cynomolgus monkeys. Fibrous plaques were rare, and there were no complicated plaques. This survey reemphasizes the necessity of using adequate controls in experimental investigation of atherosclerosis.

STRONG *et al.* [60] reviewed the subject of atherosclerosis in free-ranging primates in 1968 and estimated the comparative prevalence and severity of naturally occurring sudanophilic aortic lesions in selected primate species and their relationship to lesions in man. They emphasized that 'spontaneous' aortic arterial lesions in different primate species vary considerably in extent of intimal surface involved but are usually *much less extensive and less severe than lesions in humans*. Furthermore, they indicated that naturally occurring lipid-containing lesions are rarely found in major coronary branches of primates, even in those species in which aortic lesions are frequent.

III. Atherosclerosis in Captive Primates

The subject of arterial lesions in captive primates including chimpanzees, baboons, gorillas, and marmosets was also reviewed by STRONG *et al.* [60] and by STOUT and LEMMON [57]. While aortic lesions are present in many captive animals, interpretation of findings is difficult because of differing conditions during captivity.

MIDDLETON *et al.* [38] compared arterial lesions in squirrel monkeys in their natural environment with lesions in squirrel monkeys held in an animal supplier's acclimatizing compounds for three months to one year. The monkeys were fed low-fat monkey ration while in captivity. Aortic fatty streaks occurred frequently in the free-living squirrel monkeys but were even more frequent in the acclimatized squirrel monkeys. Experience in the author's laboratory indicates that sudanophilic aortic lesions are more extensive in rhesus monkeys fed low-fat laboratory monkey food diet for two years than in rhesus monkeys fed the same diet for only three months.

IV. Experimental Atherosclerosis in Primates

Experimental investigation of atherosclerosis in human subjects is limited by practical and ethical considerations. Much valuable information has been obtained by using experimental animals; however, scientists are constantly trying to make their experiments more relevant to the human situation. As a result of these considerations, increasing numbers of investigators have turned toward nonhuman primates as experimental models for studying atherosclerosis. The justification for this choice is the reasonable assumption that their biological behavior is more like that of man than is the behavior of other animals, and that experimental findings will be more directly applicable to man. Not only are nonhuman primates closer phylogenetically to man than other commonly used laboratory animals, but they also have been demonstrated to have many anatomic, biochemical and metabolic similarities to the human.

KAWAMURA [25] in 1927, and HUEPER [23] in 1946, were not able to produce experimental atherosclerotic lesions in rhesus monkeys. MANN *et al.*

[31] in 1953, produced hypercholesterolemia and sudanophilic aortic lesions in Cebus monkeys with high-cholesterol diets deficient in sulphur-containing amino acids. Since then, investigators have reported the production of experimental arterial lesions in many primate species by feeding diets high in fat and cholesterol but varying in other constituents.

Despite early unsuccessful attempts with the rhesus monkey, this species has become important as an experimental model for investigating atherosclerosis. MANN and ANDRUS [30] reported the occurrence of xanthomatosis and atherosclerosis associated with a high-cholesterol-high-fat diet in an old adult male rhesus monkey. COX et al. [12] and TAYLOR et al. [67] produced hypercholesterolemia and atherosclerotic lesions in rhesus monkeys by feeding a cholesterol-butter dietary supplement for periods as long as three years. One of their animals developed coronary artery lesions with stenosis, thrombosis and myocardial infarction [68]. Other investigators have since produced lipid-containing arterial lesions in the rhesus monkeys [32, 40, 52, 73]. HAMM et al. [21] have recently reported the development of myocardial infarction in three rhesus monkeys fed atherogenic diets. ARMSTRONG et al. [3], EGGEN et al. [17], DE PALMA et al. [13], STARY [53], TUCKER et al. [69], and VESSELINOVITCH et al. [70] have induced experimental arterial lesions in rhesus monkeys by feeding a high-cholesterol high-fat diet and have studied the regression of these lesions after return to a control or 'therapeutic' diet. The details of some of these regression experiments are contained in subsequent chapters of the monograph [2, 62]. The fine structure of the coronary arteries of normocholesterolemia and hypercholesterolemic rhesus monkeys has been described [54, 55].[2]

Lipid-containing arterial lesions varying in severity have also been produced in chimpanzees [1], baboons [20, 62], woolly monkeys [45], squirrel monkeys [37, 45], cynomolgus macaques [2, 26, 48], and stumptailed macaques [5, 43], and these lesions are discussed in depth in this monograph [2, 9, 10, 35].

In addition to these studies involving the production of experimental arterial lesions, there have been studies of cholesterol, lipoprotein and arterial metabolism as related to atherosclerosis in laboratory primates. These studies are also reviewed in detail in this volume [15, 46, 50].

---

[2] GOLDFISCHER et al. recently demonstrated lipid accumulation in intimal smooth muscle cell lysosomes in rhesus monkeys fed an atherogenic diet (Am. J. Path *78:* 497–504, 1975). VESSELINOVITCH et al. investigated the effects of three different food fats on atherosclerosis in rhesus monkeys and found that peanut oil was unusually atherogenic (Atherosclerosis *20:* 303–321, 1974).

The scope and volume of the most recent experimental work in the field is reflected by pertinent abstracts at national and international scientific meetings. Nine reports on the atherosclerosis and lipid metabolism in nonhuman primates were presented (and published) at the 1973 annual meeting of the Federation of American Societies of Experimental Biology and Medicine. Seventeen such abstracts were published in the program of the Third International Symposium on Atherosclerosis held in West Berlin, October 1973. Eight abstracts on nonhuman primate atherosclerosis and lipid metabolism were published and seven papers presented at the November 1973 meeting of the American Heart Association.[3] These figures do not reflect the entire scope of worldwide investigation, but they represent current activity in the field and convincingly confirm the widespread use of primates in atherosclerosis research. It seems likely that continued growth in this activity will occur.

V. Supply of Primates for Experimental Research

Judging from the increasing use of nonhuman primates in atherosclerosis research, it seems likely that investigators will choose nonhuman primates when they have the option. Because of this increasing use of primates, serious problems of supply are anticipated in the future as demand increases and destruction of the natural habitats of the free-living primates continues. For example, it has been suggested that the supply of stumptailed macaques *(Macaca arctoides)* from nature could easily become exhausted in four or five years if these animals become more popular for biomedical research. Since investigators have recently reported that a fibrous-type atherosclerotic lesion can be produced in the stumptailed macaque [5, 42] demand could increase rapidly and the problem of supply could soon become critical. It seems highly desirable to begin now to establish breeding colonies of selected primate species to meet future needs. These breeding colonies could also meet the need for disease-free primate species of known genetic stock, known age, and known dietary history.

[3] STARY reviewed all papers from the 1974 scientific session of the American Heart Association in which nonhuman primate models were used for cardiovascular research (J. med. Primat. in press). Seventeen papers were presented and 15 additional abstracts published involving primate animals. Twenty of these 32 reports were studies of atherosclerosis.

VI. Concluding Remarks

Primates as a group seem especially suited for investigating atherosclerosis; moreover, different primate species meet special requirements, such as size, availability of base line information, and varying degrees of susceptibility to induction of arterial lesions. Experimental investigations in primates which combine morphological, chemical, metabolic, genetic, behavioral and pharmacological approaches appear to be most promising in shedding light on unresolved questions concerning human atherosclerosis.

Much work needs to be done on developing experimental lesions which resemble advanced human lesions more closely. Most induced lesions have been produced by experimental regimens of less than two years, and have been similar to human fatty streaks, an early stage in the atherosclerotic process. More recent studies have indicated that fibrous lesions can be produced more easily in certain species of monkeys than in others [5]. However, factual data concerning the type of lesion which may be produced by long-term experiments are scarce at the present time. It seems likely that experiments lasting five to ten years or longer may be necessary to simulate human-like advanced arterial lesions and their clinical sequelae more closely.

The preponderance of current experimental work is related to the production of lesions by atherogenic diets which have elevated the concentration of blood lipids. The mechanisms of lesion development and lesion regression have been items of special interest and emphasis. As more and more becomes known about the relationship of diet and hyperlipoproteinemia to experimental arterial lesions, research emphasis may shift to the other risk factors for atherosclerosis and end-organ disease (coronary heart disease and stroke). It seems likely that investigators will increasingly use primates in future investigation of the relationships of hypertension, cigarette smoking, obesity, glucose intolerance, physical activity and genetic disorders to vascular disease. This shift in emphasis is just perceptible now.[4] I predict that it will be much more obvious by the end of the decade.

---

[4] Examples of this shift in emphasis include recent reports on the effect of alloxan induced diabetes mellitus on aortic atherosclerosis in squirrel monkeys (LEHNER N. D. M.: Fed. Proc. Fed. Am. Socs. exp. Biol. *34:* 876, 1975) and on the effect of hypertension on atherosclerosis in cynomolgus monkeys (HOLLANDER, W.; MADOFF, I. M.; McCOMBS, H. L.; PADDOCK, J., and KIRKPATRICK, B.: Fed. Proc. Fed. Am. Socs. exp. Biol. *34:* 236, 1975).

References

1 ANDRUS, S. B.; PORTMAN, O. W., and RIOPELLE, A. J.: Comparative studies of spontaneous and experimental atherosclerosis in primates. II. Lesions in chimpanzees including myocardial infarction and cerebral aneurysms. Prog. biochem. Pharmacol., vol. *4*, pp. 393–419 (Karger, Basel 1968).

2 ARMSTRONG, M. L.: Atherosclerosis in rhesus and cynomolgus monkeys; in STRONG Atherosclerosis in primates (this volume).

3 ARMSTRONG, M. L.; WARNER, E. D., and CONNER, W. E.: Regression of coronary atheromatosis in rhesus monkeys. Circulation Res. *57:* 59–67 (1970).

4 BULLOCK, B. C.; CLARKSON, T. B.; LEHNER, N. D. M.; LOFLAND, H. B., jr., and ST. CLAIR, R. W.: Atherosclerosis in *Cebus albifrons* monkeys. III. Clinical and pathologic studies. Expl. molec. Path. *10:* 39–60 (1969).

5 BULLOCK, B. C.; LEHNER, N. D. M.; CLARKSON, T. B., and FELDNER, M. D.: Response of nonhuman primates to an atherogenic diet. Circulation *46:* suppl. II, p. 251 (1972).

6 CHAWLA, K. K.; MURTHY, C. D. S.; CHAKRAVARTI, R. N., and CHHUTTANI, P. N.: Arteriosclerosis and thrombosis in wild rhesus monkeys. Am. Heart J. *73:* 85–91 (1967).

7 CLARKSON, T. B.: Animal models of atherosclerosis; in HARMISON Research animals in medicine (in press).

8 CLARKSON, T. B.: Spontaneous atherosclerosis in subhuman primates; in ROBERTS and STRAUS Comparative atherosclerosis, pp. 211–214 (Harper & Row, New York 1965).

9 CLARKSON, T. B.; HAMM, T. E.; BULLOCK, B. C., and LEHNER, N. D. M.: Old world monkeys in atherosclerosis research; in STRONG Atherosclerosis in primates (this volume).

10 CLARKSON, T. B.; LEHNER, N. D. M.; BULLOCK, B. C.; LOFLAND, H. D., and WAGNER, W. D.: Atherosclerosis in new world monkeys; in STRONG Atherosclerosis in primates (this volume).

11 CLARKSON, T. B.; LOFLAND, H. B.; BULLOCK, B. C.; LEHNER, N. D. M.; ST. CLAIR, R., and PRICHARD, R. W.: Atherosclerosis in some species of new world monkeys. Ann. N. Y. Acad. Sci. *162:* 103–109 (1969).

12 COX, G. E.; TAYLOR, C. B.; COX, L. G., and COUNTS M. D.: Atherosclerosis in rhesus monkeys. I. Hypercholesteremia induced by dietary fat and cholesterol. Archs Path. *66:* 32–52 (1958).

13 DEPALMA, R. G.; BELLON, E. M.; INSULL, W., jr.; ROTH, W. T., and ROBINSON, A. V.: Studies on progression and regression of experimental atherosclerosis; in Medical primatology, part III, pp. 313–323 (Karger, Basel 1972).

14 DUFF, G. L. and MCMILLAN, G. C.: Pathology of atherosclerosis. Am. J. Med. *11:* 92–108 (1951).

15 EGGEN, D. A.: Cholesterol metabolism in nonhuman primates; in STRONG Atherosclerosis in primates (this volume).

16 EGGEN, D. A.; STRONG, J. P., and NEWMAN, W. P., III: Experimental atherosclerosis in primates. A comparison of selected species. Ann. N. Y. Acad. Sci. *162:* 110–119 (1969).

17 EGGEN, D. A.; STRONG, J. P.; NEWMAN, W. P., III; CATSULIS, C.; MALCOLM, G. T., and KOKATNUR, M. G.: Regression of diet-induced fatty streaks in rhesus monkeys Lab. Invest. *31:* 294–301 (1974).

18 GEER, J. C. and HAUST, M. D.: Smooth muscle cells in atherosclerosis; in POLLAK *et al.* Monographs on atherosclerosis, vol. 2 (Karger, Basel 1972).

19 GRESHAM, G. A. and HOWARD, A. N.: Experimental atherosclerosis in baboons. Ann. N. Y. Acad. Sci. *162:* 99–102 (1969).

20 GRESHAM, G. A. and HOWARD, A. N.: Vascular lesions in primates. Ann. N. Y. Acad. Sci *127:* 694–701 (1965).

21 HAMM, T. E.; ABEE, C. R.; RIGGS, T. A., and CLARKSON, T. B.: Myocardial infarction in three rhesus monkeys *(Macaca mulatta)* fed atherogenic diets. Fed. Proc. Fed. Am. Socs exp. Biol. *33:* 236 (1974).

22 HARRIS, R. S.: Feeding and nutrition of nonhuman primates (Academic Press, London 1970).

23 HUEPER, W. C.: Experimental atherosclerosis in macacus rhesus monkeys. Am. J. Path. 22: 1287–1289 (1946).

24 JONES, R. J.: Atherosclerosis. Proc. 2nd Symp. (Springer, Berlin 1970).

25 KAWAMURA, R.: in Neue Beiträge zur Morphologie und Physiologie der Cholinesterinsteatose, p. 267 (Fischer, Jena 1927).

26 KRAMSCH, D. M. and HOLLANDER, W.: Occlusive atherosclerotic disease of the coronary arteries in monkey *(Macaca irus)* induced by diet. Expl. molec. Path. *9:* 1–22 (1968).

27 KRITCHEVSKY, D.: Experimental atherosclerosis in primates and other species. Ann. N. Y. Acad. Sci. *162:* 80–88 (1969).

28 LONG, E. R.: Development of our knowledge of arteriosclerosis; in BLUMENTHAL Cowdry's Arteriosclerosis. A survey of the problem; 4th ed., pp. 5–20 (Ch. C. Thomas, Springfield 1967).

29 MALINOW, M. R.: Atherosclerosis in subhuman primates. Folia primatol. *3:* 277–300 (1965).

30 MANN, G. V. and ANDRUS, S. B.: Xanthomatosis and atherosclerosis produced by diet in an adult rhesus monkey. J. Lab. clin. Med. *48:* 533–550 (1956).

31 MANN, G. V.; ANDRUS, S. B.; MCNALLY, A., and STARE, F. J.: Experimental atherosclerosis in cebus monkey. J. exp. Med. *98:* 195–218 (1953).

32 MANNING, P. J. and CLARKSON, T. B.: Development, distribution, and lipid content of diet-induced atherosclerotic lesions of rhesus monkeys. Expl. molec. Path. *17:* 38–54 (1972).

33 MCGILL, H. C., jr.: The geographic pathology of atherosclerosis (Williams & Wilkins, Baltimore 1968).

34 MCGILL, H. C., jr.; EGGEN, D. A., and STRONG, J. P.: Atherosclerotic lesions in the aorta and coronary arteries of man; in ROBERTS and STRAUS Comparative atherosclerosis, pp. 311–326 (Harper & Row, New York 1965).

35 MCGILL, H. C., jr.; MOTT, G. A., and BRAMBLETT, C. A.: Experimental atherosclerosis in the baboon; in STRONG Atherosclerosis in primates (this volume).

36 MCGILL, H. C., jr.; STRONG, J. P.; HOLMAN, R. L.; and WERTHESSEN, N. T.: Arterial lesions in the Kenya baboon. Circulation Res. *8:* 670–679 (1960).

37 MIDDLETON, C. C.; CLARKSON, T. B.; LOFLAND, H. B., and PRICHARD, R. W.: Diet and atherosclerosis of squirrel monkeys. Archs Path. *83:* 145–153 (1967).

38 MIDDLETON, C. C.; ROSAL, J.; CLARKSON, T. B.; NEWMAN, W. P., III, and MCGILL, H. C., jr.: Arterial lesions in squirrel monkeys. Archs Path. *83:* 352–358 (1967).

39 NAPIER, J. R. and NAPIER, P. H.: A handbook of living primates (Academic Press, London 1967).

40 NEWMAN, W. P., III; Eggen, D. A., and STRONG, J. P.: Comparison of arterial lesions and serum lipids in spider and rhesus monkeys on an egg and butter diet. Atherosclerosis *19:* 75–86 (1974).

41 NEWMAN, W. P., III.; MIDDLETON, C. C.; CLARKSON, T. B.; ROSAL, J., and STRONG, J. P.: Naturally occurring arterial lesions in five species of new world primates. Archs Path. *98:* 173–176 (1974).

42 PICK, R. and GLICK, G.: The influence of hypertension on diet-induced aortic, coronary and cerebral atherosclerosis in the stumptail macaque. Circulation *46:* suppl. II, p. 248 (1972).

43 PICK, R. and KATZ, L. N.: Cholesterol-fat induced atherosclerosis in the stumptail macaque *(Macaca speciosa)*. Circulation *39:* suppl. III, p. 20 (1969).

44 PORTMAN, O. W.: Atherosclerosis in nonhuman primates. Sequences and possible mechanisms of change in phospholipid composition and metabolism. Ann. N. Y. Acad. Sci. *162:* 120–136 (1969).

45 PORTMAN, O. W. and ANDRUS, S. B.: Comparative evaluation of three species of new world monkeys for studies of dietary factors, tissue lipids, and atherogenesis. J. Nutr. *87:* 429–438 (1965).

46 PORTMAN, O. A. and ILLINGWORTH, D. R.: Arterial metabolism in primates; in STRONG Atherosclerosis in primates (this volume).

47 PRATHAP, K.: Spontaneous aortic lesions in wild adult malaysian long-tailed monkeys *(Macaca irus)*. J. Path. *110:* 135–143 (1973).

48 PRATHAP, K. and LAU, K. S.: Spontaneous and experimental arterial lesions in the Malaysian long-tailed monkey; in GOLDSMITH and MOOR-JANKOWSKI Medical primatology, part III, pp. 343–349 (Karger, Basel 1972).

49 ROBERTS, J. C. and STRAUS, R.: Comparative atherosclerosis (Harper & Row, New York 1965).

50 RUDEL, L. L. and LOFLAND, H. B.: Circulating lipoproteins in nonhuman primates; in STRONG Atherosclerosis in primates (this volume).

51 SCHETTLER, G. and WEIZEL, A.: Atherosclerosis III, Proc. 3rd Int. Symp. Atherosclerosis (Springer, Berlin 1974).

52 SCOTT, R. F.; MORRISON, E. S.; JARMOLYCH, J.; NAM, S. C.; KROMS, M., and COULSTON, F.: Experimental atherosclerosis in rhesus monkeys. I. Gross and microscopic features and lipid values in serum and aorta. Expl molec. Path. *7:* 11–33 (1967).

53 STARY, H. C.: Progression and regression of experimental atherosclerosis in rhesus monkeys; in GOLDSMITH and MOOR-JANKOWSKI Medical primatology, part III, pp. 356–367 (Karger, Basel 1972).

54 STARY, H. C.: Coronary artery fine structure in rhesus monkeys: The early atherosclerotic lesion and its progression; in STRONG Atherosclerosis in primates (this volume).

55 STARY, H. C. and STRONG, J. P.: Coronary artery fine structure in rhesus monkeys: Nonatherosclerotic intimal thickening; in STRONG Atherosclerosis in primates (this volume).

56 STOUT, C. and GROOVER, M. E., jr.: Spontaneous versus experimental atherosclerosis. Ann. N. Y. Acad. Sci. *162:* 89–98 (1969).

57 STOUT, C. and LEMMON, W. B.: Predominant coronary and cerebral atherosclerosis in captive nonhuman primates. Expl molec. Path. *10:* 312–322 (1969).

58 STRONG, J. P.: Arterial lesions in primates; in ROBERTS and STRAUS Comparative atherosclerosis, pp. 244–252 (Harper & Row, New York 1965).

59 STRONG, J. P.: Atherosclerosis in primates. J. med. Primat *3:* 89–94 (1974).

60 STRONG, J. P.; EGGEN, D. A.; NEWMAN, W. P., III, and MARTINEZ, R. D.: Naturally occurring and experimental atherosclerosis in primates. Ann. N. Y. Acad. Sci. *149:* 882–894 (1968).

61 STRONG, J. P.; EGGEN, D. A., and OALMANN, M. C.: The natural history, geographic pathology, and epidemiology of atherosclerosis; in WISSLER and GEER The pathogenesis of atherosclerosis, pp. 20–40 (Williams & Wilkins, Baltimore 1972).

62 STRONG, J. P.; EGGEN, D. A., and STARY, H. C.: Reversibility of experimental fatty streaks in rhesus monkeys; in STRONG Atherosclerosis in primates (this volume).

63 STRONG, J. P. and McGILL, H. C., jr.: Diet and experimental atherosclerosis in baboons. Am. J. Path. *50:* 669–690 (1967).

64 STRONG, J. P.; MILLER, J. H., and McGILL, H. C., jr.: Naturally occurring parasitic lesions in baboons; in VAGTBORG The baboon in medical research, pp. 503–512 (Univ. Texas Press, Austin 1965).

65 STRONG, J. P. and TAPPEN, N. C.: Naturally occurring arterial lesions in african monkeys. Archs Path. *79:* 199–205 (1965).

66 TAYLOR, C. B.: Experimentally induced arteriosclerosis in nonhuman primates; in ROBERTS and STRAUS Comparative atherosclerosis, pp. 215–243 (Harper & Row, New York 1965).

67 TAYLOR, C. B.; COX, G. E.; MANALO-ESTRELLA, P.; SOUTHWORTH, J.; PATTON, D. E., and CATHCART, C.: Atherosclerosis in rhesus monkeys. II. Arterial lesions associated with hypercholesterolemia induced by dietary fat and cholesterol. Archs Path. *74:* 16–34 (1962).

68 TAYLOR, C. B.; PATTON, D. E., and COX, G. E.: Atherosclerosis in rhesus monkeys. VI. Fatal myocardial infarction in a monkey fed fat and cholesterol. Archs Path. *76:* 404–412 (1963).

69 TUCKER, C. F.; CATSULIS, C.; STRONG, J. P. and EGGEN, D. A.: Regression of early cholesterol-induced aortic lesions in rhesus monkeys. Am. J. Path. *65:* 493–502 (1971).

70 VESSELINOVITCH, D.; HUGHES, R.; FRAZIER, L., and WISSLER, R. W.: Studies of the reversal of advanced atherosclerosis in the rhesus monkey. Am. J. Path. *70:* 41a (1973).

71 WISSLER, R. W.; FRAZIER, L. E.; HUGHES, R. H., and RASMUSSEN, R. A.: Atherogenesis in the cebus monkey. I. A comparison of three food fats under controlled dietary conditions. Archs Path. *74:* 312–322 (1962).

72 WISSLER, R. W. and GEER, J. C.: The pathogenesis of atherosclerosis (Williams & Wilkins, Baltimore 1972).

73 Wissler, R. W.; Hughes, R. H.; Frazier, L. E.; Getz, G. S., and Turner, D.:
Aortic lesions and blood lipids in rhesus monkeys fed 'table prepared' human diets.
Circulation *32:* suppl. II, p. 220 (1965).

Jack P. Strong, M.D., Department of Pathology, Louisiana State University Medical
Center, 1542 Tulane Avenue, *New Orleans, LA 70112* (USA)

Prim. Med. vol. 9, pp. 16–40 (Karger, Basel 1976)

Atherosclerosis in Rhesus and Cynomolgus Monkeys[1]

MARK L. ARMSTRONG

Department of Internal Medicine, University of Iowa, Iowa City, Iowa

Contents

[1] Supported by USPHS, NIH grants HL 14320 and HL 014388.

I. Introduction

Among nonhuman primates used in the research of various aspects of experimental atherosclerosis, the rhesus monkey *(M. mulatta)* has been studied by more investigators than any other species during the past quarter century. Other macaques, less common or familiar to medical investigators, have been relatively neglected until recently. Among these, the cynomolgus *(M. fascicularis)*, has been found very suitable for studies of induced atherosclerosis. With a relatively short span marking the phylogenetic branching of macaques and the pongid-hominid line, it is reasonable to hope that a comprehensive evaluation of the metabolic factors and the morphologic features of induced atherosclerosis in the members of the macaque genus may eventually advance our understanding of evolutionary aspects of human atherosclerosis.

Meanwhile, the more immediate justification for the use of nonhuman primates in atherosclerosis research rests on the general presupposition that the experimental lesions produced in monkeys belonging to the Old World radiation have a closer parallelism, aside from purely morphologic resemblances, to the human lesion than do atherosclerotic lesions found or induced in species outside the primate order. This chapter, describing induced atherosclerosis in two species, the most familiar macaque and the lightest of the macaques, is written from the view that the stated proposition has considerable evidence to support it.

II. Atherosclerosis in Rhesus Monkeys

A. Failures and Successes in the Induction of Atherosclerosis

The arterial tree of the rhesus in the wild has, as a rule, remarkably little arteriosclerotic change in the intima as detected by meticulous gross examination [12]. This is also true of captive animals under controlled conditions. At the magnification used in light microscopy, arterial sections often or even typically show the endothelium closely applied to the internal elastic lamina [8, 52, 58]. However, old animals may show some intimal thickening, especially in the aorta, in which connective tissue characteristically predominates, but lipid is present to a variable degree [31, 30]. Even in younger animals, SCOTT *et al.* [47] found small proliferative intimal lesions in one-third of the sections from the aortic intima and none from the coronaries. VLODAVER *et al.* [60] described proliferative changes in the intima of coronary arteries that varied with the duration of captivity, and STARY and STRONG [50] in this volume describe coronary thickening in the form of cushions and diffuse thickening in normocholesterolemic rhesus monkeys.

Although these cited departures from a somewhat idealized normal arterial architecture indicate the need for careful matching of control and experimental animals in studies of atherosclerosis, the ordinary experience of the investigator will be the discovery of minimal atherosclerotic change in his controls. This fortunate circumstance greatly simplifies the evaluation of atherogenic regimens. It is the more noteworthy therefore that initial efforts to induce atherosclerosis by dietary means were apparently unsuccessful [24, 25]. MANN and ANDRUS [3] documented the course of an adult animal that developed hypercholesterolemia with preponderant β-lipoproteinemia, xanthomatosis, and atherosclerosis as a result of a nutritionally complete diet enriched cholesterol and fat. In a series of reports from 1954 to 1963, TAYLOR and co-workers [15, 52–54] showed a wide spectrum of experimental atherosclerosis and hypercholesterolemia induced in rhesus monkeys by diets of high cholesterol and fat content. Subsequent studies of lesion induction [8, 34, 47, 63, 67] have uniformly shown unequivocal atherosclerotic change, the degree of which is in part a reflection of various experimental designs and diets. Some form of dietary induction of atherosclerosis has been almost uniformly employed, an apparent exception being the induction of fatty streaks and fibrous plaques in the aorta by hypertension [38].

B. Dietary Factors

The *cholesterol intake* of macaques in the wild is both a matter of surmise and of some knowledge. Inference from several types of data permits the operational judgment that cholesterol intake must be quite low. In the experimental situation the use of dietary cholesterol is a *sine qua non* in many nutritionally complete diets if the investigator hopes to induce more than minimal lesions.

Dietary cholesterol may not be as well absorbed without added fat [10]. In unpublished studies we found that radiocholesterol was readily absorbed when no fat was added to commercial chow containing 4% fat; nonetheless, the quantitative aspects of the issue remain of interest because the dietary fat requirements for maximal cholesterol absorption in the rhesus have not been established. Commonly cholesterol is added to diets in which the fat content is in the range of 10–25% by weight. However, the amount of cholesterol added has varied greatly, with additions of 0.37% [58], 0.5% [34], 2.0% [63], 2.6% [15], 6.5% [33], and even 10% [67]. In cholesterol balance studies, we found that most of the dietary cholesterol was excreted when 1.2% cholesterol was given [2]. The hypercholesterolemic effect of added cholesterol, on which the induction of significant atherosclerosis is dependent, apparently conforms to a classical dose-response curve [3], in which the asymptotic limb of the rectangular hyperbola is approached or reached at all of the more frequently used amounts of dietary cholesterol in atherogenic diets. These cholesterol supplements to the diets of nonhuman primates exceed the average human level of cholesterol consumption in all countries where a considerable amount of the diet is from animal origin (high in cholesterol content) and coronary heart disease rates are high [14]. An attempt to obtain arterial lesions in rhesus monkeys at lower cholesterol intakes for 18 months resulted in small intimal lesions in which atherogenic factors were regarded as operative but at a low level [7]. Rhesus monkeys fed 'table-prepared diets' that compared to the average American diet and a recommended prophylactic diet showed a difference in the extent of lesion formation, depending on the levels of cholesterol intakes, but with additional variables in the composition of the diets [63].

Dietary fat does not *per se* have a very significant atherogenic effect with the possible exception of coconut oil. Differences are noted in the type and extent of lesion formation, however, that seem to depend on the type of fat that is combined with cholesterol in the diet. In combination with 2% cholesterol, peanut oil caused more hyperplasia of intimal connective tissue than corn oil [65]. Butterfat and coconut oil combined with cholesterol resulted in quite advanced atherosclerotic lesions in the rhesus aorta after 18 months [59]. In comparison, egg yolk fat with 0.4% added cholesterol, yielding a total cholesterol content of 1.2%, may have caused less advanced aortic lesions after 17 months [8]. Lard at 25% by weight caused some elevation of plasma cholesterol [35] and with 0.5% cholesterol added caused marked hypercholesterolemia and widespread lesions in the aorta and branch arteries after 18 months [34].

The relative atherogenicity of the various fats in combination with dietary cholesterol cannot be quantified at present. Current evidence suggests that for a given amount of absorbed cholesterol (including that contained in some fats), the more saturated fats (coconut oil, lard, butterfat) together with peanut oil may accelerate the occurrence of widespread and more complex lesions, and that less saturated fats (corn oil, egg yolk) may be less potently atherogenic. It is important to recognize that this provisional statement is made more to raise a significant question that requires an answer, than to voice a conclusion. The attempt to obtain substantial evidence on this point is severely hampered by the numerous variables among the reported investigations. Whether the degree of saturation of a fat is a distinguishing feature in the combined atherogenic effect of fat and high cholesterol intakes has been usually confounded by considerable variation in the chain lengths of the fatty acids of the compared fats, as well as by other biochemical differences.

The type and amount of *carbohydrate intake* have been inadequately studied for atherogenic effect. Recently, MURAKAMI *et al.* [37] reported that sucrose caused elevation of very low density lipoproteins (VLDL) and some evidence suggesting atherosclerotic change. In unpublished studies we also found that sucrose as the carbohydrate source tended to cause variable but sustained elevation of VLDL; sucrose did not appear to be atherogenic in the absence of dietary cholesterol.

Dietary deficiencies have been noted to cause or exacerbate atherosclerotic lesions. Pyridoxine deficiency caused a sclerotic type of lesion without much lipid [4]. Whether ascorbic acid deficiency exacerbates atherosclerosis as it does intimal lesions in guinea pigs [55] is not known. The amount of protein intake above deficiency levels does not seem to be a factor in the degree of atherosclerosis induced. The use of soy protein has been reported to cause iron deficiency and anemia in young rhesus monkeys [18], but after feeding a diet containing soy protein to adult animals [2] we found no such effect [9].

C. Plasma Lipid and Lipoprotein Changes

The key changes in lipids and lipoproteins of rhesus monkeys fed large amounts of cholesterol and fat are a rise in cholesterol and low-density lipoprotein (LDL) [2, 15, 33]. Changes in triglyceride concentration are minor during the usual diet but a rise has been noted when the diet contained

---

Figures 1–6 are of lesions found in rhesus after 17 months of atherogenic diet.

Fig. 1. Abdominal aorta: fibromuscular cap, intermediate layers of fibrous material and extracellular lipid, basal concentration of sterol clefts in area of increased cellular breakdown and phagocytosis. HE. × 50.

Fig. 2. Carotid artery: fibrosis of intima predominates, but with basal lipid pools and numerous areas of medial involvement. Verhoeff-Van Gieson. × 35.

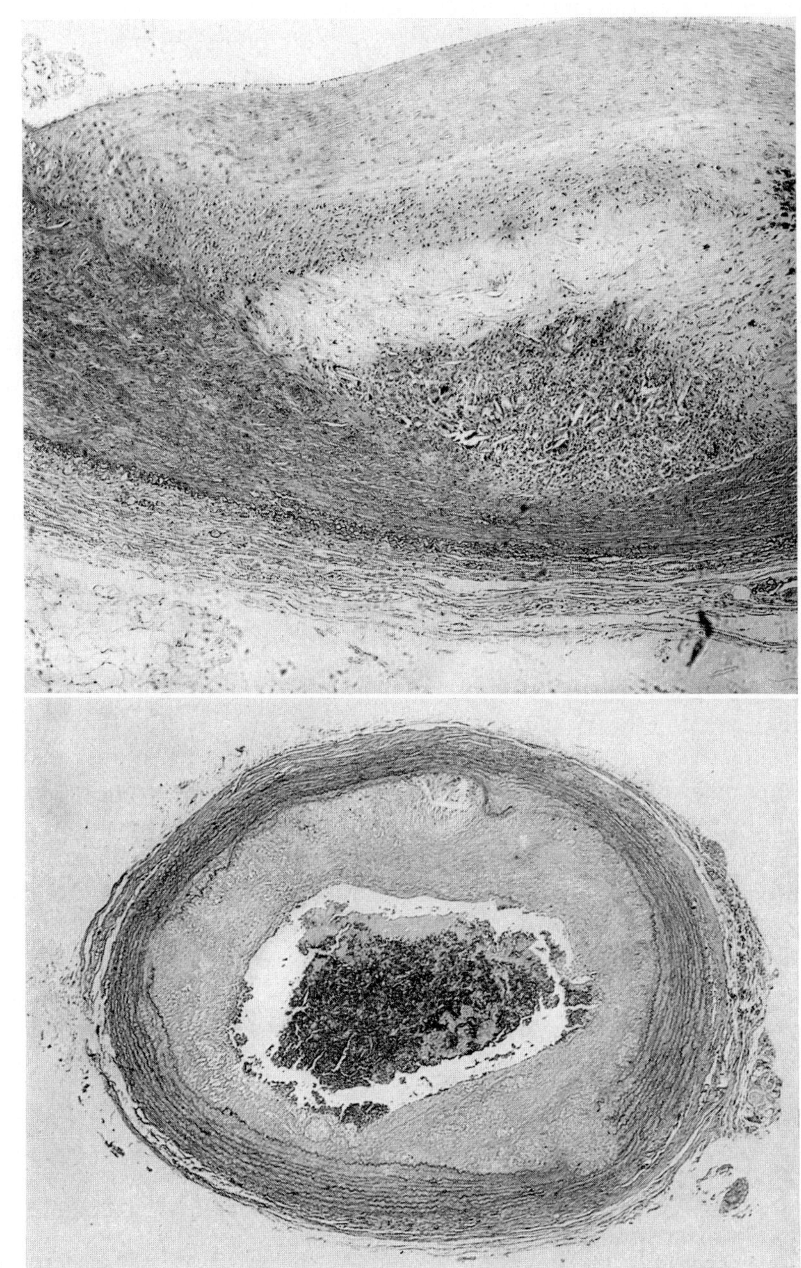

1

2

only 10% fat by weight [29]. The use of agents and procedures that cause
significant triglyceridemia and VLDL elevation [22, 37] needs further elucida-
tion. Studies along these lines may allow a more precise delineation of the
comparative atherogenicity of LDL and VLDL in primates.

D. Characteristics of the Atherosclerotic Lesion

1. Time Course of Lesion Development in Different Arterial Beds

TAYLOR *et al.* [52] estimated from a series of studies that lesion formation
occurred in the order: aorta, coronary, and then mesenteric and cerebral
arteries. We have expressed concurrence with this opinion [8], but to a
considerable degree our impression of the presumed sequence was an extrapolation from the morphologic appearance of the observed lesions. Subsequent ongoing observations have not dissuaded us from the view that atherosclerotic lesions in the aorta precede those in the coronary artery, but MAN-
NING and CLARKSON [34] found raised plaques in coronary arteries sooner
than in the aorta. They also noted earlier lesions of the fatty streak type
in the thoracic aorta, but found late plaques preponderant in the abdominal
aorta, in concurrence with most investigators.

2. Distribution of Atherosclerotic Lesions in Long-Term Studies

The aorta and the coronary arteries become extensive sites of fatty streaks
and plaques as an atherogenic diet is continued. The carotid bulbs, the iliac
and subclavian arteries, and progressively more distal portions of the extramural coronary tree may become extensively involved. The intracranial
arteries are quite spared in most studies. This fact may be related to the
duration of the challenge diet. Cerebral atherosclerosis has been observed
after four or more years [34, 53]. There is some variability in the reported
extent of lesion development in peripheral arteries beyond the subclavian
and iliac vessels. We have commonly found lesions in the femoral and even
in the tibial vessels of the lower limb and to a lesser degree in the brachial
and proximal portions of the radial and ulnar arteries of the upper limb [8].

Fig. 3. Subclavian artery: a similar lesion showing minor mineralization. Verhoeff-Van
Gieson. × 50.

Fig. 4. Brachial artery: eccentric lesion (T-shaped lumen top left). Fibromuscular cap
contains numerous lipid-bearing cells. Subjacent palisading and some basal necrosis. Van
Gieson. × 50.

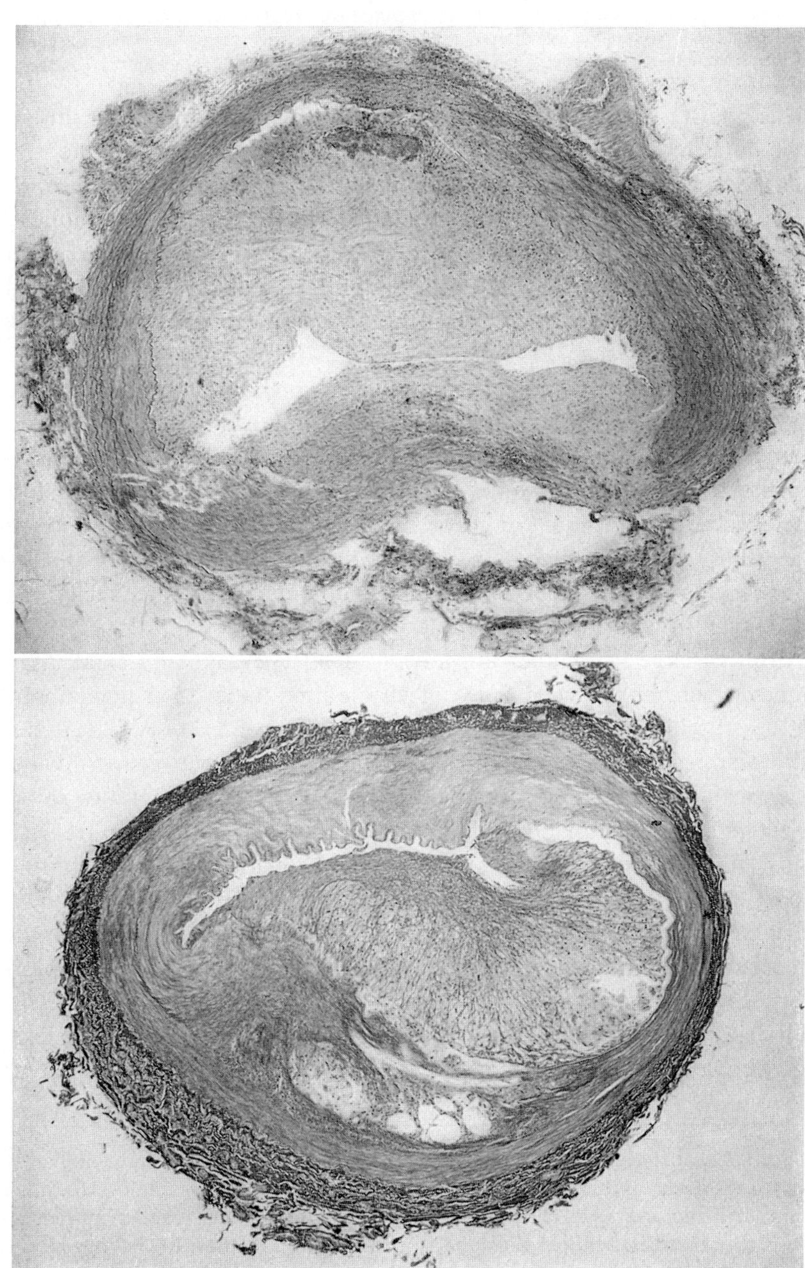

The mesenteric arteries are often less involved, and the celiac axis and renal arteries frequently show only an intermediate degree of atherosclerosis.

3. Morphology. Variations in Different Arteries

The relative frequency of lesion types in a given artery is a fair indicator of the duration and degree of hypercholesterolemia. After many months of atherogenic diet, the investigator can distinguish, in addition to typical fatty streaks, two types of larger lesions: (1) uncomplicated fibrous plaques i.e. collections of extracellular and intracellular lipid beneath a hallmark feature of quite discrete fibromuscular proliferation [1, 66], and (2) complicated plaques that exhibit more advanced necrosis and less frequently calcification and hemorrhage.

Fatty streaks are exceedingly common. The nearly total involvement of the intimal surface of the rhesus aorta after relatively short challenges of atherogenic diet is the result of the formation and coalescence of innumerable lesions that are chiefly of the fatty streak type [58]; in long-term studies these are the background for more advanced lesions at sites of increased vulnerability related to hemodynamic and mechanical factors [19, 20, 43, 44, 56, 57] and to poorly understood regional differences in the capacity of the arterial wall to withstand hypercholesterolemia [21].

More variable in number are advanced fatty streaks and small fibrous plaques which show a wide range of visible lipid under their fibromuscular caps. Extracellular lipid is not as prominent in either lesion type as it is in a well-developed fibrous plaque; however, the distinction between an advanced fatty streak and a small uncomplicated fibrous plaque sometimes rests on subjective evaluation of borderline phenomena and is quite arbitrary. The foam cells found in all of these lesions are thought to be principally fat-laden smooth muscle cells [46]. Fibrous plaques are quite common, and clearly necrotic plaques will often be found at sites where maximal predilection occurs in man, the most notable exception being the intracranial arteries.

Some lesions that have a necrotic base have very thin subendothelial caps. Sometimes caps contain many foam cells in their interstices. Such lesions resist simple classification; they are the inevitable consequence of the great

---

Fig. 5. Coronary artery (left anterior descending branch): lesion is predominantly fibrous except for basal third, which is largely lipid. Verhoeff-Van Gieson. × 110.

Fig. 6. Femoral artery: prominent periluminal fibromuscular cap. Lipid-rich areas show patchy necrosis. Marked gaps and splits in internal elastica. Medial atrophy. HE. × 60.

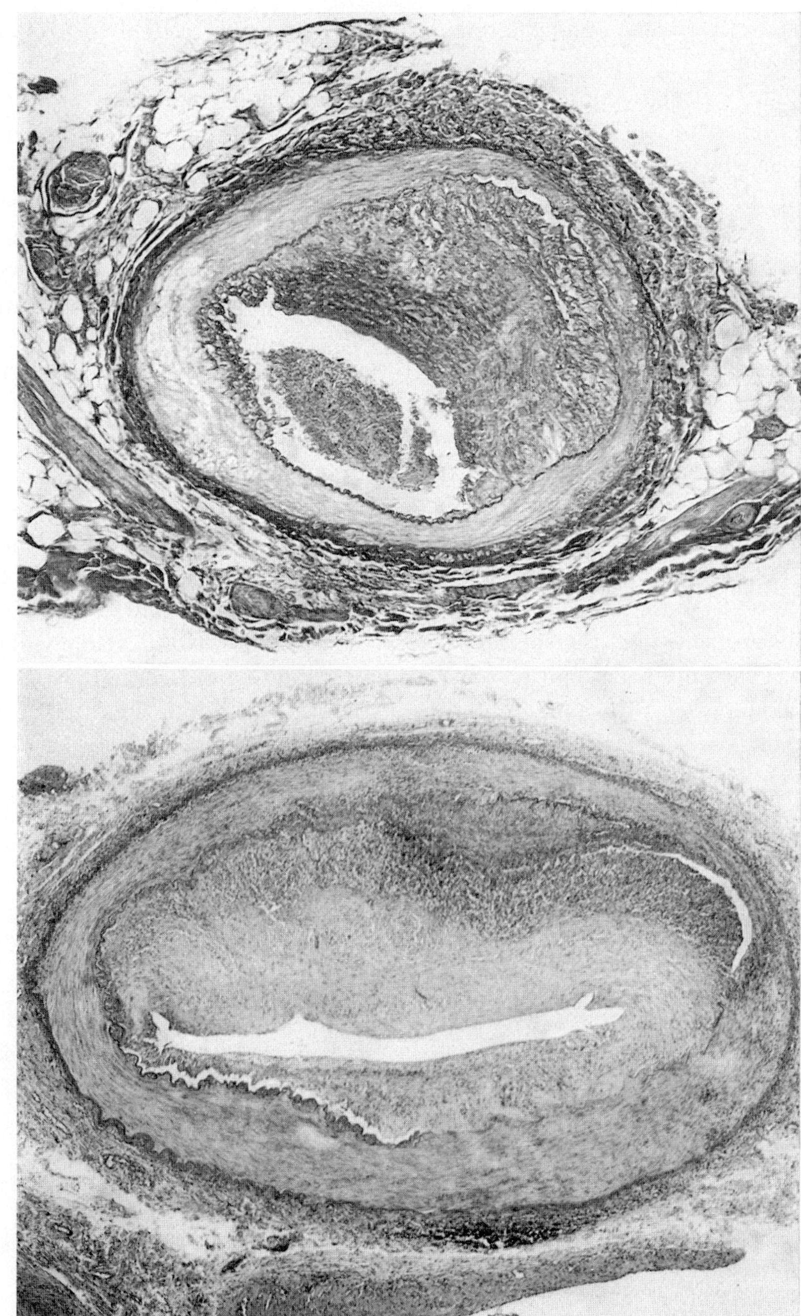

5

6

hypercholesterolemia and the relatively short time course used to produce the lesion.

Some vascularization of the intima is common. Hemorrhage, although seen, tends to be quite minor. Mineralization is found but its relative frequency in atheromas is not great. Thrombosis must be rare; microscopically evident platelet thrombi are infrequent enough to suggest that their ordinary fate is dissolution within the lumen without their making a significant contribution to lesion formation. Whether this opinion would be more challengeable for lesions created in a milieu that comprised a broader range of hyperlipidemia (e.g. excess of VLDL as well as LDL) remains to be explored. The lesions have apparently not progressed beyond superficial ulceration in any investigation to date.

The frequency of lesion types varies in different arteries in some correspondence to the time course of lesion development in the various arterial beds, as noted in section II. D. 1. Generally the aorta is regarded as the site of more advanced lesions. We thought, in terms of cell-thickness equivalence, that lesions in numerous branch arteries were at least equal to those in aorta, and speculated that the age of the animals may be an important determinant of this propensity [8]. Subspecies differences in the pattern of development of lesions have not been reported, and we have found no difference thus far in rhesus imported from India and a distinctly different rhesus imported from Pakistan.

III. Atherosclerosis in Cynomolgus Monkeys

A. Resumé of Investigations to Date

MALMROS and WIGNAND [32] reported on the hyperlipidemic aspects of an atherogenic diet, but the effect on arteries was not described. KRAMSCH and HOLLANDER [26] appear to have been the first to report the vascular consequences of an atherogenic diet in this species. PRATHAP and LAU [40] confirmed several aspects of these positive findings. Our recent report [5] also parallels the findings of KRAMSCH and HOLLANDER.

Figures 7–10 are of residual lesions in rhesus after 40 months of regression diet.

Fig. 7. Abdominal aorta: intima remains moderately thickened by fibromuscular layers. HE. × 150.

Fig. 8. Carotid artery: only a narrow rim of intimal fibroplasia persists in this regression specimen. Verhoeff-Van Gieson. × 50.

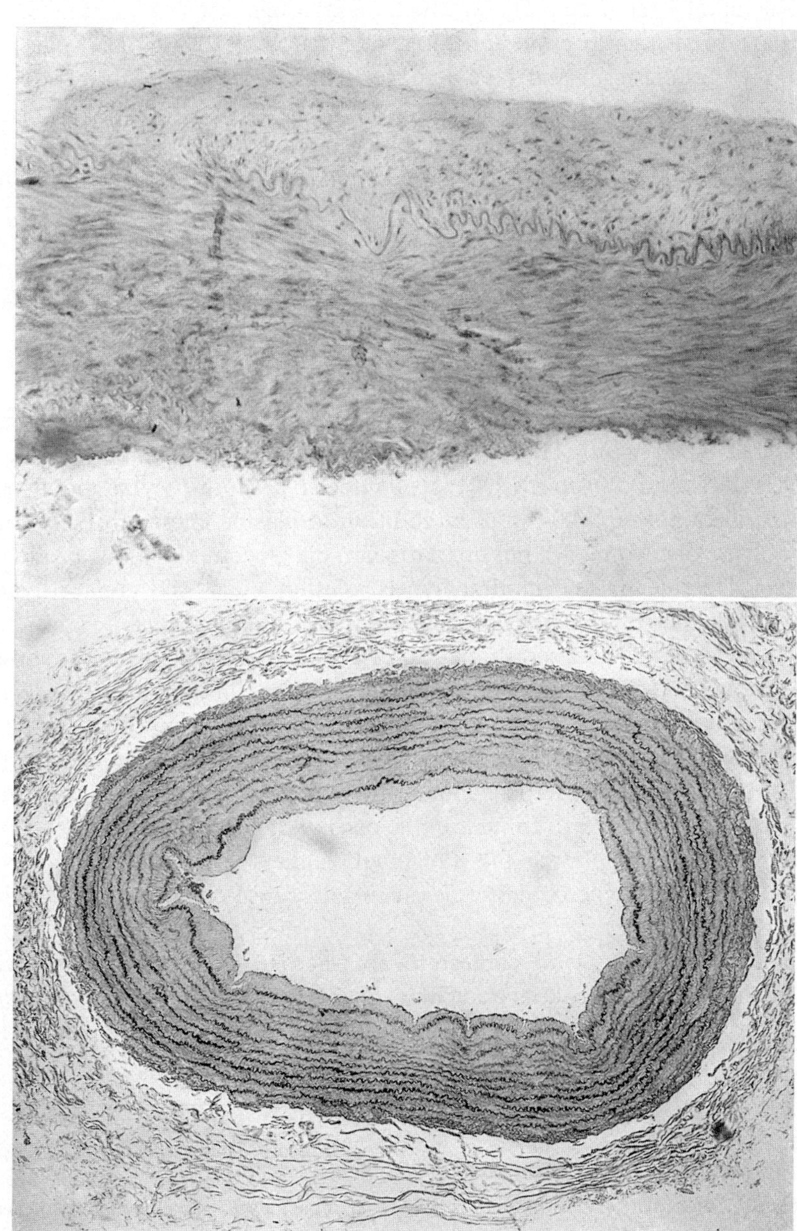

B. Dietary Induction of Atherosclerosis

The initially reported diets, in which an atherogenic effect was unequivocal, have been high in cholesterol and fat. KRAMSCH and HOLLANDER [26] used 1.4% cholesterol and less than 10% fat by weight in their original studies. PRATHRAP and LAU [40] fed 0–15% cholesterol with high fat enrichment. We used 1.2% cholesterol and 18% egg yolk fat as in our original rhesus studies [9]. More recently, however, KRAMSCH *et al.* [28] have found significant atherosclerosis with less hypercholesterolemia after using 0.1% cholesterol together with 10% butterfat.

C. Plasma Lipid and Lipoprotein Changes

All investigations show a moderate to marked hypercholesterolemia, averaging 400–800 mg/dl. Triglycerides typically show little change. The study of MALMROS and WIGNAND [32] is of interest because they found hypertriglyceridemia up to 400 mg/dl in addition to plasma cholesterol elevation with the use of coconut oil, but not with corn oil. Coconut oil alone produced hypertriglyceridemia and modest hypercholesterolemia, and greater hypercholesterolemia and hypertriglyceridemia when cholesterol was fed in addition to coconut oil. KRAMSCH's diet of 0.1% cholesterol and butterfat caused hypercholesterolemia of 400–500 mg/dl. The lipoprotein changes after atherogenic diets are apparently similar to those observed in rhesus. KRAMSCH and HOLLANDER [26], as well as our laboratory, have found a striking rise in the LDL band by electrophoretic evaluation. There was a correspondingly great increase of the cholesterol content in the density-defined fraction 1.019–1.063 of ultracentrifuged plasma [5]. Presumably, diets causing both hypertriglyceridemia and hypercholesterolemia would cause combined VLDL and LDL elevation.

The response of plasma cholesterol to atherogenic diets is prompt, with marked elevation to peak levels achieved within a few months. In our comparison of the plasma cholesterol changes in cynomolgus versus rhesus fed

---

Fig. 9. Coronary artery: intimal thickening with greatly decreased evidence of lipid. HE. × 120.

Fig. 10. Femoral artery: intima has receded to compact layers of connective tissue, leaving a large lumen. HE. × 60.

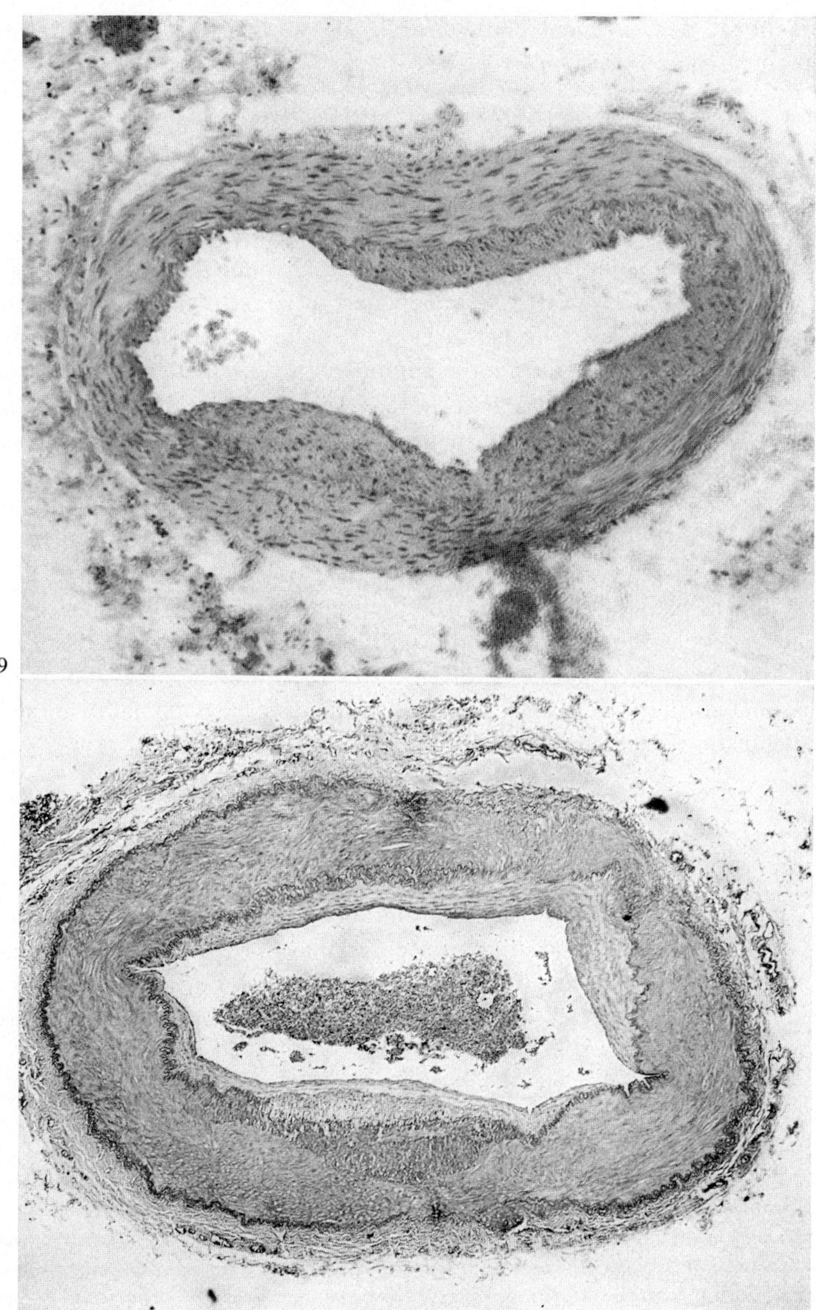

9

10

an identical diet, average cholesterol levels were significantly lower in cynomolgus than rhesus [5].

D. Characteristics of the Experimental Lesion

Somewhat similar to rhesus, cynomolgus control animals typically have grossly quite minor intimal lesions [26]. After 12 months of atherosclerotic diet KRAMSCH and HOLLANDER found gross atherosclerotic lesions in the main coronary arteries; all other arteries, including the aorta, were grossly normal except for tiny lesions in the pulmonary arteries. After 18 months of atherosclerotic diet, however, gross lesions were found in the aorta and its other branch arteries with the coronary lesions more severe and widespread. The intracranial vessels were spared. PRATHRAP and LAU [40] found massive fatty streaking of the aorta and fibrous plaques in their experimental animals after two to four years and lesions were seen also in the coronary and other branch arteries. KRAMSCH and HOLLANDER, as well as PRATHRAP and LAU found fibrous plaques to a greater extent in the abdominal aorta than in the thoracic aorta at 18 or more months. Our observations are not in disagreement, but the atherosclerotic involvement of the lower thoracic and upper abdominal aorta in our material varied considerably from animal to animal, and the distinction was neither absolute nor uniform.

E. Comparison of Atherosclerotic Lesions in Cynomolgus and Rhesus

After 17 months of identical atherogenic diet we found that atherosclerosis in cynomolgus equaled and usually exceeded that in rhesus in the aorta and the subclavian, carotid, femoral and coronary arteries. Furthermore, several features of atherosclerosis in cynomolgus, aside from a tendency toward

---

Figures 11–12 are of lesions in carotid arteries of cynomolgus.

Fig. 11. Atherosclerosis after 17 months of atherogenic diet. Marked intimal fibroplasia with extracellular lipid more prominent than in rhesus. Verhoeff-Van Gieson. × 25.

Fig. 12. Lesion appearance after 20 months of regression diet. Densely fibrous intima, residual lipid pools, and replacement fibrosis of medial damage that is presumed to have occurred during atherogenic period. Enlargement of lumen areas by regression regimen was less marked than in rhesus. Verhoeff-Van Gieson. × 45.

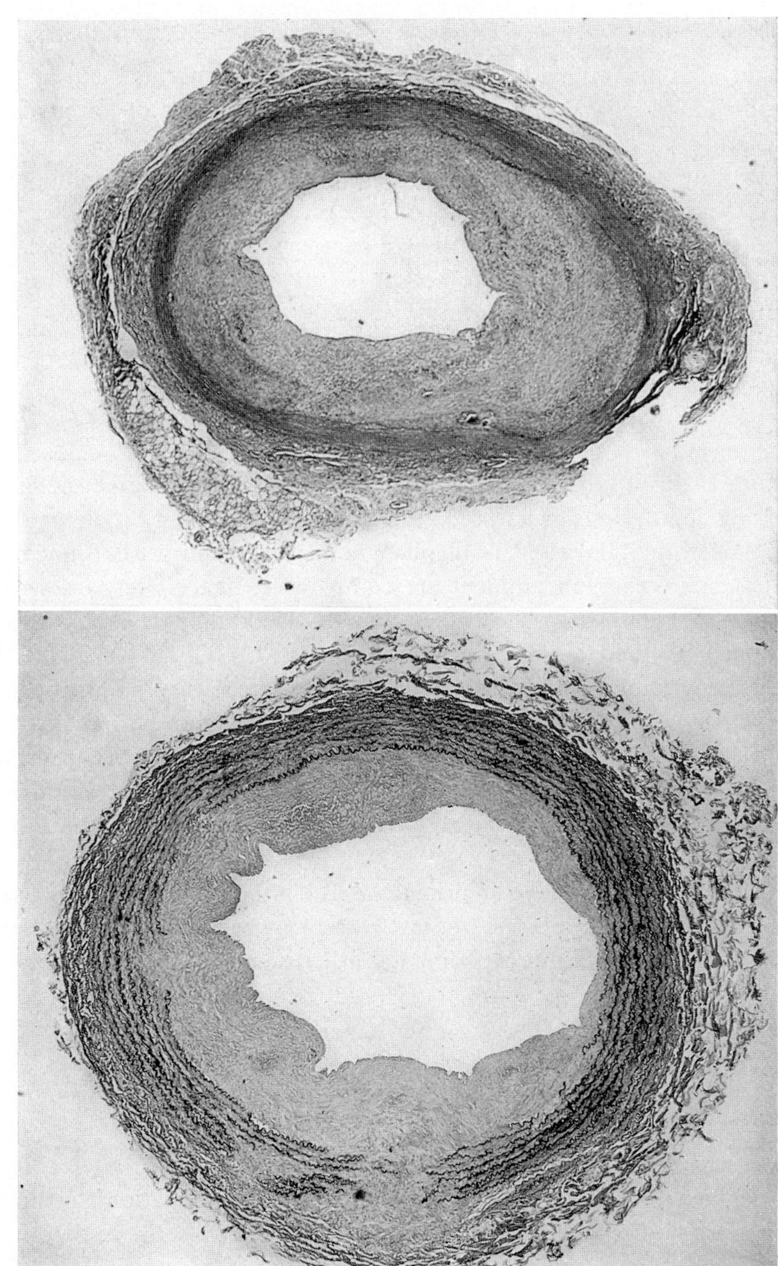

11

12

more intimal thickening, distinguished the lesions from those of the rhesus: medial damage in elastic arteries was clearly greater, more sterol clefts were observed, visible extracellular lipid was, in general, more extensive in cynomolgus, and the intimal fibrotic reaction to the atherogenic diet was more prominent. We have proposed that the divergent findings of lower plasma cholesterol and greater atherosclerosis under the conditions of our study may indicate a greater susceptibility of the arterial wall of cynomolgus to atherosclerosis [5]. This opinion is of necessity tentative until further work is reported, but we believe that these differences in induced atherosclerosis are valid for the Thailand cynomolgus versus rhesus (in which subspecies differences have not been found).

F. Cynomolgus Selection for Atherosclerosis Study

More than 20 subspecies of cynomolgus have been described [39]. The preceding descriptions of induced atherosclerosis are based on observations in animals from Malaya and Thailand. Whether those subspecies found in the southeastern Asian subcontinent are comparable, in their responses to an atherogenic stimulus, to those that inhabit the Philippine archipelago is questioned [27] and a matter of current study [13]. One issue involved is whether geographically isolated subspecific groups may vary in the propensity with which coronary versus aortic atherosclerosis may be induced. The clear identification of a primate model with highly selective resistance in one of these arterial beds and susceptibility in the other would have obvious value for studies of metabolic differences that may underlie atherogenesis.

The role of cynomolgus as a primate useful in the study of atherosclerosis has been clearly defined in the last few years. Less certain is whether the full potential of this species in experimental atherosclerosis has been exploited.

IV. Comparison of Experimental Lesions in Macaques to Human Atherosclerosis

A common desideratum among investigators has been the creation of lesions which resemble those of adult man in human populations susceptible to the morbid and lethal complications of atherosclerosis. Success from this perspective has been somewhat controversial. Severe atherosclerotic disease

was achieved in the very long-term studies of TAYLOR *et al.* [53, 54] and of
MANN and ANDRUS [33]. Numerous examples of necrotic and calcified
plaques occurred in the shorter studies of MANNING and CLARKSON [34] and
VESSELINOVITCH *et al.* [59]. In our own material we have found lesions from
macaque arteries fed atherogenic diets that were histologically similar to
lesions from human atherosclerotic arteries.

Simulations of relatively severe human lesions, having significant amounts
of necrosis, have thus to some extent been induced in macaques. The limits
of this simulation should likewise be noted. Although there has been some
evidence of hemorrhage, extravasation at the slopes of elevated lesions has
not been observed in the experimental lesions we have seen, perhaps in part
because the connective tissue in the subendothelial cap has not yet become
unyieldingly rigid. Connective tissue stains indicate that the fibrous elements
of the subendothelial cap in the experimental lesion are obviously younger
than in most human material. Deep ulceration of the plaque and thrombosis
are typically missing in the experimental setting, the important questions
being to what extent they occur at all by current manipulative measures,
and what means exist (other than injury for thrombosis [54]) to induce these
complications more adequately in primate investigations.

Since the primate lesion is induced more rapidly than the human lesion
develops, except in extraordinary circumstances, the two lesions tend to
reflect this difference in yet another way. The experimental lesion, evoked
under the forced conditions of a degree of hyperbetalipoproteinemia encountered in only a small fraction of human population, has florid features
that several workers have emphasized in some cases of human hyperlipidemia
[42, 61, 62]. These, in brief, consist of foam cell abundance and an increased
number of round cell infiltrates in the adventitia and outer media. These
distinctions are not absolute, nor do they occur in all patients with marked
hyperlipidemia, but we have suggested that these features may form a
bridging characteristic between the atherosclerotic macaque and the arteries
of hyperlipidemic man [8].

Because regression of the experimental lesion has been reported, an
appreciation of morphologic and metabolic differences between induced
macaque lesions and human lesions becomes an important exercise in perspective; experimental manipulations that cause favorable changes in the
macaque lesions are inherently exaggerations of the possibility of favorably
modifying human lesions. However, such magnified versions give the investigator a remarkable opportunity to study tendencies toward change that
might have implications for the more indolently evolving human lesions.

V. Ischemic and Necrotic Changes in Macaques

A. Myocardial Infarction

The study of TAYLOR *et al.* [53] on the occurrence of myocardial infarction in a female rhesus after 40 months of high-cholesterol, high-fat diet is the classical document describing the induction of unequivocal and fatal ischemic heart disease in a primate animal. The lesions affecting the coronary arteries at this very late stage of atherogenic diet show many resemblances to human material. Recently, HAMM *et al.* [23] found evidence by electrocardiogram or autopsy of myocardial infarction in three animals in a large group made hypercholesterolemic by diet.

B. Sudden Death

KRAMSCH and HOLLANDER [26] noted sudden death to be a complication of atherogenic diet in a much smaller number of cynomolgus monkeys. One animal had some histologic evidence suggestive of myocardial ischemia. Whether this species has an unperceived difference in distribution of coronary atherosclerotic lesions leading to more electrical instability and sudden death than is found in rhesus is a noteworthy question that has not been resolved by adequate comparative study. We believe the small size of the coronary arteries in the cynomolgus may predispose to induced ischemic heart disease, with sudden death as its common presentation.

C. Gangrene of an Extremity

TAYLOR *et al.* [54] also succeeded in inducing gangrene of the leg by atherosclerotic occlusive changes. The compromising lesion was an atheroma complicated by thrombosis in the femoral artery. We have suggested that the peripheral atherosclerosis that is evident in older animals may enlarge the experimental approaches to the pathophysiology of diseased peripheral arteries [8].

VI. Regression of Induced Atherosclerotic Lesions

A. Rhesus

1. Morphologic Changes

We found that lesions induced by 17 months of atherogenic diet had partial regression of the atherosclerotic state after 40 months of cholesterol-free diet whether the diet was low in fat or enriched by corn oil to 40% of total calories. Atherosclerotic lesions throughout the arterial tree showed this phenomenon, and the decrease in intimal thickness and the increase in lumen size in the coronary arteries have been described in detail [9]. Two animals autopsied at 20 months (one following a short illness) showed regression changes much like those seen at 40 months. Whether 20 months or an even shorter period would have been adequate for the overall regression obtained cannot be deduced from this study. VESSELINOVITCH *et al.* [59] however, have recently established that quite severe aortic lesions regress very perceptibly after 18 months of cholesterol-free diet. DE PALMA *et al.* [16] found angiographic and autopsy evidence of regression of stenotic lesions of the superior mesenteric artery after nine months of internal biliary diversion combined with a cholesterol-free diet, and regression of gross yellow lesions seen at abdominal aortotomy after twelve months of cholesterol-free diet. EGGEN *et al.* [17] and STARY [48] have shown that early atherosclerotic lesions (fatty streaks) induced in rhesus monkeys by dietary fat and cholesterol also regress when the dietary stimulus is removed. STARY [49] has compared cell proliferation and ultrastructural changes in these early lesions during progression and regression. The gross histological, chemical, ultrastructural and cell proliferation changes in the early lesions studied by EGGEN *et al.* and STARY are reviewed by STRONG *et al.* [51] in another chapter of this volume.

More systematic time course studies of regression regimens for rhesus after long-term atherogenic challenges would be of great value in indicating the rate of change of the various types of induced lesions.

2. Biochemical Changes in the Regression Lesion

The lipid concentration in the coronary arteries has been reported for control, atherosclerotic, and regression animals [4]. The levels of major lipid classes in induced atherosclerosis in the rhesus were similar to those reported in human atherosclerosis [11, 36, 45]. In regression these values were lower. The most mobile lipid was cholesterol in both its free and ester forms. Whether cholesterol ester was depleted from the arterial wall before hydrolysis to free cholesterol was not shown, however. The fall in free cholesterol concentration in the intima-media was sufficiently marked to indicate that depletion of lipid was from extracellular as well as intracellular sites.

The connective tissue so prominent in regression lesions cannot be estimated adequately by morphologic means as to whether (1) it is increased beyond the amount present in atherosclerosis, (2) it is unchanged in amount but is condensed as lipid is depleted from the arterial wall, or (3) it is somewhat decreased from atherosclerotic levels. WISSLER [64] has recently found a decrease in aortic collagen after a regression period of 18 months. In another chapter of this volume, STRONG *et al.* [51] discuss chemical changes occurring during regression of the early diet induced aortic fatty streaks in rhesus monkeys.

B. Cynomolgus

1. Morphologic Changes

In our study, regression in cynomolgus monkeys, defined as a decrease in intimal thickness and an increase in lumen size, was distinctly less after 20 months than was found in our rhesus material. Two factors may have contributed to this difference: the lesions tended to be more severe in cynomolgus and the plasma cholesterol did not return to control levels during regression until after the first year, in contrast to the more prompt drop in cholesterol values seen in rhesus [9, 48].

The histologic features of the regression lesion in cynomolgus differed somewhat from the rhesus lesion. In elastic arteries transmural replacement fibrosis was seen occasionally rather than rarely, and 25–50% fibrous replacement, measured from internal to external elastica, was common. Intimal collagen fibres were more densely compacted, and residual visible extracellular lipid was greater. Fibrohyaline changes in the intimal connective tissue were more frequent, occasionally assuming a chondroid appearance [5]. Rarely, bone formation occurred. Thus, the atherosclerotic arteries of cynomolgus showed more striking mesenchymal changes during regression. We believe that cynomolgus may be valuable in the investigation of repair processes in atherosclerosis.

2. Biochemical Changes in the Regression Lesion

Whether arteries showing marked mesenchymal reaction have a net accrual of connective tissue, particularly collagen, beyond that found in induced atherosclerosis is a significant issue. KRAMSCH *et al.* [28] found that elastin content decreases in regression. We have also noted that elastin is decreased from atherosclerotic levels in regression and that collagen content is apparently decreased in late regression [6]. There is an urgent need for information regarding fibrous protein synthesis and turnover in addition to mass data in the normal, atherosclerotic, and regression states. Knowledge of the *in vivo* kinetics of connective tissue formation and loss is important in the continuing effort to understand the interactions between fibroplasia and other factors in the experimental lesion. Knowledge of this sort will be of keystone significance in a comprehensive approach to the ultimate goal of experimental regression studies: favorable change of the fate and, conceivably, of the actual size of the clinically significant human lesion.

References

1 ADAMS, C. W. M.: Vascular histochemistry, pp. 36–37 (Year Book Med. Publishers Chicago 1967).

2 ARMSTRONG, M. L.; CONNOR, W. E., and WARNER, E. D.: Xanthomatosis in rhesus monkeys fed a hypercholesterolemic diet. Archs Path. *84:* 227–237 (1967).

3 ARMSTRONG, M. L. and MEGAN, M. B.: Plasma carcass cholesterol in rhesus monkeys after low and intermediate levels of dietary cholesterol. Circulation *44:* Suppl. II, p. 3 (1971).

4 ARMSTRONG, M. L. and MEGAN, M. B.: Lipid depletion in atheromatous coronary arteries in rhesus monkeys after regresssion diets. Circulation Res. *30:* 675–680 (1972).

5 ARMSTRONG, M. L. and MEGAN, M. B.: Responses of two macaque species to atherogenic diet and its withdrawal; in SCHETTLER and SCHLIERF Proc. 3rd Int. Symp. Atheroscler., Berlin (Springer, Berlin, in press).

6 ARMSTRONG, M. L. and MEGAN, M. B.: Arterial fibrous proteins in cynomolgus monkeys after atherogenic and regression regimens. Circulation *48:* Suppl. IV, p. 41 (1973).

7 ARMSTRONG, M. L. and MEGAN, M. B.: Intimal thickening in normocholesterolemic rhesus monkeys fed low supplements of dietary cholesterol. Circulation Res. (in press).

8 ARMSTRONG, M. L. and WARNER, E. D.: Morphology and distribution of diet-induced atherosclerosis in rhesus monkeys. Archs Path. *92:* 395–401 (1971).

9 ARMSTRONG, M. L.; WARNER, E. D., and CONNOR, W. E.: Regression of coronary atheromatosis in rhesus monkeys. Circulation Res. *27:* 59–67 (1970).

10 BANERJEE, S.; KUMAN, K. S., and BANDYOPADHJAY, A.: Effect of vegetable oils on plasma lipids of rhesus monkeys. Proc. Soc. exp. Biol. Med. *109:* 313–317 (1962).

11 BÖTTCHER, C. J. F.; BOELSMA-VAN HOUTE, E.; TER HAAR ROMENY-WACHTER, C. C.; WOODFORD, F. P., and VAN GENT, C. M.: Lipid and fatty acid composition of coronary and cerebral arteries at different stages of atherosclerosis. Lancet *ii:* 1162–1166 (1960).

12 CHAWLA, K. K.; MURTHY, C. D. S.; CHAKRAVARTI, R. N., and CHUTTANI, P. N.: Arteriosclerosis and thrombosis in wild rhesus monkeys. Am. Heart J. *73:* 85–91 (1967).

13 CLARKSON, T. B.: Personal commun. (1973).

14 CONNOR, W. E.: Dietary cholesterol and the pathogenesis of atherosclerosis. Geriatrics *16:* 407–415 (1961).

15 COX, G. E.; TAYLOR, C. B.; COX, L. G., and COUNTS, M. A.: Atherosclerosis in rhesus monkeys. I. Hypercholesterolemia induced by dietary fat and cholesterol. Archs Path. *66:* 32–52 (1958).

16 DE PALMA, R. G.; INSULL, W., jr.; BELLON, E. M.; ROTH, W. T., and ROBINSON,
    A. V.: Animal models for the study of progression and regression of atherosclerosis.
    Surgery *72:* 268–278 (1972).

17 EGGEN, D. A.; STRONG, J. P.; NEWMAN, W. P , III; CATSULIS, C.; MALCOM, G. T.,
    and KOKATNUR, M. G.: Regression of diet-induced fatty streaks in rhesus monkeys.
    Lab. Invest. (in press).

18 FITCH, C. D.; HARVILLE, W. E.; DINNING, J. S., and PORTER. S. F.: Iron deficiency in
    monkeys fed diets containing soybean protein. Proc. Soc. exp. Biol. Med. *116:* 128–133
    (1964).

19 FRY, D. L.: Acute vascular endothelial changes associated with increased velocity
    gradients. Circulation Res. *22:* 165–197 (1968).

20 FRY, D. L.: Certain histological and chemical responses of the vascular interface to
    acutely induced mechanical stress in the aorta of the dog. Circulation Res. *24:*
    93–108 (1969).

21 HAIMOVICI, H. and MAIER, N.: Role of arterial tissue susceptibility in experimental
    canine atherosclerosis. J. Atheroscler. Res. *6:* 62–74 (1966).

22 HAMILTON, C. L.; FENG, L., and KUO, P. T.: Response of spontaneous and hypo-
    thalamic-lesioned hyperlipidemic and obese monkeys to sucrose. Circulation *44:*
    Suppl. II, p. 16 (1971).

23 HAMM, T. E.; ABEE, C. R.; Riggs, T. A., and CLARKSON, T. B.: Myocardial infarction
    in three rhesus monkeys *(Macaca mulatta)* fed atherogenic diets. I. Fed. Proc. Fed.
    Am. Socs exp. Biol. *33:* 236 (1974).

24 HUEPER, W. C.: Experimental studies in cardiovascular pathology. XIV. Experimen-
    tal atheromatosis in macacus rhesus monkeys. Am. J. Path. *22:* 1287–1291 (1946).

25 KAWAMURA, R.: Neue Beiträge zur Morphologie und Physiologie der Cholesterin-
    steatose, p. 159 (Fischer, Jena 1927).

26 KRAMSCH, D. M. and HOLLANDER, W.: Occlusive atherosclerotic disease of the cor-
    onary arteries in monkey *(Macaca irus)* induced by diet. Expl molec. Path. *9:* 1–22
    (1968).

27 KRAMSCH. D. M. and HOLLANDER, W.: Personal commun. (1971).

28 KRAMSCH, D. M.; HOLLANDER, W., and RENAUD, S.: Induction of fibrous plaques
    versus foam cell lesions in *Macaca fascicularis* by varying the composition of dietary
    fats. Circulation *48:* suppl. IV, p. 41 (1973).

29 LANG, C. M. and BARTHEL, C. H.: Effects of simple and complex carbohydrates on
    serum lipids and atherosclerosis in nonhuman primates. Am. J. clin. Nutr. *25:*
    470–475 (1972).

30 LAPIN, B. A. and YAKOVLEVA, L. A.: Comparative pathology of monkeys, pp.
    147–162 (Ch. C. Thomas, Springfield 1963).

31 LINDSAY, S. and CHAIKOFF, I. L.: Naturally occurring atherosclerosis in nonhuman
    primates. J. Atheroscler. Res. *6:* 36–61 (1966).

32 MALMROS, H. and WIGNAND, G.: Experimental hypercholesterolemia and hypertri-
    glyceridemia in cynomolgus monkeys fed saturated fat and cholesterol. J. Athero-
    scler. Res. *5:* 474–482 (1965).

33 MANN, C. V. and ANDRUS, S. B.: Xanthomatosis and atherosclerosis produced by
    diet in adult rhesus monkey. J. Lab. clin. Med. *48:* 533–550 (1956).

34 MANNING, P. J. and CLARKSON, T. B.: Development, distribution and lipid content of diet-induced atherosclerotic lesions of rhesus monkeys. Expl molec. Path. *17:* 38–54 (1972).

35 MANNING, P. J.; CLARKSON, T. B., and LOFLAND, H. B.: Cholesterol absorption, turnover and excretion rates in hypercholesterolemic rhesus monkeys. Expl molec. Path. *14:* 75–89 (1971).

36 MEYER, B. J.; MEYER, A. C.; PEPLER, W. J., and THERON, J. J.: Chemical composition of the aorta, coronary arteries and cerebral arteries of Europeans and Bantu. Am. Heart J. *71:* 68–78 (1966).

37 MURAKAMI, M.; SEKIMOTO, H.; NAKADA, I.; NAKAMURA, K., and TAKAORI, M.: On pathogenic factors concerned with vascular lesions induced by high sucrose diet in monkey; in SCHETTLER and SCHLIERF Proc. 3rd Int. Symp. Atheroscler., Berlin (Springer, Berlin, in press).

38 McGILL, H. C., jr.; FRANK, M. H., and GEER, J. C.: Aortic lesions in hypertensive monkeys. Archs Path. *71:* 96–102 (1961).

39 NAPIER, J. R. and NAPIER, P. H.: A handbook of living primates, p. 207 (Academic Press, New York 1967).

40 PRATHAP, K. and LAU, K. S.: Spontaneous and experimental arterial lesions in the Malaysian long-tailed monkey; in GOLDSMITH and MOOR-JANKOWSKI Medical primatology 1972, part III, pp. 343–349 (Karger, Basel 1972).

41 RINEHART, J. F. and GREENBERG, L. D.: Arteriosclerotic lesions in pyridoxine-deficient monkeys. Am. J. Path. *25:* 481–485 (1949).

42 ROBERTS, W. C.; LEVY, R. I., and FREDERICKSON, D. S.: Hyperlipoproteinemia. Archs Path. *90:* 46–56 (1970).

43 RODBARD, S.: Physical factors in the progression of stenotic vascular lesions. Circulation *17:* 410–417 (1958).

44 RODBARD, S.: Dynamics of blood flow in stenotic vascular lesions; in BREST and MOYER Atherosclerotic vascular diseases, chap. 45, pp. 470–479 (Appleton Century Crofts, New York 1967).

45 SCOTT, R. F.; DAOUD, A. S.; WORTMAN, B.; MORRISON, E. S., and JARMOLYCH, J.: Proliferation and necrosis in coronary and cerebral arteries. J. Atheroscler. Res. *6:* 499–509 (1966).

46 SCOTT, R. F.; JONES, R.; DAOUD, A. S.; ZUMBO, O.; COULSTON, F., and THOMAS, W. A.: Experimental atherosclerosis in rhesus monkeys. II. Cellular elements of proliferative lesions and possible role of cytoplasmic degeneration in pathogenesis as studied by electron microscopy. Expl molec. Path. *7:* 34–57 (1967).

47 SCOTT, R. F.; MORRISON, E. S.; JARMOLYCH, J.; NAM, S. C.; KROMS, M., and COULSTON, F.: Experimental atherosclerosis in rhesus monkeys. I. Gross and light microscopy features and lipid values in serum and aorta. Expl molec. Path. *7:* 11–33 (1967).

48 STARY, H. C.: Progression and regression of experimental atherosclerosis in rhesus monkeys; in GOLDSMITH and MOOR-JANKOWSKI Medical primatology 1972, part III, pp. 356–367 (Karger, Basel 1972).

49 STARY, H. C.: Cell proliferation and ultrastructural changes in regressing atherosclerotic lesions after reduction of serum cholesterol, in SCHETTLER and WEIZEL Atherosclerosis III, pp. 187–190 (Springer, Berlin 1974).

50 STARY, H. C. and STRONG, J. P.: Coronary artery fine structure in rhesus monkeys. I Nonatherosclerotic intimal thickening, in STRONG Atherosclerosis in primates (this volume).

51 STRONG, J. P.; EGGEN, D. A., and STARY, H. C.: Reversibility of experimental fatty streaks in rhesus monkeys; in STRONG Atherosclerosis in primates (this volume).

52 TAYLOR, C. B.; COX, G. E.; MANALO-ESTRELLA, P., and SOUTHWORTH, J.: Atherosclerosis in rhesus monkeys. II. Arterial lesions associated with hypercholesterolemia induced by dietary fat and cholesterol. Archs Path. *74:* 16–34 (1962).

53 TAYLOR, C. B.; MANALO-ESTRELLA, P., and COX, G. E.: Atherosclerosis in rhesus monkeys. V. Marked diet-induced hypercholesterolemia with xanthomatosis and severe atherosclerosis. Archs Path. *76:* 239–249 (1963).

54 TAYLOR, C. B.; PATTON, D. E., and COX, G. E.: Atherosclerosis in rhesus monkeys. VI. Fatal myocardial infarction in a monkey fed fat and cholesterol. Archs Path. *76:* 404–412 (1963).

55 TAKAO, F. and KOTA, O.: Experimental atherosclerosis with ascorbic acid deficiency. Proc. 3rd Int. Symp. Atheroscler., Berlin 1973 (Springer, Berlin, in press).

56 TEXON, M.: Mechanical factors involved in atherosclerosis in; in BREST and MOYER Atherosclerotic vascular disease, chap. 3, pp. 23–42 (Appleton Century Crofts, New York 1967).

57 TEXON, M.; IMPARATO, A. M., and HELPERN, M.: The role of vascular dynamics in the development of atherosclerosis. J. am. med. Ass. *194:* 1226–1230 (1965).

58 TUCKER, C. F.; CATSULIS, C.; STRONG, J. P., and EGGEN, D. A.: Regression of early cholesterol-induced aortic lesions in the rhesus monkey. Am. J. Path. *65:* 493–514 (1971).

59 VESSELINOVITCH, D.; HUGHES, R.; FRAZIER, L., and WISSLER, R. W.: Studies of reversal of advanced atherosclerosis in rhesus monkeys. Am. Ass. Path. Bact. 17th Ann. Meet. Washington 1973.

60 VLODAVER, Z.; MEDALIE, J., and NEUFELD, H. N.: Coronary arteries in immature monkeys. Preliminary report of the relationships to activity and diet. J. Atheroscler. Res. *8:* 923–933 (1968).

61 WARNER. E. D.: Personal commun. (1971).

62 WATANABE, T.; TANAKA, K., and YANAI, N.: Essential familial hypercholesterolemic xanthomatosis. An autopsy case with special reference to the pathogenesis of its cardiovascular lipidosis. Acta path. jap. *18:* 319–331 (1968).

63 WISSLER, R. W.: Recent progress in studies of experimental primate atherosclerosis. Prog. biochem. Pharmacol., vol. 4, pp. 378–392 (Karger, Basel 1968).

64 WISSLER, R. W.: Personal commun. (1973).

65 WISSLER, R. W.; VESSELINOVITCH, D.; GETZ, G. S., and HUGHES, R. H.: Aortic lesions and blood lipids in rhesus monkeys fed three food fats. Fed. Proc. Fed. Am. Socs exp. Biol. *26:* 317 (1967).

66 Classification of atherosclerotic lesions. Tech. Rep. Ser. Wld Hlth Org. No 14 (1958).

67 YOUNGER, R. K.; SCOTT, W. H., jr.; BUTTS, W. H., and STEPHENSON, S. E., jr: Rapid production of experimental hypercholesterolemia and atherosclerosis in the rhesus monkey. Comparison of five dietary regimens. J. surg. Res. *9:* 263–271 (1969).

MARK L. ARMSTRONG, M.D., Division of Cardiovascular Diseases, Department of Internal Medicine, University of Iowa College of Medicine, *Iowa City, IA 52242* (USA)

Prim. Med., vol. 9, pp. 41–65 (Karger, Basel 1976)

Experimental Atherosclerosis in the Baboon[1]

Henry C. McGill, jr., Glen E. Mott and Claud A. Bramblett

Departments of Pathology and Biochemistry, The University of Texas Health Center at San Antonio, and Department of Anthropology, The University of Texas at Austin, and Southwest Foundation for Research and Education, San Antonio, Tex.

Contents

[1] Supported in part by USPHS, NIH grant HL-15914.

I. Introduction

Limitations in human experimentation have made it necessary to utilize nonhuman primates in research, particularly in studying diseases such as atherosclerosis that result from interactions of multiple environmental agents and polygenic traits. Economic, ecologic, and other practical considerations together with similarity to man in anatomy, physiology, biochemistry, behavior, and social organization are constraints which limit the choice of a suitable animal. Standing phylogenetically just below the great apes, the baboon ranks high among the available and acceptable species.

It is the most numerous and widespread African nonhuman primate, ranging in habitat from Ethiopia in the north to the Cape in the south, and across the breadth of Africa south of the Sahara. In some areas baboons are so numerous they are considered vermin and many are killed annually to control their depredations of crops. The relationship of man and the baboon was perhaps a happier one in ancient Egypt, where the hamadryas or 'sacred' baboon was considered a representative of the god Thoth. Drawings and statutes of baboons are frequent in material from Egyptian excavations.

Baboons are captured with relative ease in baited traps from the wild. They are hardy, breed readily in captivity, and are relatively resistant to tuberculosis. These characteristics make them excellent experimental subjects, particularly when a medium-sized to large nonhuman primate is desirable. They have been used more extensively by biomedical scientists in South Africa and Russia than in the US or Europe.

Interest in baboons as experimental animals for the investigation of atherosclerosis began in the late 1950's as a result of several independent observations that baboons spontaneously develop arterial lesions resembling those of human atherosclerosis. Subsequently, research programs at several institutions tested the response of the baboon to atherogenic diets. The Southwest Foundation for Research and Education in San Antonio, Texas, began an extensive program to develop background information on the care, management, and breeding of baboons, and on their anatomic, physiological, and biochemical characteristics. These data have confirmed the humanoid nature of the baboon in many respects, and have provided a basis for many types of investigation. Much of the information is contained in the published proceedings of two symposia on the baboon [87, 88], and in other publications of visiting and staff scientists of the Southwest Foundation. These include reports on reproductive physiology [30], embryology [29,

30], dentition [85], management [40], feeding [41], blood chemical and hematologic normal values [62–64], and microflora [1, 44, 50, 66].

The following sections summarize existing knowledge concerning the baboon in areas that are most relevant to the investigation of experimental atherosclerosis with this species.[2]

II. General Characteristics of the Baboon

A. Reproduction, Growth, and Development

The adult female baboon ovulates the year around with an average menstrual cycle of 33 days [30]. The state of ovarian activity can be identified readily by observation of the sex skin, which becomes turgescent for 8 to 10

Fig. 1. Juvenile baboons about 1.5 years old reared at the Southwest Foundation from East African stocks. There is no difference in size between the sexes at this age.

[1] For most up-to-date review on the baboon see S. S. Kalter: The baboon. Microbiology, clinical chemistry and some hematological aspects. Prim. Med. vol. 8 (Karger, Basel 1973).

Fig. 2. Adult male and female East African baboons. Males average in weight about twice the weight of females. Note the strong, well-muscled shoulders of the male, the quadrupedal stance, and the stiff tail.

days prior to ovulation and deturgesces approximately 3 days after ovulation. The female is receptive to sexual advances by the male during the period preceding ovulation. The gestation period averages 175 days. Newborn baboons of both sexes weigh, on the average, about 750 g. Infants of both sexes grow at the same rate until 2.5 years of age, at which time they weigh about 6.5 kg. Thereafter, females grow slowly to attain a maximum size of about 12.5 kg, and males continue to grow to attain an average adult weight of about 22 kg [74]. Puberty occurs at about the age of 3 to 3.5 years in both males and females.

Newborn baboons can be raised readily on commercial milk formulas prepared for human infant feeding [89]. Under these conditions, they grow somewhat more rapidly in weight during the suckling period than do breast fed infants.

The life-span of the baboon in the wild is probably about 15 years. In a skeletal study of 60 wild baboons from the Darajani area of Kenya, Bram-

BLETT [6] found the oldest male to be 17 years and the oldest female to be 15 years (age estimates based on dental attrition rates, cranial suture closure, and postcranial epiphyseal union). The life span in captivity is about the same, with survival to 20 years occasionally reported.

Juvenile and adult baboons in outdoor cages at the Southwest Foundation for Research and Education are shown in figures 1 and 2.

B. Evolutionary History

Fossil histories of baboons are poorly documented at present. Presumably, the radiation which resulted in recent genera occurred in the African Pliocene [43]. Large-bodied terrestrial cercopithecoids appear in early Pleistocene in Africa, Europe, and Asia. Specimens which morphologically resemble modern macaques are described from the Pliocene of Algeria *(Macaca)*, Pleistocene of China *(Procynocephalus)*, and Pleistocene of India *(Cynocephalus)*. Pleistocene African deposits contain *Papio*-like forms *(Parapapio* and *Gorgopithecus)*. The modern genus *Theropithecus* is represented by the Pleistocene from *Simopithecus*. PATTERSON [61] found *Parapapio* and *Simopithecus* in deposits at least four-million years old in Kenya. The time separation between these baboon-like groups must be at least five-million years. Some species of *Simopithecus* appear more terrestrial in anatomy than recent forms. One well-preserved female specimen from Olduvai IV has the body size of a gorilla [72].

The evolutionary history of baboons prior to the late Pliocene is uncertain. Our knowledge of Oligocene Old World anthropoids comes almost exclusively from Egypt. Although there is a possibility that separation between the subfamily *Colobinae* and subfamily *Cercopithecinae* may have occurred 19 to 22-million years ago [65], fossil representation is minimal. At this time, ape fossils are about 20 times more common than monkey fossils in the African Miocene. *Prohylobates* and *Victoriapithecus* appear to be Miocene *Cercopithecinae* with unclear affinities to the later forms [72].

Any common ancestor between baboons and man would seem to be (at the very latest) near the Oligocene/Miocene boundary in time (26 million years). If dental cusp morphology is a reliable indicator of phylogenetic history, dryopithecine and bilophodont molars are clearly distinguishable in later Miocene deposits. Estimates from quantitative immunochemistry [71] propose a 22 ± 2-million year divergence between *Cercopithecoidea* and *Hominoidea*. Fossil materials suggest that this date, if in error, is too recent.

C. Taxonomy

Subsequent to the development and widespread usage of species names for many geographic forms of baboons, field evidence clearly indicates that all savanna baboons represent a single polytypic species with interbreeding subspecies. Confusion in the literature produces a situation where common names are sometimes more useful than scientifically correct species names since the former have been used consistently to designate geographic distributions. Bᴜᴇᴛᴛɴᴇʀ-Jᴀɴᴜsᴄʜ [9] has argued that these evolutionary relationships can be represented best by combining all baboons into a single genus, *Papio,* and retaining both common names and an older version of species nomenclature in interests of clarity.

There are three kinds of baboons. The common baboon (genus *Papio*) is found throughout most of Africa south of the Sahara. Mandrills (genus *Mandrillus*) are restricted to the thick and secondary forests of West Africa. Gelada baboons (genus *Theropithecus*) are arid country forms which inhabit highland regions of Ethiopia.

Common baboons occupy such extensive areas in both forest and savanna that they are represented by at least five different forms. The Guinea baboon *(Papio papio)* is found in savanna woodlands in a very small part of extreme western Africa. Olive baboons *(Papio anubis)* are found across the southern fringe of the Sahara, from coast to coast in central Africa. Their habitat ranges from arid regions of Ethiopia to the great forests of central and western Africa. Hamadryas baboons *(Papio hamadryas)* occur only in savanna and arid parts of Ethiopia. Yellow baboons *(Papio cynocephalus)* are found in savanna and riverine forests in a region between the olive baboons in the central part of the continent and the chacma baboons *(Papio ursinus)* of South Africa. The correct species name for these five forms is probably *Papio hamadryas* Linnaeus, 1758 [86], but its use would result in undesirable bibliographic confusion.

D. Social Organization and Activities in the Natural Habitat

Baboons have been well-studied in their natural habitat since Wᴀsʜʙᴜʀɴ and Dᴇ Vᴏʀᴇ [90] completed their widely publicized observations in the Nairobi Game Park. Most important have been the works of Hᴀʟʟ [25] and of Sᴛᴏʟᴛᴢ and Sᴀᴀʏᴍᴀɴ [79] on chacma baboons; Rᴏᴡᴇʟʟ [69] and Rᴀɴsᴏᴍ [67] on olive baboons; Aʟᴛᴍᴀɴɴ and Aʟᴛᴍᴀɴɴ [2] on yellow baboons in

Kenya; CROOK [12] on gelada baboons; KUMMER [49] on hamadryas; and STRUHSAKER [84], GARTLAN [17], and SABATER Pí [70] on mandrills.

Except for patas monkeys, baboons have made the most extensive adaptations toward a terrestrial mode of life among the Old World monkeys. All baboons still depend upon trees or rock cliffs for safe refuge at night. Much of the day is spent on the ground, often moving far from trees which might provide safety. Security is found in the proximity of one or more powerful males. Several such males acting as a team are substantial deterrents to predators. Heavy body weight, quadrupedal stance, shortened digits, and an inflexible tail handicap a baboon's climbing ability, but are consistent with movements on the ground.

Among common baboons, the 'alpha role' [4], is shared by several males ('central hierarchy') who act as a political alliance to police the group. Characteristic features of baboon social life are the formidable organized structure of the troop and exaggerated behavioral dimorphism. Males of a troop have a stable priority to social space and to resources encountered in a social context. Male behavior principally involves play and aggression. Females also are arranged in a dominance hierarchy, but behaviors which involve females are affected by their changes from estrus to pregnancy and motherhood. Fights among females are more frequent than among males.

Baboons use almost any staple food item a local area may offer. Their ability to dig, chase, and climb enables them to exploit a wide variety of foods and a wide variety of habitats. Their diet is basically vegetable, supplemented by substantial predation on insects, birds, reptiles, and small mammals.

There appear to be few differences in these patterns among the many forms of *Papio* except for *Papio hamadryas*. Hamadryas baboon troops are subdivided into several social units (unfortunately called harems) which frequently contain only one fully adult male. This male herds females in his one-male group with neck bites, and prevents any female from straying more than a few feet from him. Several such adult males may cooperate with each other against outsiders, but the one-male group's rigid social spacing is maintained and respected within a troop. A narrow zone of hybridization occurs between olive baboons and hamadryas baboons in Ethiopia in spite of the differences in social behaviors.

Gelada baboons are physically different, with contrasting face, posture, and skin. Gelada troops can be quite large and have a loose organization that has been described as a 'herd'. Within the herd, the bisexual units are hamadryas-like one-male groups. The gestures used are different, but the pattern of 'herding' the females is just as extreme.

Generally among Old World monkeys, status hierarchies are stable, and female status is influenced by kinship and social traditions in the group. Groups may have extensive learned traditions which affect movements and behaviors. Reproduction is a seasonal event with elaborate roles being performed by both sexes in the supervision and socialization of infants. Baboons differ in that the characteristic cercopithecoid pattern of male defense of the groups by vigilance, display, and attack is greatly exaggerated and shared by several animals in each group.

III. Special Characteristics of the Baboon Relevant to Experimental Atherosclerosis

A. Spontaneous Arterial Lesions

The finding that baboons develop atherosclerosis-like arterial fatty streaks without dietary manipulation first directed attention to the baboon as an experimental animal for atherosclerosis. Since young humans develop aortic fatty streaks without high fat, high cholesterol diets, and without hyperlipidemia [55], this characteristic appeared to provide an important parallel to the human. On the other hand, detection of treatment effects in the presence of spontaneous lesions requires rigorous controls in all experiments.

Fox [16] first described an atheromatous plaque at the orifice of a coronary artery in a single adult baboon. Lindsay and Chaikoff [51] described fatty streaks and fibromuscular plaques in the aortas and common iliac arteries of two male baboons, each about 20 years old, that had lived continuously in the San Diego zoo on diets low in fat and probably free of cholesterol. The coronary arteries contained fibromuscular intimal plaques with little lipid.

Gillman and Gilbert [20] described extensive aortic lesions in 85 adult baboons *(Papio ursinus)*, many of which had been subjected to experiments involving removal of thyroid, adrenal, pituitary, and pancreatic glands, and removal of ovaries and testes. These baboons were from 2.75 to 15 years of age and had been in captivity for varying periods on a diet low in cholesterol. Thirty-eight of 59 females and 14 of 26 males had some degree of aortic fatty streaks. The severity of lesions was not associated with endocrine gland ablation, but severity was greater in older animals. Fibromuscular intimal plaques also occurred in a smaller percentage of the animals, and some of these were associated with lipid deposits. A few of the animals that received cortisone showed elevations of serum cholesterol concentration (up to 405 mg/dl) for a short time, but there was no consistent association of fatty streaks with serum cholesterol concentration. Most of the fatty streaks occurred in animals that had no elevation of serum cholesterol.

GILBERT and GILLMAN [19] reported a survey of histologic changes in the anterior descending branch of the left coronary artery of 133 baboons ranging in age from newborn to 18 years. Many of these were the same animals whose aortic lesions had been described by GILLMAN and GILBERT [20]. After one year of age, approximately three quarters of the animals showed a fibrous and cellular thickening of the coronary artery intima that occasionally contained small amounts of lipid. There was no relationship to age in animals older than one year. None of these lesions appreciably reduced the lumen and none resembled the typical human fibrous plaque with a core of extracellular lipid and a fibromuscular cap. The plaques resembled the fibromuscular intimal thickening that is almost universal in young humans, and appeared identical to the fibromuscular plaques described in normal rhesus monkeys by STARY and STRONG [77].

McGILL et al. [53] found lesions in the aorta of a 16-year-old female zoo baboon similar to those described by LINDSAY and CHAIKOFF [51] and by GILLMAN and GILBERT [20], and reported the results of a systematic survey of 163 baboons *(Papio anubis)* examined immediately after being trapped from their natural habitat in Kenya, East Africa. Infants showed no fatty streaks. Among 40 prepubertal and pubertal juvenile males, five had appreciable fatty streaks; and among 39 juvenile females, 12 had even more extensive fatty streaks. One young adult female baboon had approximately 80% of the thoracic aorta and 50% of the abdominal aorta involved by fatty streaks. The greater extent of fatty streaks in juvenile females than in juvenile males is similar to the situation in young humans with regard to aortic fatty streaks [55]. Among adults, 26 out of 41 males and 24 out of 26 females had aortic fatty streaks.

Both grossly and histologically, two distinct types of fatty deposits could be distinguished. A frequent type was a small area of sudan-stained lipid in fibromuscular intimal cushions at the orifices of aortic branches. The other type involved more extensive lipid deposition away from orifices, and microscopically showed both extracellular and intracellular lipid lying just beneath the endothelium and between the inner two or three elastic laminae.

The Kenya baboons occasionally showed fibromuscular intimal plaques in the aorta, often located at the orifices of branches. These plaques were not similar to typical human fibrous plaques, but appeared more likely to represent focal intimal hyperplasia of smooth muscle with occasional minimal lipid deposits.

Coronary arteries from the same animals showed no appreciable lipid deposits but did contain numerous fibromuscular intimal cushions similar to those described by GILBERT and GILLMAN [19].

HOWARD *et al.* [37] found fatty streaks in the aortas of 14 of 23 adult baboons from Kenya, but found no lesions of any kind in the aortas of 50 young baboons that had been in captivity for one to two years. Their findings were similar to those of previous studies.

Some years later, VAN DER WATT *et al.* [91, 92] surveyed 77 newly captured baboons in South Africa. This species of baboon *(Papio ursinus)* is identical to that studied earlier in captivity by GILLMAN and GILBERT [20]. Sudan IV staining disclosed that all animals had aortic fatty streaks covering from 2 to 63% of the aortic intimal surface. The most frequently occuring extent among 44 males was 6–10%, and the most frequent extent among 31 females was 11–20%. As with other groups, the frequency and extent of fatty streaks was greater in females than in males. Aortic fibromuscular intimal plaques were more numerous in the South African baboon than in the Kenya baboon. Microscopically, both fatty streaks and fibromuscular plaques were identical to those in the Kenya baboons. Coronary arteries of the South African baboons also had fibromuscular intimal cushions similar to those in the Kenya baboons.

B. Serum Cholesterol Concentrations in Baboons

The serum cholesterol concentrations of baboons trapped from the wild show considerable differences depending on the stage of development. Newborn baboons in South Africa *(Papio ursinus)* had an average serum cholesterol concentration of 75 mg/dl; those that were nursing had a mean serum cholesterol of 224 mg/dl; and adult animals had a mean serum cholesterol of 85 mg/dl [92]. This adult value was very similar to that found in Kenya baboons, 78 mg/dl [53].

Baboons maintained on conventional diets (i.e. low in fat and cholesterol) in captivity show higher levels of serum cholesterol that do those in the wild. Among young adult males, STRONG *et al.* [80] and McMAHAN and STRONG

Table I. Serum cholesterol concentrations of baboons by age, sex, and state of pregnancy
and lactation at the Southwest Foundation for Research and Education
San Antonio, Texas, 1972–74

	Number of animals	Number of observations	Mean, mg/dl	Range, mg/dl	Standard deviation[1] between animals	within animals
Males, adult	17	17	109	58–153	26.2	
Females						
Non-pregnant						
3–6 years	18	18	119	84–140	14.5	
8–11 years	26	26	119	73–201	27.4	
12–15 years	36	36	133	68–212	30.5	
Pregnant, all ages	16	16	63	33–151	29.5	
Lactating, all ages	12	111	118	59–194	15.9	15.6
Infants, newborn	23	23	81	55–161	27.1	
Infants, 1–4 weeks						
Breast fed	12	94	125	64–273	21.8	31.8
Formula fed	6	47	105	63–152	14.0	18.2

[1] Standard deviations computed by analysis of variance.

[57] found that animal means (mean of 4 consecutive days) ranged from a
low of 108 to a high of 169 mg/dl with a standard deviation of 21 and an
overall mean of 130 mg/dl. The variance among animals was greater than the
day-to-day variance in the same animal. These values were based on the
analytical technique of SPERRY [75]. In a larger sample of 40 young adult
males under similar conditions, the overall mean was 114 mg/dl with a range
from 88 to 166 mg/dl [81]. This value compares well with the control values
of 116 mg/dl found by BLATON *et al.* [5]; 118 mg/dl by HOWARD *et al.* [39];
116 mg/dl for males and 126 mg/dl for females by KRITCHEVSKY *et al.* [47],
and 142 mg/dl by VAN ZYL [97]. In all these studies, there was a similar
degree of variability among animals.

Table I summarizes serum cholesterol values obtained from several pilot
studies on baboons at various ages and stages of pregnancy at the Southwest
Foundation for Research and Education in San Antonio. The low value for
newborn infants is similar to that found in infants trapped from the wild by
VAN DER WATT [92]. VAN ZYL [98] found the mean concentration of serum
cholesterol in suckling infants to be 200 mg/dl, in contrast to the 125 mg/dl
that we found. As with humans, there is a tendency for the average serum

cholesterol levels to rise with age, but not markedly. A remarkable feature of serum cholesterol in the baboon, also noted by van Zyl [98], is the drop early in pregnancy to levels about half those in the preconception period. Soon after delivery, they rise to normal levels and remain so during lactation.

C. Hemostatic Mechanism

Hampton and Matthews [26, 27] investigated the coagulation mechanism of 16 healthy adult baboons of both sexes. They studied clot initiation, prothrombin activation, production of thromboplastin, and fibrin formation and stability, and concluded that the clotting mechanism of baboons was similar to that of humans with only two exceptions: the mean silicone clotting time of baboons was shorter, and baboon fibrinolysin could not be activated by streptokinase. Similarities to the human procoagulant concentrations were confirmed by Wessler and Yin [94].

McKee *et al.* [56] found human, chimpanzee, and baboon plasminogens, in that order, decreasingly susceptible to activation by streptokinase. All three plasminogens were activated readily by human urokinase. They suggested that the reactor site on plasminogen for streptokinase had undergone evolutionary changes, while that for urokinase had not.

There are few studies of the characteristics of baboon platelets. Hawkey and Symons [28] compared ADP aggregation tests on several primate species, including baboons and humans, and found a wide variation in response. At a final concentration of 1.25 µg of ADP, baboon platelets first aggregated as did those of humans, but rapidly disaggregated whereas human platelets remained aggregated. They presented evidence to indicate that baboon plasma contained a higher level of ADP-splitting enzymes.

Despite the similarity of the procoagulant system of baboons to humans, no experiments to test the relationship of thrombosis to atherosclerosis or its complications in baboons have been reported. The hemostatic mechanism of experimental endotoxin shock in baboons has been studied by Cavanagh *et al.* [10].

D. Heart Weight and Myocardial Lesions

Herrmann *et al.* [32] analyzed heart weights of 162 Kenya baboons and found that the heart averaged 0.479 % of body weight with a standard deviation of ± 0.076. In another smaller series of 12 baboons, the heart averaged

0.54% of body weight, and left ventricle to right ventricle weight ratios averaged 1.62 [33]. In a third series of 72 baboons, HERRMANN *et al.* [31] found the heart to be 0.40% of body weight in mature males and 0.48% of body weight in mature females. BRINK *et al.* [8] found the heart to be 0.55% of the body weight in 28 adults. These variations may be due to differences in technique, or to true differences among animals of different backgrounds. On the average, it appears that the heart weight is about 0.50% of body weight.

Myocardial scarring has been observed in a number of baboons without demonstrable occlusion of a coronary artery and without evidence of any other obvious etiologic agent [24, 53]. It seems unlikely that these represent true infarcts, and almost certainly do not represent infarcts with a pathogenesis like that of human coronary heart disease.

E. Coronary Artery Anatomy

BRINK *et al.* [8] surveyed the anatomy of the coronary arteries in 28 adult baboons *(Papio ursinus)* by the injection of radioopaque material and radiography. Right preponderance was found in 48%; left preponderance in 18%; and a balanced circulation in 34%. In all hearts, the septal blood supply came from multiple branches of the anterior descending branch of the left coronary artery, but 67.5% of the hearts showed an additional larger single septal artery. Seventy-eight percent of the left coronary arteries had a third primary division, which would provide greater collateral blood supply in the distribution of the left coronary artery and would modify the distribution of ischemic necrosis if a major branch were occluded.

F. Electrocardiogram

KAMINER [45] performed 12 lead electrocardiograms on 15 lightly anesthetized baboons and compared these with human electrocardiograms. The amplitude of the P wave, the P-R interval, and the amplitude of the QRS complex were similar to the human. There were differences in heart rate, form of the P wave, Q-S duration, and several other features. Discordancy of the T wave was frequent, but the cause of discordancy was not determined.

HERRMANN and WILLIAMS [34] studied electrocardiograms on 170 baboons, some under sernylan and some under pentobarbital anesthesia. Most of these were similar to the normal human electrocardiogram, but about 10%

showed ST and T waves that would be considered abnormal. They were unable to explain these apparent abnormalities. Herrmann *et al.* [35] later took electrocardiograms on 72 more baboons that had been in captivity for a year or more, and found only two abnormal tracings. They attributed the decrease in abnormal electrocardiograms to the longer period of stabilization in captivity.

IV. Experimental Atherosclerosis and Related Findings in the Baboon

A. Experimentally Induced Atherosclerotic Lesions

Most experimental investigations of atherosclerosis using baboons have concentrated on the effects of dietary cholesterol and type of fat on serum lipids and on arterial lesions. In general, the baboon is only moderately sensitive to dietary cholesterol, and does not develop the very high serum cholesterol levels seen in rhesus monkeys.

Strong and McGill [82] observed 40 young adult male baboons for two years on semisynthetic diets designed to provide each of three dietary variables at two levels. The variables were cholesterol (0.5 and 0.01 %), type of fat (highly saturated and highly unsaturated), and protein (25 and 10 %). At the termination of the experiment, five baboons had serum cholesterol concentrations greater than 180 mg/dl, and all five were on high cholesterol diets. The highest serum cholesterol concentration was 262 mg/dl. Dietary cholesterol significantly increased serum cholesterol. Type of fat and level of protein had no main effect on serum cholesterol, but there was an interaction between dietary cholesterol, protein, and type of fat in the effect on serum cholesterol.

Arterial lesions in the baboons autopsied at the end of two years were almost exclusively aortic fatty streaks. Six animals showed more than 10% of the aortic intimal surface covered by fatty streaks, and five of these six had received the high cholesterol diet. Just as with the serum cholesterol elevation, dietary cholesterol was the only dietary variable associated with the extent of aortic fatty streaks.

The character of the aortic lesions induced in these baboons after two years of dietary cholesterol and mild hypercholesterolemia was similar to that of human juvenile fatty streaks and of fatty streaks induced in other species [18]. Fine sudanophilic particles lay beneath the endothelium and inside the internal elastic lamina, and occasionally extended into the media. By electron microscopy, there was abundant electron-dense material in the interstitial spaces, presumably representing extracellular lipid. Among lipid containing cells, smooth muscle cells were the most numerous. These appeared in all varieties, from typical smooth muscle cells with minimal lipid inclusions to scarcely recognizable smooth muscle cells distended with lipid particles. Lesions also contained monocytes or lymphocytes with no lipid particles in their cytoplasm, and foam cells distorted with large numbers of lipid particles so that they were not identifiable as either of smooth muscle or monocytic origin. Occasional lipid-filled cells also appeared to have swollen mitochondria, disrupted plasma membranes, and electron-dense inclusions suggesting degeneration. The lesions also contained elastic and collagen fibres, but not in greatly increased amount. None of these features would distinguish the experimentally induced fatty streaks in baboons from juvenile human fatty streaks, nor from experimentally induced fatty streaks in other species.

HOWARD *et al.* [38] achieved substantially similar results with young baboons fed an egg yolk and butter-enriched diet with a cholesterol content of 0.55%, and a butter- and cholesterol-enriched diet with a cholesterol content of 2%. These two groups reached mean serum cholesterol levels of 204 and 302 mg/dl and developed aortic lesions similar to those described by GEER *et al.* [18].

KRITCHEVSKY *et al.* [47] studied the effects of semisynthetic diets free of cholesterol but high in saturated fat and with different carbohydrate sources. After 12 months, serum cholesterol levels rose by about 35% in all groups, and reached levels of 145–167 mg/dl. Aortas showed up to 35% intimal sudanophilic fatty streaks, with an average of approximately 8%. There was a high degree of variability in aortic lesions and little correlation between aortic lesions and serum cholesterol concentration as noted by STRONG and McGILL [82]. There were no significant differences in either serum lipids or lesions among the four different carbohydrate sources.

In all of the above-described experiments, coronary artery lesions were rare or nonexistent, as were lesions in the arteries of the neck, cranium, and legs. Furthermore, except for fibromuscular plaques similar to those seen in control animals, the lesions were all of the fatty streak type, and no lesions resembling human fibrous plaques appeared. STRONG [83] maintained one adult male baboon on a high cholesterol diet for four years, and this animal developed both aortic and coronary artery fatty streaks and fibrous plaques remarkably like uncomplicated human fibrous plaques. This isolated example suggests that truly chronic experiments, in which serum cholesterol levels in the human range (200–300 mg/dl) are induced over periods of several years, would lead to more realistic facsimiles of human lesions. both in distribution and in characteristics of lesions.

Howard *et al.* [39] combined cholesterol feeding for six months with injections of bovine serum albumin (BSA) to determine the interaction of hypercholesterolemia with immunological injury. The serum cholesterol of cholesterol-fed baboons rose to an average of approximately 250 mg/dl, or about 2.5 times the control level. BSA injections were not associated with elevations in serum cholesterol concentration or with changes in other serum lipids. However, BSA-injected animals had an average of nearly 50% of their aortic intimal surface covered by fatty streaks, whereas the non-BSA-injected animals had no fatty streaks at all. BSA alone produced no lesions. The authors do not comment on the character of the fatty streaks, but it would be of interest to know if they had any of the characteristics of fibrous plaques. This combination of experimental manipulations demonstrates a remarkable acceleration in the production of experimental atherosclerosis in baboons, even more than was demonstrated by BSA injection in rabbits [60]·

B. Lipoproteins

The response of lipoproteins of the baboon on a high cholesterol, high fat diet is also similar to that of man [5, 36]. Baboons fed an atherogenic diet high in cholesterol for two months had elevated free and esterified serum cholesterol, and increased cholesterol-phospholipid ratios chiefly in the β-lipoproteins, and also in the α-lipoproteins. Pre-β-lipoproteins or triglycerides were not altered. A similar lipid composition has been observed in serum lipoproteins from humans with type II hyperlipoproteinemia. Baboon serum high-density lipoprotein (HDL) is structurally very similar to human HDL, with considerable cross reactivity with antibodies to human HDL polypeptides [23]. Determination of the structure and characteristics of baboon lipoproteins, which currently is under way in several laboratories, will be valuable in the study of experimental atherosclerosis in this species.

C. Cholesterol Synthesis, Absorption, and Turnover

The liver is the principal source of endogenous cholesterol in baboons on low cholesterol diets. However, hepatic cholesterogenesis is markedly suppressed by dietary cholesterol, but not until cholesterol absorption is approximately equal to the synthesis of cholesterol by liver on a low cholesterol intake [95]. Relatively low absorption of dietary cholesterol is a characteristic of baboons as it is of humans. Baboons fed diets containing 20% fat and 1 mg cholesterol/kcal absorb only 25–50% of the cholesterol ingested [13]. The half-life of cholesterol in the miscible pool is approximately 50 days with only slight changes observed due to dietary manipulations. Details regarding techniques and results of sterol balance studies in baboons are discussed in this volume by EGGEN [14].

V. Unexploited Research Potential with the Baboon

A. Behavior

The association of coronary heart disease with behavior patterns and environmental, social, and psychologic stresses has been debated widely [42], and is based largely on epidemiological studies in humans. However, there has been practically no experimental investigation of the relationship of behavior to atherosclerosis, despite the rich potential of such investigations in nonhuman primates. The baboon offers a particularly valuable research resource in this regard, even when the elaborate social organization is modified by captivity.

Studies which focus on social behaviors of baboons in captivity are rare. ZUCKERMAN [96], KUMMER [48], SPIVAK [76], and FEDIGAN [15] studied the behavior of zoo baboons. Several species have been studied at the Yerkes Regional Primate Center [3, 7]. ROWELL [68, 69] compared caged baboons to free-ranging animals in Uganda and observed that the same social patterns were seen in both places. Social interactions were four times more frequent in the caged group, and caged animals spent much less time foraging and moving. When one considers the limited number of desirable sitting positions in the cage, restricted area, limited novelty, and restriction in foraging, behaviors were remarkably consistent with those observed in the wild.

Behavioral studies involve recording behaviors which may occur many thousand times, but behaviors that are also of biological significance are relatively rare. Despite this limitation, the investigator can document reliably many behavioral characteristics which reflect the 'personality' of the individual baboon or nonhuman primate [11]. The observer has access to many behaviors and communicative acts that structure or express interactions between animals in the group. Events, such as exercise, that have metabolic consequences can be quantified readily. The animal's pattern of social communication is a potential source of data about the group's social organization and status hierarchies. Sampling and quantitation of such behaviors in a systematic way will specify the behavioral characteristics of individuals in a particular group and context. Experiments in which behavioral analysis is used to test for correlations between behavior and experimental atherosclerosis may help to test the hypotheses generated from human epidemiological studies regarding the effects of behavior on atherosclerosis and its sequelae.

B. Cigarette Smoking

Cigarette smoking is one of the best documented and strongest risk factors for coronary heart disease, and is associated with increased aortic and coronary atherosclerosis. However, there is no reported experimental investigation of the effect of cigarette smoking on atherosclerosis, and there are only a few reports of experiments with isolated cigarette smoke components, principally carbon monoxide and nicotine. Most of these studies have been made in rabbits and one study of carbon monoxide effects [93] involved squirrel monkeys. Since cigarette smoke is a chemically complex substance and since the process of cigarette smoking is more complex than inhalation of a gas or injection of a drug, there is no assurance that these experimental regimens simulate human cigarette smoking.

Preliminary work indicates that baboons can be taught to smoke cigarettes with smoking patterns similar to those of humans [54, 78]. This has been most efficiently accomplished with a specially constructed training device that rewards a water-deprived animal with water when he puffs on a lighted cigarette with sufficient duration and negative pressure to simulate human smoking. The size of the baboon is an asset for such studies because its large size makes it easier to approximate human puff duration and volume. This technique promises to be useful in studying the mechanism of the cigarette smoking effect on atherosclerosis and possibly on other cardiovascular and pulmonary diseases.

C. Heritability of Response to Dietary Cholesterol

A feature of all experiments with baboons has been the high degree of variability among animals on the same diet in both serum cholesterol elevation and arterial lesions. LOFLAND *et al.* [52] demonstrated the heritability of response to dietary cholesterol in squirrel monkeys, and renewed interest has developed in the genetic aspects of human atherosclerosis [22] with the identification of several genetically determined hyperlipidemias. Therefore, variability in responses among experimental animals has provided opportunity to investigate the genetic basis of hyperlipidemia. Analysis of limited preliminary data on sires, dams, and progeny in the baboon colony at the Southwest Foundation has indicated that in the baboon, as in other animals, both basal serum cholesterol concentration and response to dietary cholesterol have a high degree of heritability, approaching 0.50 [46]. The mechanism by which this trait is expressed – i.e., whether by regulating absorption, synthesis, or excretion of cholesterol – is not known in the baboon. Variability in the extent of lesions among animals with identical serum cholesterol levels also has been observed frequently, and it appears likely that this phenomenon is also under genetic control.

D. Cholelithiasis

The baboon is one of the few species that frequently produces cholesterol gallstones [21]. Bile secreted by the baboon is similar in composition to that of man. The primary bile acids synthesized by the liver, chenodeoxycholic and cholic acids, are secreted in the bile with phospholipids and high concentrations of cholesterol [58]. The principal bacterial metabolite of bile acids in both species also is deoxycholic acid. As shown by SMALL [73], the ratios of these three bile acids in human bile appear to be important in solubilization of cholesterol. Chemical analysis of normal baboon hepatic bile, MC SHERRY *et al.* [59], demonstrated that 50 % of the animals examined were secreting bile supersaturated with cholesterol. These similarities indicate that the baboon is an ideal model in which to study the pathogenesis of cholelithiasis, and is likely to be a source of advances in the understanding of cholesterol and bile acid metabolism in man.

References

1 AL-DOORY, Y.: Microbiological parameters of the baboon (*Papio* sp.): Mycology; in
  VAGTBORG The baboon in medical research, vol. 2, pp. 731–739 (Univ. of Texas Press,
  Austin 1967).

2 Altmann, S. A. and Altmann, J.: Baboon ecology (Karger, Basel 1970).

3 Bernstein, I. S.: Primate status hierarchies; in Rosenblum Primate behavior, vol. 1, pp. 71–109 (Academic Press, New York 1970).

4 Bernstein, I. S. and Sharpe, L. G.: Social roles in a rhesus monkey group. Behaviour *26:* 91–104 (1966).

5 Blaton, V.; Howard, A. N.; Gresham, G. A.; Vandamme, D., and Peeters, H.: Lipid changes in the plasma lipoproteins of baboons given an atherogenic diet. I. Changes in the lipids of total plasma and of α- and β-lipoproteins. Atherosclerosis *11:* 497–507 (1970).

6 Bramblett, C. A.: Non-metric skeletal age changes in the Darajani baboon. Am. J. phys. Anthrop. *30:* 161–171 (1969).

7 Bramblett, C. A.: Social organization as an expression of role behavior among Old World monkeys. Primates *14:* 101–112 (1973).

8 Brink, A. J.; Lewis, C. M.; Bosman, A. R., and Lochner, A.: The baboon *(Papio ursinus)* heart (coronary blood supply, muscle function and metabolism). Folia primatol. *13:* 11–22 (1970)

9 Buettner-Janusch, J.: A problem in evolutionary systematics: nomenclature and classification of baboons, genus *Papio.* Folia primatol. *4:* 288–308 (1966).

10 Cavanagh, D., Rao, P. S.; Sutton, D. M. C.; Bhagat, B. C. D., and Bachmann, F: Pathophysiology of endotoxin shock in the primate. Am. J. Obstet. Gynec. *108:* 705–722 (1970).

11 Chamove, A. S.; Eysenck, H. J., and Harlow, H. F.: Personality in monkeys. Factor analyses of rhesus social behaviour. Q. J. exp. Psychol. *24:* 496–504 (1972).

12 Crook, J. H.: Gelada baboon herd structure and movement, a comparative report. Symp. zool. Soc. Lond. *18:* 237–258 (1966).

13 Eggen, D. E.: Cholesterol metabolism in rhesus monkey, squirrel monkey, and baboon. J. Lipid Res. *15:* 139–145 (1974).

14 Eggen D. E. A.: Cholesterol metabolism in nonhuman primates; in Strong Atherosclerosis in primates (this volume).

15 Fedigan, L. M.: Roles and activities of male geladas *(Theropithecus gelada).* Behaviour *41:* 82–90 (1972).

16 Fox, H.: Arteriosclerosis in lower mammals and birds; in Cowdry Arteriosclerosis. A survey of the problem, chap. 21 (Macmillan, New York 1933).

17 Gartlan, J. S.: Preliminary notes on the ecology and behavior of the drill, *Mandrillus leucophaeus* Ritgen, 1824; in Napier and Napier Old World monkeys, pp. 445–480 (Academic Press, New York 1970).

18 Geer, J. C.; Catsulis C.; McGill, H. C. jr., and Strong, J. P.: Fine structure of the baboon aortic fatty streak. Am. J. Path. *52:* 265–286 (1968).

19 Gilbert, C. and Gillman, J.: Structural modifications in the coronary artery of the baboon *(Papio ursinus)* with special reference to age and endocrine status. S. Afr. J. med. Sci. *25:* 59–70 (1960).

20 Gillman, J. and Gilbert, C.: Atherosis in the baboon *(Papio ursinus).* Its pathogenesis and etiology. Expl Med. Surg. *15:* 181–221 (1957).

21 Glenn, F. and McSherry, C. K.: The baboon and experimental cholelithiasis. Archs Surg., Chicago *100:* 105–108 (1970).

22 GOLDSTEIN, J. L.; HAZZARD, W. R.; SCHROTT, H. G.; BIERMAN, E. L., and MOTULSKY, A. G.: Genetics of hyperlipidemia in coronary heart disease. Trans. Ass. Am. Physns *85:* 120–128 (1972).

23 GOLDSWORTHY, P. D.; LIM, G.; GLOMSET, J. A., and VOLWILER, W.: Polypeptides of baboon serum high density lipoprotein. Fed. Proc. *32:* 547, abstr. 1856 (1973).

24 GROOVER, M. E., jr.; SELJESKOG, E. L.; HAGLIN, J. J., and HITCHCOCK, C. R.: Myocardial infarction in the Kenya baboon without demonstrable atherosclerosis. Angiology, Baltimore *14:* 409–416 (1963).

25 HALL, K. R. L.: Variations in the ecology of the chacma baboon, *Papio ursinus.* Symp. zool. Soc. Lond. *10:* 1–28 (1963).

26 HAMPTON, J. W. and MATTHEWS, C.: Similarities between baboon and human blood clotting. J. appl. Physiol. *21:* 1713–1716 (1966).

27 HAMPTON, J. W. and MATTHEWS, C.: Observations on the clotting mechanism in the baboon; in VAGTBORG The baboon in medical research, vol. 2, pp. 659–665 (Univ. of Texas Press, Austin 1967).

28 HAWKEY, C. M. and SYMONS, C.: Variation in ADP-induced platelet aggregation *in vitro* in primates as a result of differences in plasma ADP inhibitor levels. Thromb. Diath. haemorrh. *19:* 29–35 (1968).

29 HENDRICKX, A. G.: Studies in the development of the baboon; in VAGTBORG The baboon in medical research, vol. 2, pp. 283–307 (Univ. of Texas Press, Austin 1967).

30 HENDRICKX, A. G.: Embryology of the baboon (Univ. of Chicago Press, Chicago 1971).

31 HERRMANN, G. R.; OVERALL, J. E.; CLABORN, L.; KRIEWALDT, F. H., and MAPLES, W. R.: Body, heart, left and right ventricular, left ventricular free wall, right ventricular wall, and septum weights and ratios in the baboon; in VAGTBORG The baboon in medical research, vol. 2, pp. 535-544 (Univ. of Texas Press, Austin 1967).

32 HERRMANN, G. R.; VICE, T. E.; McGILL, H. C., jr.; STRONG, J., and WERTHESSEN, N. T.: The heart of the wild Kenya African baboon. Normal control series of heart weight to body weight ratios; in VAGTBORG The baboon in medical research, vol. 1, pp. 265–268 (Univ. of Texas Press, Austin 1965).

33 HERRMANN, G. R., VICE, T. E.; RODRIGUEZ, A. R., and HERRMANN, A. W.: Heart weight to body weight, left to right ventricular weights, and ratios of baboon hearts; in VAGTBORG The baboon in medical research, vol. 1, pp. 269–274 (Univ. of Texas Press, Austin 1965).

34 HERRMANN, G. R. and WILLIAMS DE H., A. H.: The electrocardiographic patterns in 170 baboons in the domestic and African colonies at the Primate Center of the Southwest Foundation for Research and Education; in VAGTBORG The baboon in medical research, vol. 1, pp. 251–264 (Univ. of Texas Press, Austin 1965).

35 HERRMANN, G. R.; WILLIAMS DE H., A. H.; OVERALL, J. E.; KRIEWALDT, F. H., and CLABORN, L.: Electrocardiographic data on *Papio cynocephalus* baboons from Darijani and from Nairobi; in VAGTBORG The baboon in medical research, vol. 2, pp. 523–533 (Univ. of Texas Press, Austin 1967).

36 HOWARD, A. N.; BLATON, V.; VANDAMME, D.; VAN LANDSCHOOT, N., and PEETERS, H.: Lipid changes in the plasma lipoproteins of baboons given an atherogenic diet. III. A comparison between lipid changes in the plasma of the baboon and chimpanzee

given atherogenic diets and those in human plasma lipoproteins of type II hyperlipo-proteinaemia. Atherosclerosis *16:* 257–272 (1972).

37   Howard, A. N.; Gresham, G. A.; Bowyer, D. E., and Lindgren, F. T.: Aortic and coronary atherosclerosis in baboons. Prog. biochem. Pharmacol., vol. 4, pp. 438–444 (Karger, Basel 1968).

38   Howard, A. N.; Gresham, G. A.; Richards, C., and Bowyer, D. E.: Serum proteins, lipoproteins, and lipids in baboons given normal and atherogenic diets; in Vagtborg The baboon in medical research, vol. 1, pp. 283–299 (Univ. of Texas Press, Austin 1965).

39   Howard, A. N.; Patelski, J.; Bowyer, D. E., and Gresham, G. A.: Atherosclerosis induced in hypercholesterolaemic baboons by immunological injury; and the effects of intravenous polyunsaturated phosphatidyl choline. Atherosclerosis *14:* 17–29 (1971).

40   Hummer, R. L.: Preventive medicine practices in baboon colony management; in Vagtborg The baboon in medical research, vol. 2, pp. 51–68 (Univ. of Texas Press, Austin 1967).

41   Hummer, R. L.: Observations on the feeding of baboons; in Feeding and nutrition of nonhuman primates, pp. 183–203 (Academic Press, New York 1970).

42   Jenkins, C. D.: Psychologic and social precursors of coronary disease. New Engl. J. Med. *248:* 244–255, 307–317 (1971).

43   Jolly C. J.: The evolution of the baboons; in Vagtborg The baboon in medical research vol. 2, pp. 23–50 (Univ. of Texas Press, Austin 1967).

44   Kalter, S. S.; Ratner, J. J.; Rodriguez, A. R., and Kalter, G. V.: Microbiological parameters of the baboon (*Papio* sp.). Virology; in Vagtborg The baboon in medical research, vol. 2, pp. 757–773 (Univ. of Texas Press, Austin 1967).

45   Kaminer, B.: The electrocardiogram of the baboon *(Papio ursinus)*. S. Afr. J. med. Sci. *23:* 231–240 (1958).

46   Kraemer, D.: Personal commun. (1972).

47   Kritchevsky, D.; Davidson, L. M.; Shapiro, I. L.; Kim, H. K.; Kitagawa, M.; Malhotra, S.; Nair, P. O.; Clarkson, T. B.; Bersohn, I., and Winter, P. A. D.: Lipid metabolism and experimental atherosclerosis in baboons. Influence of cholesterol-free, semisynthetic diets. Am. J. clin. Nutr. *27:* 29–50 (1974).

48   Kummer, H.: Soziales Verhalten einer Mantelpavian-Gruppe (Hans Huber, Bern 1957).

49   Kummer, H.: Social organization of hamadryas baboons (Karger, Basel 1968).

50   Kuntz, R. E. and Myers, B. J.: Microbiological parameters of the baboon (*Papio* sp.) Parasitology; in Vagtborg The baboon in medical research, vol. 2, pp. 741–755 (Univ. of Texas Press, Austin 1967).

51   Lindsay, S. and Chaikoff, I. L.: Arteriosclerosis in the baboon. Naturally occuring lesions in the aorta and coronary and iliac arteries. Archs Path. *63:* 460–471 (1957).

52   Lofland, H. B., jr.; Clarkson, T. B.; Clair, R. W. St., and Lehner, N. D. M.: Studies on the regulation of plasma cholesterol levels in squirrel monkeys of two genotypes. J. Lipid Res. *13:* 39–47 (1972).

53   McGill, H. C., jr.; Strong, J. P.; Holman, R. L., and Werthessen, N. T.: Arterial lesions in the Kenya baboon. Circulation Res. *8:* 670–679 (1960).

54 McGill, H. C., jr.; Strong, J. P.; Newman, W. P., III, and Eggen, D. A.: The baboon in atherosclerosis research; in Vagtborg The baboon in medical research, vol. 2, pp. 351–363 (Univ. of Texas Press, Austin 1967).

55 McGill, H. C., jr.: Fatty streaks in the coronary arteries and aorta. Lab. Invest. *18:* 560–564 (1968).

56 McKee, P. A.; Lemmon, W. B., and Hampton, J. W.: Streptokinase and urokinase activation of human, chimpanzee and baboon plasminogen. Thromb. Diath. haemorrh. *26:* 512–522 (1971).

57 McMahan, C. A. and Strong, J. P.: Factorial experiments for pilot studies of serum cholesterol in baboons; in Vagtborg The baboon in medical research, vol. 1, pp. 485–501 (Univ. of Texas Press, Austin 1965).

58 McSherry, C. K.; Javitt, N. B.; Corvalho, J. M. de, and Glenn, F.: Cholesterol gallstones and the chemical composition of bile in baboons. Ann. Surg. *173:* 569–577 (1971).

59 McSherry, C. K.; Glenn, F., and Javitt, N. B.: Composition of basal and stimulated hepatic bile in baboons, and the formation of cholesterol gallstones. Proc. natn. Acad. Sci., Wash. *68:* 1564–1568 (1971).

60 Minick, C. R.; Murphy, G. E., and Campbell, W. G. jr.: Experimental induction of athero-arteriosclerosis by the synergy of allergic injury to arteries and lipid-rich diet. J. exp. Med. *124:* 635–651 (1966).

61 Patterson, B.: The extinct baboon *Parapapio jonesi* in the early Pleistocene of Northwestern Kenya. Breviora *282:* 1–4 (1968).

62 Pena, A. de la and Goldzieher, J. W.: Clinical parameters of the normal baboon; in Vagtborg The baboon in medical research, vol. 2, pp. 379–389 (Univ. of Texas Press, Austin 1967).

63 Pena, A. de la; Matthijssen, C., and Goldzieher, J. W.: Normal values for blood constituents of the baboon (*Papio* species). Lab. Anim. Care *20:* 251–261 (1970).

64 Pena, A. de la; Matthijssen, C., and Goldzieher, J. W. Normal values for blood constituents of the baboon. II. Lab. Anim. Sci. *22:* 249–257 (1972).

65 Pilbeam, D. and Walker, A.: Fossil monkeys from the miocene of Napak, North East Uganda. Nature, Lond. *220:* 657–660 (1968).

66 Pinkerton, M. E.; Boncyk, L. H., and Cline, J. A.: Microbiological parameters of the baboon (*Papio* sp.): Bacteriology; in Vagtborg The baboon in medical research, vol. 2, pp. 717–730 (Univ. of Texas Press, Austin 1967).

67 Ransom, T. W.: Ecology and behaviour of the baboon, *Papio anubis*, at the Gombe Stream National Park, Tanzania, Ph. D. thesis, Berkeley (1971).

68 Rowell, T. E.: A quantitative comparison of the behaviour of a wild and a caged baboon group. Anim. Behav. *15:* 499–509 (1967).

69 Rowell, T.: The social behaviour of monkeys (Penguin, Harmondsworth 1972).

70 Sabater Pí, J.: Contribution to the ecology of *Mandrillus sphinx* Linnaeus 1758 of Rio Muni (Republic of Equatorial Guinea). Folia primatol. *17:* 304–319 (1972).

71 Sarich, V.: Quantitative immunochemistry and the evolution of the anthropoidea; in Chiarelli Taxonomy and phylogeny of Old World primates with references to the origin of man (Rosenberg & Sellier, Torino 1968).

72 Simons, E. L.: The deployment and history of Old World monkeys (Cercopithecidae, primates); in Napier and Napier Old World monkeys, pp. 97–137 (Academic Press, New York 1970).

73 Small, D. M.: Size and structure of bile salt micelles; Influence of structure, concentration, counterion concentration, pH and temperature. Adv. Chem. Ser., vol. 84, pp. 31–52 (Am. Chemical Society, Washington 1968).

74 Snow, C. C.: Some observations on the growth and development of the baboon; in Vagtborg The baboon in medical research, vol. 2, pp. 187–199 (Univ. of Texas Press, Austin 1967).

75 Sperry, W. M.: A micro method for the determination of total and free cholesterol. Am. J. clin. Pathol. Techn. Suppl. *81:* 91–99 (1938).

76 Spivak, H.: Ausdrucksformen und soziale Beziehungen in einer Dschelada-Gruppe *(Theropithecus gelada)* im Zoo. Z. Tierpsychol. *28:* 279–296 (1971).

77 Stary, H. C. and Strong, J. P.: Coronary artery fine structure in rhesus monkeys: Nonatherosclerotic intimal thickening: in Strong Atherosclerosis in primates (this volume).

78 Stevens, R. E., III; Wilbur, R. L., and Johnson, D. E.: Training nonhuman primates to inhale cigarette smoke (unpublished data).

79 Stoltz, L. P. and Saayman, G. S.: Ecology and behaviour of baboons in the Northern Transvaal. Ann. Transvaal Museum *26:* 99–143 (1970).

80 Strong, J. P.; Radelat, P. B.; Guidry, M. A., and McMahan, C. A.: Variability of serum cholesterol values in baboons. Circulation. Res. *14:* 367–372 (1964).

81 Strong, J. P.; Rosal, J.; Deupree, R. H., and McGill, H. C., jr.: Diet and serum cholesterol levels in baboons. Expl molec. Path. *5:* 82–91 (1966).

82 Strong, J. P. and McGill, H. C., jr.: Diet and experimental atherosclerosis in baboons. Am. J. Path. *50:* 669–690 (1967).

83 Strong, J. P.: Personal commun. (1972).

84 Struhsaker, T. T.: Correlates of ecology and social organization among African cercopithecines. Folia primatol. *11:* 80–118 (1969).

85 Swindler, D. R.; McCoy, H. A., and Hornbeck, P. V.: The dentition of the baboon *(Papio anubis)*; in Vagtborg The baboon in medical research, vol. 2, pp. 133–150 (Univ. of Texas Press, Austin 1967).

86 Thorington, R. W., jr. and Groves, C. P.: An annotated classification of the Cercopithecoidea; in Napier and Napier Old World monkeys, pp. 629–647 (Academic Press, New York 1970).

87 Vagtborg, H.: The baboon in medical research, vol. 1 (Univ. of Texas Press, Austin 1965).

88 Vagtborg, H.: The baboon in medical research, vol. 2 (Univ. of Texas Press, Austin 1967).

89 Vice, T. E.; Britton, H. A.; Ratner, I. A., and Kalter, S. S.: Care and raising of newborn baboons. Lab. Anim. Care *16:* 12–22 (1966).

90 Washburn, S. L. and Vore, I. de: The social life of baboons. Sci. Am. *204:* 62–71 (1961).

91 Watt, J. J. van der: Naturally occurring vascular lesions in the major arteries of the Baboon *(Papio ursinus)*; thesis for degree of Master of Veterinary Medicine, Pretoria (1972).

92 WATT, J. J. VAN DER; KOTZHE, J. P.; KEMPFF, P. G.; PLESSIS, J. P. DU, and LAUBSCHER, N. F.: Aortic intimal lesions and serum lipids in wild baboons. J. med. Primatol. *2:* 25–38 (1973).

93 WEBSTER, W. S.; CLARKSON, T. B., and LOFLAND, H. B.: Carbon monoxide-aggravated atherosclerosis in the squirrel monkey. Expl molec. Path. *13:* 36–50 (1970).

94 WESSLER, S. and YIN, E. T.: The role of animal models in thrombosis. A point of departure. Thrombosis et diathesis haemorrhagica, Suppl. 59, pp. 83–90, 1974.

95 WILSON, J. D.: The relation between cholesterol absorption and cholesterol synthesis in the baboon. J. clin. Invest. *51:* 1450–1458 (1972).

96 ZUCKERMAN, S.: The social life of monkeys and apes (Harcourt, New York 1932).

97 ZYL, A. VAN: The serum lipids and age in the baboon *(Papio ursinus)*. S. Afr. J. med. Sci. *20:* 97–117 (1955).

98 ZYL, A. VAN: The lipid and iodine content in blood and milk of the lactating baboon: their influence on the serum lipids of the suckling infant. S. Afr. J. med. Sci. *20:* 119–128 (1955).

HENRY C. MCGILL, jr., M.D. and GLEN E. MOTT, Ph. D., Department of Pathology, The University of Texas Health Science Center, 7703 Floyd Curl Drive, *San Antonio, TX 78284;* CLAUD A. BRAMBLETT, Ph. D., Department of Anthropology, The University of Texas at Austin, *Austin, TX 78712* (USA)

Prim. Med., vol. 9, pp. 66–89 (Karger, Basel 1976)

Atherosclerosis in Old World Monkeys[1]

Thomas B. Clarkson, Thomas E. Hamm, Bill C. Bullock
and Noel D. M. Lehner

Arteriosclerosis Research Center and Department of Comparative Medicine,
Bowman Gray School of Medicine, Wake Forest University, Winston-Salem, N. C.

Contents

I. Introduction

Old World monkeys are indigenous to Africa and Asia. Their classification has been revised frequently and for this discussion the system established by Thorington and Groves [30] will be used with a few exceptions.

[1] Supported in part by USPHS NIH grants RR00236 and HL 14164.

Evaluation of data collected from primates, especially the less common ones, requires precise and accurate identification of the species used. THORINGTON [29] proposed that investigators identify their animals as precisely as possible taxonomically and indicate the geographic origin. Voucher specimens should be placed in museums or photographs of the animals stored with the records and the existence of these registered specimens or photographs should be mentioned in the publications.

The Old World monkeys are all members of the family Cercopithecidae, which is divided into two subfamilies. The subfamiliy Cercopithecinae is composed of six genera (*Macaca, Cercocebus, Papio, Theropithecus, Erythocebus*, and *Cercopithecus* which will include *Miopithecus*, and *Allenopithecus*, for this discussion). The subfamily Colobinae is composed of five genera (*Colobus, Procolobus, Pygathrix, Nasalis*, and *Presbytis*), none of which has been used in research on arteriosclerosis.

Arteriosclerosis in *Papio* sp. (baboons) and in *Macaca mulatta* (rhesus monkey) has been studied much more extensively than in other Old World monkeys, and information on them is covered in other chapters of this monograph. In this chapter we shall discuss the data available from six other species of the genus *Macaca*, two species of the genus *Cercopithecus*, one species of *Cercocebus*, and one species of *Erythrocebus*.

II. Old World Monkeys Used in Atherosclerosis Research

A. Macaca

The genus *Macaca* (commonly called the macaques) consists of 13 species of which seven have been investigated as models for atherosclerosis research. *M. mulatta* (rhesus monkey) will be discussed in another chapter. Here we shall report information available on *M. fascicularis* (*M. irus*, crab-eating monkey) *M. arctoides* (*M. speciosa*, stumptailed macaque), *M. nigra* (*Cynopithecus niger*, Celebes black ape), *M. radiata* (bonnet monkey). *M. fuscata* (Japanese macaque), and *M. nemestrina* (pigtailed macaque).

1. Macaca fascicularis (M. irus, Cynomolgus Monkeys,
Crab-Eating Macaque, Java Monkey)
ARMSTRONG [1] in this volume summarized his work with MEGAN [2, 3] on progression and regression of experimentally induced atherosclerosis in

the cynomolgus monkey. He also reviewed the work by MALMROS *et al.* [20], KRAMSCH and HOLLANDER [16], KRAMSCH *et al.* [17], and PRATHAP and LAU [25] on experimental atherosclerosis in this species. STRONG [27], in the introductory chapter of this monograph, referred to the work of PRATHAP [24] on spontaneous arterial lesions in this species. We shall provide additional information on this species as a basis for comparison with the other macaques, including some unpublished data on *M. fascicularis* from our own laboratory.

M. fascicularis is the smallest species of the genus *Macaca* (mature males weigh 3.5–8.5 kg and females 2.5–6 kg). They are widely distributed in Southeast Asia ranging from Thailand and the Malayan Peninsula throughout Indonesia to the Philippines, typically inhabiting coastal areas arond tidal creeks and mangrove swamps, and are quite numerous in their native habitats. They are readily available, and are exported mainly from Manila, Philippine Islands, Kuala Lumpur, Malaysia and Bangkok, Thailand [29]. Twenty-one subspecies have been described [22]. *M. fascicularis* may warrant serious consideration for studies on arteriosclerosis, especially if the availability of rhesus monkeys does not improve or become seven more critical.

DOYLE *et al.* [9] used four *M. fascicularis* to study the effect of atherosclerosis on periodontal disease. A 5.6-kg male and a 4.65-kg female were fed 100 g of canned banana mash poured over 300 g of commercial monkey chow. Two males, weighing 5.8 kg and 5.1 kg, were fed the same diet with the addition of 7 g of cholesterol in 20 ml cottonseed oil. Oranges and apples were cut and mixed with monkey chow and fed on random days since the monkeys reportedly did not like the cholesterol and cottonseed oil-containing diet. The diets were fed for slightly over one year. One male animal from each group underwent surgery to 'produce coarctation of the aorta with the hope of inducing hypertension'. When the animals fed the diet without added cholesterol were necropsied, the female had gained 1.3 kg and had a serum cholesterol concentration of 157 mg/dl, the male had gained 0.4 kg and had serum cholesterol concentration of 102 mg/dl. Neither animal had atherosclerosis. The two cholesterol-fed animals had serum cholesterol concentrations of 439 and 218 mg/dl (male with coarctation) and had gained 0.4 and 2.6 kg, respectively. Both animals had extensive atherosclerosis of the coronary arteries, focal atherosclerotic lesions in the pulmonary arteries, and collections of foamy macrophages in the bone marrow. The authors asserted that the number of animals was too small to contribute to an understanding of the relationship between atherosclerosis and periodontal disease. The effect of coarctation of the aorta on blood pressure was not stated.

The plasma cholesterol and triglyceride concentration changes of juvenile cynomolgus monkeys fed saturated or unsaturated fat with and without dietary cholesterol have been reported by COREY et al. [8]. They used 12 juveniles (14.6 $\pm$ 3.2 months old) born and raised in captivity. The monkeys were divided into groups of five and seven animals. Base line serum cholesterol concentrations averaged about 180 mg/dl when the animals were fed a control diet containing cottonseed-soybean oil. Diets were then fed that provided 17% of calories from protein (lactalbumin), 62% of calories from carbohydrates (dextrin and sucrose), 21% of calories from fat (safflower or coconut oil), 0.6 mg/Cal cholesterol, and a high or low level of vitamin E. The effect of the level of vitamin E was not significant and the data were pooled according to type of dietary fat (safflower oil or coconut oil). The monkeys fed coconut oil tended to have higher but not statistically significant serum cholesterol levels than those fed safflower oil. After one year the serum cholesterol levels of the coconut oil-fed animals were 506 $\pm$ 137 mg/dl (mean $\pm$ SD) and those of the safflower oil-fed animals were 426 $\pm$ 55 mg/dl. Plasma triglycerides were highly variable, ranging from 24 $\pm$ 7 to 77 $\pm$ 37 mg/dl. The cholesterol-containing diets were fed for 48 weeks when cholesterol was removed from the diets and the feeding continued for 59 more weeks. During this last feeding period the serum cholesterol concentrations of the coconut oil-fed monkeys ranged between 177 and 230 mg/dl, compared to a range of 130-174 mg/dl for the animals fed the safflower oil diet.

Our experience with *M. fascicularis* is limited to observations on monkeys that are a part of two long-range experiments. The first experiment includes 24 adult males (12 from Thailand and 12 from the Philippine Islands), half of which have consumed an atherogenic diet (table I, 1 mg/cal cholesterol, lard as source of fat) for more than 30 months. The other half of the animals have consumed the same diet without added cholesterol for the same period. At termination of the experiment all of the monkeys will be necropsied to determine if there are racial differences between the two groups in their development of atherosclerosis. Additionally, they will be compared to rhesus monkeys that have been fed the same diet for a comparable amount of time.

Base line serum cholesterol concentrations obtained while the monkeys consumed monkey chow were not different between groups and averaged 118 $\pm$ 4 mg/dl (mean $\pm$ standard ; error of the mean). After initiation of the experimental diets, the serum cholesterol concentrations of the various groups

Table I. Composition (G/kg) of the test diet[1]

	Control[2]	Atherogenic[3]
Nonfat dry milk solids	300.0	300.0
Wheat flour	200.0	200.0
Casein, USP	130.0	130.0
Lard	250.0	250.0
Apple sauce	78.0	73.0
Complete Vitamin Mix[4]	22.0	22.0
USP XIV salts mixture	20.0	20.0
Cholesterol	–	5.0
Total, g	1,000.0	1,000.0

[1] Adapted from BULLOCK *et al.* [7].

[2] Approximately 0.05 mg cholesterol/Cal primarily from lard.

[3] Approximately 1 mg cholesterol/Cal.

[4] Nutritional Biochemicals Corp., Cleveland, Ohio: vitamin D_3 substituted for D_2 to provide 2.5 IU/g of diet.

Table II. Composition of the test diet fed to *M. fascicularis.*

	%	Cal/g	% Cal
Protein			
Lactalbumin	5		
Casein USP	11,2	72	15
Gelatin	1.8		
Carbohydrate			
Sucrose	23	184	38
Dextrin	23		
Fat			
Crisco	8		
Corn oil	8	216	45
Lard	8		
Hegsted's salt mix	4		
Vitamin mix	2.5	10	
D_3 in corn oil	(6.25 cc)		
Alphacel	5.3		
Cholesterol	0.27[1]		
Total	100	482	

[1] Cholesterol approximately 0.56 mg/Cal.

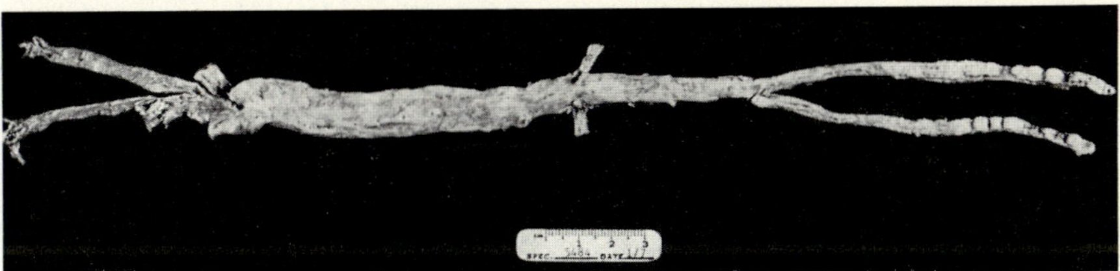

Fig. 1. Intimal surface of the major arteries from an adult male *M. fascicularis* fed an atherogenic diet. Approximately 90–95 % of the surface of all the arteries was covered by atherosclerotic plaque. Twenty to 30 % of the plaques had a glistening white surface while the rest were yellow. The plaques in the femoral arteries are separated by transverse areas of normal intima.

of monkeys have averaged the following: control Thailand monkeys, 210 $\pm$ 15 mg/dl; cholesterol-fed Thailand monkeys, 829 $\pm$ 41 mg/dl; control Philippine monkeys, 188 $\pm$ 11 mg/dl; and cholesterol-fed Philippine monkeys, 707 $\pm$ 26 mg/dl.

Two Thailand cynomolgus monkeys have died, one after eating the test diet for nine months, and one after test diet of 24 months. The last animal that died had atherosclerotic plaques covering 90–95 % of the intimal surface of the aorta, carotid, and femoral arteries (figure 1). Plaques that were 'pearly'·in appearance covered 30 % of the intimal surface of the aortic arch, 25 % of the thoracic aorta, and 20 % of the abdominal aorta. The microscopic appearance of these lesions from the aorta is shown in figure 2 and from a coronary artery in figure 3.

The second group of cynomolgus monkeys are part of an experiment designed to investigate the relationship of atherosclerosis and sex in Malaysian *M. fascicularis*. Females of this species have a regular monthly menstrual cycle [21]. Twenty adult males and 20 adult females have been fed a diet with 15 % of calories from protein, 38 % from carbohydrate, and 45 % from fat (table II). For a period of 10 weeks, 16 members of each sex also were fed this diet with the addition of 0.2 % cholesterol (0.42 mg/Cal). Then the cholesterol content of the diet for these animals was increased to 0.27 % (0.56 mg/Cal). The serum cholesterol concentration five weeks later was 462 $\pm$ 47 mg/dl (mean $\pm$ SEM) for the test females, 509 $\pm$ 41 mg/dl for the test males, 129 $\pm$ 14 mg/dl for the control females, and 132 $\pm$ 17 mg/dl for the control males. The increase in serum cholesterol concentration has been in the β plus pre-β lipoprotein fractions, and the α lipoprotein fraction has decreased from the control level.

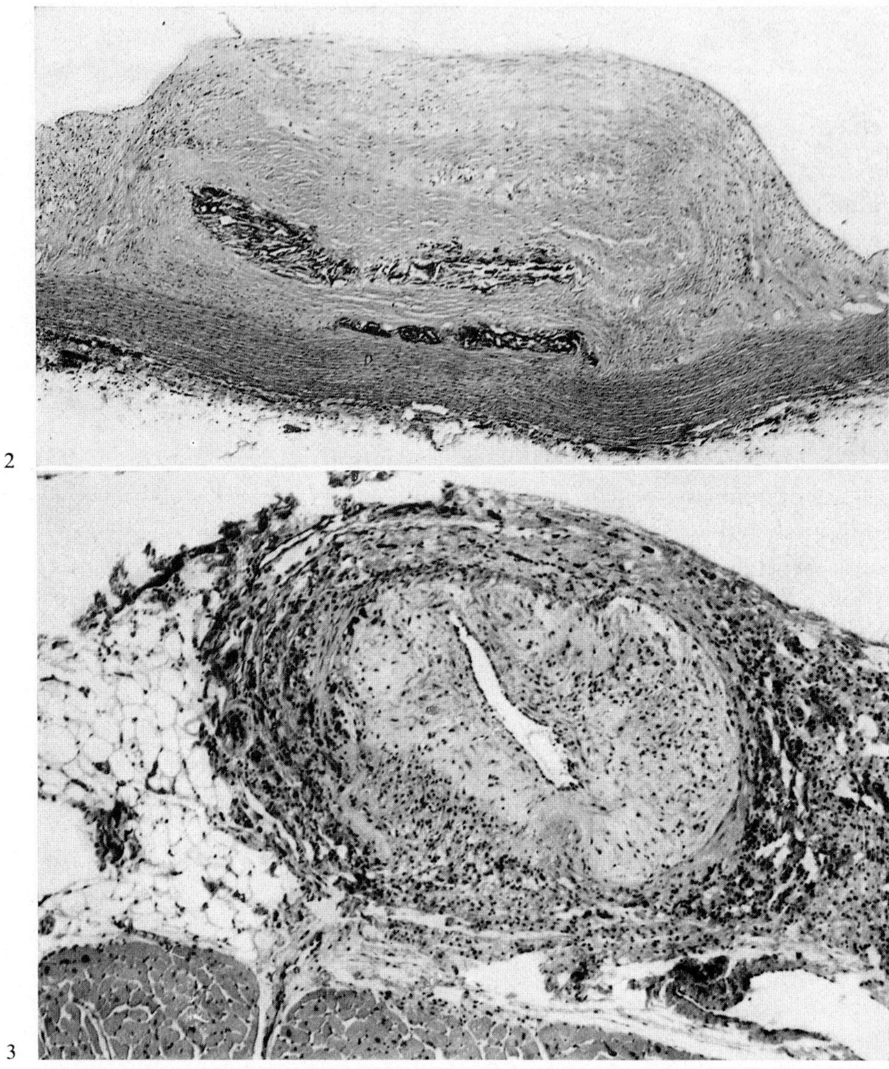

Fig. 2. Photomicrograph of a plaque from the aorta shown in figure 1. Note the large amount of mineral and lipid clefts below a fibromuscular cap. H.E. × 18.

Fig. 3. Photomicrograph of a cross-section of the left anterior descending coronary artery from the same animal whose aorta is shown in figures 1 and 2. All major coronary arteries and their extramural branches were similarly affected along their entire lengths. The media and the adventitia contain numerous foam cells. H.E. × 73.

2. *Macaca arctoides* (*M. speciosa*, Stumptailed Macaque)

Four subspecies of *M. arctoides* have been described [21]. These monkeys, especially juveniles, are remarkably more docile than other species of macaques. Adult male stumptailed macaques may weigh up to 14 kg. These monkeys are presently found in the region of southern China, Thailand, and Vietnam and are exported from Bangkok, Thailand [2]. *M. arctoides* are not thought to be numerous in their native habitats and should their use in research become extensive they would probably have to be bred and raised in domestic colonies.

PICK and KATZ [23] and PICK and GLICK [22] have published two brief reports on dietary induced atherosclerosis in stumptailed macaques. The initial report was on eight young animals (3–4 kg) that were fed a diet containing 2% cholesterol and 25% fat as an equal mixture of coconut oil and peanut oil added to a fat-free monkey mash. Plasma cholesterol concentration was 144 ± 20 mg/dl prior to feeding the atherogenic diet and varied between 322 and 594 mg/dl thereafter. Six of the monkeys died accidentally after three months and had aortic atherosclerosis but no coronary lesions. The two remaining monkeys were killed after having been fed the atherogenic diet for nine months and had aortic and coronary artery atherosclerosis.

In the second study, mild renal hypertension (170/100 mm Hg) was induced in seven of 15 *M. arctoides* that had been fed an atherogenic diet for 17 months. These as well as the eight normotensive monkeys (125/95 mm Hg) were killed and examined after being fed the atherogenic diet for an additional eight months. The average extent of aortic atherosclerosis and the percent occlusion of the coronary arteries were greater in the hypertensive monkeys. The atherosclerotic lesions in the hypertensive animals contained more fibrous and elastic elements and those in the normotensive monkeys comparatively more lipid.

The effects of simple and complex carbohydrates on serum lipids and atherosclerosis in *M. arctoides* were investigated by LANG and BARTHEL [19]. Twelve *M. arctoides* were divided into two equal groups and fed diets containing 0.5% cholesterol, 10% fat consisting of equal parts of butter and peanut oil, 17.5% casein, and 66% sucrose or dextrin. The average serum triglyceride and cholesterol concentrations were not different between the two groups of monkeys and the serum cholesterol concentrations averaged between 300 and 400 mg/dl for the 16 month duration of the experiment. The diets were equally atherogenic and both groups of monkeys had numerous and extensive atherosclerotic lesions in the coronary arteries as well as in arteries of the kidney, testes, spleen, urinary bladder, pancreas, lung, and

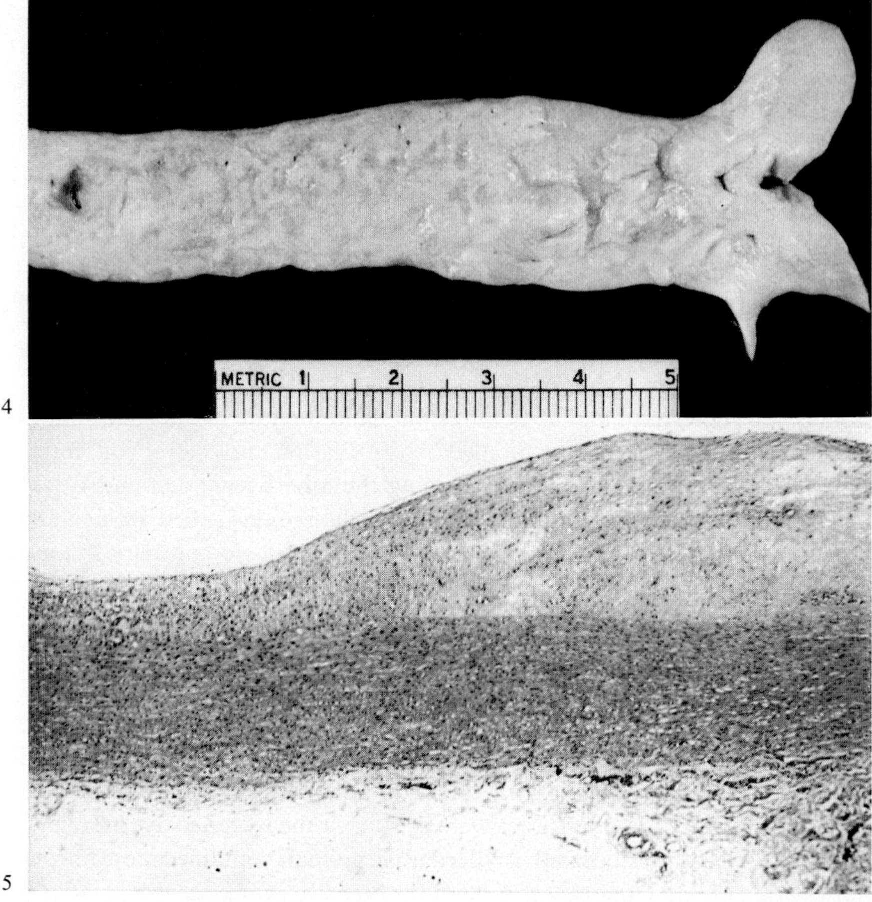

Fig. 4. Intimal surface of the aortic arch and thoracic aorta from an adult male *M. arctoides* almost completely covered by plaques. Some of the plaques were yellow and some had glistening white surfaces.

Fig. 5. Photomicrograph of a plaque from the aorta shown in figure 4. The clear areas contained lipid before the tissue was processed for paraffin sections. The relatively acellular areas are collagenous. H.E. × 18.

brain. Aortic cholesterol concentrations determined after the vessels were fixed in formalin averaged about 0.8 mg of cholesterol/gram for both groups of monkeys.

BULLOCK *et al.* [5, 6] fed four adult male stumptailed macaques a diet containing 1 mg of cholesterol/Cal and 45 % of calories as lard for 42 months (table I). The average serum cholesterol concentration during this period

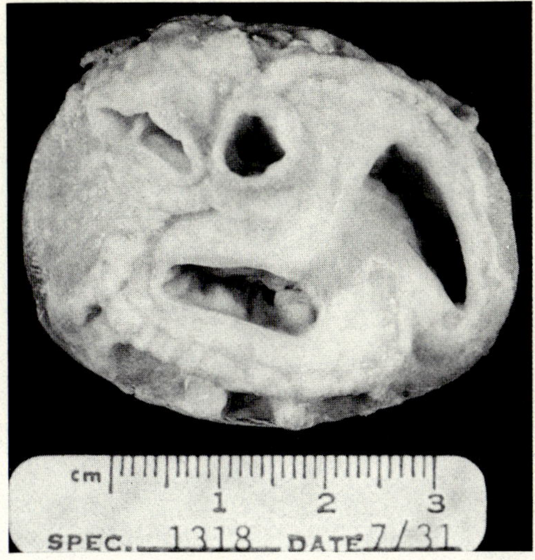

Fig. 6. Heart from a *M. arctoides* fed an atherogenic diet. The atria and epicoronary adipose tissue have been dissected away. The extramural coronary arteries are enlarged and nodular.

was 735 ± 29 mg/dl (mean ± SEM). Cutaneous xanthomata and xanthelasma were present in all animals. All of these monkeys had extensive and severe atherosclerotic lesions in the aorta, coronary, carotid, and femoral arteries.

Most of the aortic intimal surface was affected with white or yellow raised plaques (thoracic aorta, 77 ± 13%; abdominal aorta, 93 ± 4%; fig. 4). The carotid and femoral arteries were also markedly affected with severe lesions, having 95 ± 4% and 58 ± 18% of their intimal surfaces covered with raised lesions. A striking histological characteristic of the lesions in these animals was the preponderance of connective tissue elements as compared to lipid. The lesions, in general, had a multilayered interlacing appearance probably caused by the confluence of adjacent lesions. Many lesions had a thin layer of foam cells at some point on the luminal surface, overlying well-organized layers of connective tissue. Pools of lipid were usually found at the intima-media junction. Mineralization, vascularization, medial thinning, and lipid clefts were characteristics of lesions found in all of these arteries (fig. 5).

The stumptailed macaques fed the atherogenic diet also had extensive and severe coronary artery atherosclerosis. The major coronary arteries were sclerotic throughout their extramural course, from the aorta to the point where they entered the myocardium. These arteries were yellow, nodular,

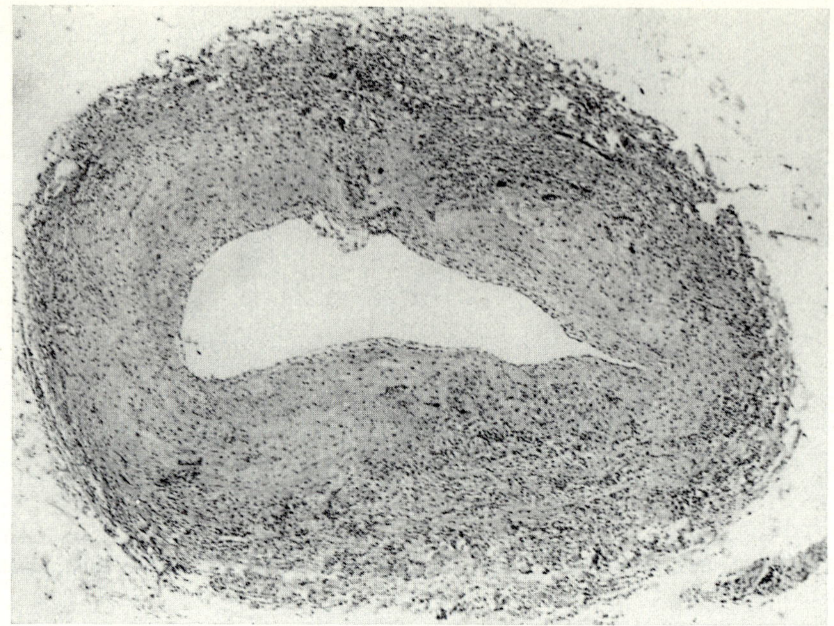

Fig. 7. Photomicrograph of the left anterior descending coronary artery from the heart shown in figure 6. Note the lumen reduced by the thickened arterial wall. The original boundaries of the media are not readily apparent. The adventitia contains numerous foamy macrophages. H.E. × 73.

and had rigid walls (fig. 6). The lesions caused significant narrowing of the lumen of all major coronary arteries with an average stenosis of $78 \pm 7\%$ for the main left, $75 \pm 3\%$ for the left anterior descending, $67 \pm 4\%$ for the left circumflex, and $80 \pm 3\%$ for the right coronary artery. The atherosclerotic lesions often involved the media of the arteries. Smooth muscle cells of the media were decreased in number and separated or replaced by collagenous tissue (fig. 7).

Intracranial arteries were the least affected of the arteries examined. Lesions were found in three of four monkeys, but compared to lesions in other vascular beds they were relatively small. The largest plaque occupied less than 20% of the vascular lumen.

In addition to the extensive and severe vascular disease, the monkeys fed the atherogenic diet had markedly increased cholesterol concentrations in several organs and tissues when compared to the control animals. The cholesterol concentration in the liver, skin, tendon, and eye were increased from 4- to 13-fold in these monkeys (table III).

Table III. Effect of cholesterol containing diet on organ and tissue cholesterol concentration of stumptailed macaques [1, 2]

	Control diet (N= 5)	Atherogenic diet (N = 4)
Brain	16.90 ±0.96	13.94 ±1.57
Myocardium	1.47 ±0.33	1.81 ±0.18
Lung	2.74 ±0.60	3.06 ±0.45
Liver	6.73 ±3.05	27.81 ±2.72
Spleen	4.12 ±0.32	5.99 ±0.27
Jejunum	2.66 ±0.28	2.97 ±0.39
Ileum	2.20 ±0.10	2.98 ±0.25
Pancreas	2.39 ±0.35	2.36 ±0.14
Kidneys	3.53 ±0.26	3.83 ±0.19
Skin		
Eyelids	1.81 ±0.28	7.05 ±3.21
Ear	2.18 ±0.31	14.31 ±4.72
Thoracic	1.96 ±0.16	10.92 ±5.01
Forearm	2.24 ±0.35	13.97 ±3.84
Tendon	0.48 ±0.05	6.52 ±3.31
Muscle	0.66 ±0.09	0.61 ±0.02
Fascia	0.76 ±1.06	3.81 ±1.61
Adipose	1.09 ±0.12	1.58 ±0.14
Eye	0.49 ±0.03	2.72 ±2.09
Cornea	0.88 ±0.12	0.79 ±0.16

[1] All data are expressed as mg/g of wet weight, mean ± SEM.
[2] Adapted from BULLOCK *et al.* [6].

Atherosclerotic lesions in monkeys fed the control diet (some were fed liquid diets containing up to 1 mg of cholesterol/Cal during the last year for metabolic studies) or monkey chow, consisted of fatty streaks in the aorta and various other vascular beds.

3. *Macaca nigra* (*Cynopithecus niger*, Celebes Black Ape)

NAPIER and NAPIER [21] list two subspecies and THORINGTON and CROVES [30] include *M. maurus* (the Celebes or Moor macaque with four subspecies) in this species. Their natural habitat is the northern peninsula of the Celebes Island and some small adjacent islands. Exportation of these animals is restricted and there are no more than about 200 outside the Celebes Island. They are sexually mature at five years and live about 20–25 years. Mature males weigh 11–12 kg and females 5–7 kg. They are said to be easier to handle than other macaques [12]. Despite their common name they are not apes.

HOWARD [12] has reported finding naturally occurring diabetes mellitus in a closed breeding colony of Celebes apes maintained at the Oregon Regional Primate Research Center. Diabetic signs in these animals have included abnormal intravenous and oral glucose tolerance tests, hyperglycemia impaired insulin response, hypertriglyceridemia, increased prebeta lipoproteinemia, retinal vascular aberrations, and abnormal islets of Langerhans. Atherosclerotic lesions have been observed in some of these diabetic monkeys fed monkey chow [13]. Mature diabetic and borderline diabetic monkeys had aortic sudanophilia that correlated well with the clinical severity of diabetes. Mature diabetic monkeys with a long history of diabetes had extensive aortic sudanophilia accompanied by visible raised fibrotic and lipid lesions. Normal adult monkeys had very little lesion development in the aorta, including sudanophilia. Based on a small number of observations juvenile diabetic monkeys had no aortic lesions.

4. *Macaca radiata* (Bonnet Monkey)

There are two subspecies of *M. radiata* and they are indigenous to peninsular India. Mature males weigh 5.7–8.9 kg and females 2.9–4.4 kg.

The effect of parenterally administered adrenaline on serum cholesterol concentration and aortic atherosclerosis in *M. radiata* was investigated by JAGANNATHAN *et al*. [14]. Twenty-three young male bonnet monkeys were fed a diet consisting of whole wheat flour, 90 g; calcium phosphate, 2 g; ground nut (peanut) oil, 7.9 ml; vanitin, 0.1 ml; and sprouted grain 35 g. The monkeys were divided into five groups and given intramuscular injections of distilled water or various doses of adrenaline, up to 1 ml of 1:1,000 adrenaline twice daily for 21 weeks. None of the animals had macroscopic plaques in the aorta, but one control monkey and one adrenaline-treated monkey had a few aortic fatty streaks. The aortas of control monkeys had diffuse intimal sudanophilia when examined microscopically. The aortic intima of eight of 11 adrenaline-treated monkeys and one control monkey had a beaded appearance associated with focal accumulations of intracellular and extracellular lipid and mucopolysaccharides. The group given the highest dose of adrenaline had an average decrease in serum cholesterol concentration of 21.5 mg/dl which was evident within two weeks after the injections began. The serum cholesterol concentrations of the monkeys given lower doses of adrenaline were highly variable while those of the control monkeys remained at base line levels.

5. *Macaca fuscata* (Japanese Macaque)

Two subspecies of *M. fuscata* are listed by NAPIER and NAPIER [21]. The natural habitat of these monkeys is Japan. They are the heaviest of the macaques, with adult males weighing 11.1–18 kg and females 8.3–16.3 kg.

KUZUYA [18] reported that arteriosclerotic lesions were produced in immature Japanese macaques by a pyridoxine-deficient diet without feeding excess fat. The monkeys eating the deficient diet had a gradual continued loss of weight. Abnormal ST segments of abnormal T waves were present in the electrocardiograms and arteriosclerotic lesions were found in the coronary arteries and the aorta as well as the arteries of the kidney, pancreas, testicle, and mesentery. The 'lipid level' in the serum was not different between test and control animals.

6. *Macaca nemestrina* (Pigtailed Macaque)

There are four subspecies of the pigtailed macaque [21]. They are exported from Kuala Lumpur, Malaysia; Bangkok, Thailand; Djakarta, Indonesia [29]. Adult males weigh 6.2–14.5 kg and females weigh 4.7–10.9 kg.

BLAKELY *et al.* [4] studied the serum cholesterol concentration of large numbers of pigtailed macaques from Malaysia and Sumatra that were fed a commercial monkey chow. They found that males from each location had significantly lower serum cholesterol concentrations than did females from the same location and pigtailed macaques from Sumatra of both sexes had significantly higher total serum cholesterol concentrations than did those from Malaysia.

STEMERMAN and Ross [26] studied eight monkeys of each sex that were 10–12 months old and weighed from 1.1 to 1.8 kg. They cannulated the right or left femoral artery with a Fogarty arterial catheter. The balloon was inflated at a pressure of 700 mm Hg, rapidly pushed up the aorta to the diaphragm, pulled back, and removed. The animals were killed at 10 min, 1 h, 24 h, 2, 3, 4, 5, 7, 14, 28 days, 3 and 6 months and their vessels examined by light and electron microscopy. An unmanipulated artery from another animal or the femoral artery from the opposite side of the same animal was used as a control. Platelets were found to cover the denuded area immediately after de-endotheliation. The thrombosis gradually disappeared and by seven days the vessel appeared re-endotheliazed. Beginning day 4, smooth muscle cells migrated into the intima and proliferated. By 28 days the lesion consisted of multiple layers of smooth muscle cells surrounded by fibrous tissue elements. With no further injury the lesion decreased after six months. The authors speculated that injury to the endothelium exposing the medial smooth muscle cells to plasma constituents, may be the principal factor in the migration and proliferation of the smooth muscle cells.

We currently are studying 12 young adult male *M. nemestrina* of Malaysian origin. One half the animals have been fed a diet containing 1 mg of cholesterol/Cal and 45% of calories as lard and the other half the same diet without added cholesterol (complete diet table I). Over a two-year period the serum cholesterol concentration has averaged 176 ± 8 mg/dl (mean ±SEM) for the control monkeys and 926 ± 30 mg/dl for the cholesterol fed animals.

An interesting observation has been the development of micronodular cirrhosis in the cholesterol fed monkeys. Wedge biopsies of the liver were obtained from three monkeys of each group after they had been fed their respective diets for 18 months. The morphologic characteristics of the liver samples obtained from the three control monkeys were normal. The liver sample from one cholesterol-fed animal had ceroid pigment in the portal areas with short collagenous septae and ductule formation, the second had complete septae and ductule formation, and that from the third animal had micronodular cirrhosis. This species is the only one of seven we have studied, that developed micronodular cirrhosis when fed this diet [7].

B. Cercopithecus

The genus *Cercopithecus* contains a large number of morphologically similar African monkeys and consists of 19 species when the genera *Miopithecus* and *Allenopithecus* are included. Only three of these species have been studied for atherosclerosis research, *C. aethiops* (grivets, vervets, African green monkeys), *C. talapoin* (*Miopithecus talapoin*, talapoin monkey, mangrove monkey), and *C. ascanius* (redtailed monkeys).

1. Cercopithecus aethiops (African Green Monkeys, Grivets, Vervets)

This species is comprised of the following recognizably different subspecies which are used in biomedical research. *C. aethiops sabeus*, the callitriche exported from Nairobi, Kenya; *C. aethiops hilgerti*, the grivet exported from Addis Ababa, and Asmara, Ethiopia; *C. aethiops arenarius* exported from Mogadishu, Somaliland; and *C. aethiops callidus* or *C. aethiops johnstoni* closely related vervets also exported from Nairobi, Kenya [29]. They are abundant and an agricultural pest in some areas.

GRESHAM *et al.* [10] examined the aortas from 82 *C. aethiops* which had been in captivity for an average of eight to nine weeks and had been fed 'rat pellets'. The animals were described as being mature or immature based on the appearance of their teeth. Half of the mature monkeys (29/61) and 15% of the immature (3/21) had macroscopic sudanophilic lesions of the aorta. These were mainly small fatty streaks situated near the orifices of the renal, superior mesenteric, and coeliac arteries. The sudanophilic areas varied in size from minute spots to lesions up to 2 mm diameter.

We have recently reported on a study of dietary induced atherosclerosis in *C. aethiops* [5, 6]. Four adult male monkeys of Kenyan origin were fed a diet containing 1 mg of cholesterol/Cal and 45% of the calories as lard for 42 months (table I). Their serum cholesterol concentrations averaged about 450 mg/dl throughout this period. The distribution and microscopic appearance of their atherosclerotic lesions is similar to that seen in human arteries. Plaques were present in the abdominal aorta, major coronary arteries, femoral arteries and at the carotid bifurcation.

The percentages of thoracic and abdominal aortic surfaces covered by fatty streaks were about the same (35%), but there was about twice as much plaque in the abdominal (40 $\pm$ 7%, mean $\pm$ SEM) as in the thoracic (26 $\pm$ 9) segment. Microscopically, the plaques usually had a core of lipid near the intima-media junction which was covered by fibromuscular elements of five or more cells in thickness (fig. 8). Extensive mineralization with some hemosiderin was present in the larger lesions.

All four of the monkeys fed the atherogenic diet had focal plaques in the left main coronary artery and smaller plaques at the origins of branches from the major extramural arteries. Two of the monkeys had average serum cholesterol concentrations (432 and 415 mg/dl) that were approximately 100 mg/dl lower than the other animals (533 and 512 mg/dl). Their plaques in the main left coronary arteries caused luminal reduction of 10–15%, to be compared to 70% stenosis in the monkeys with higher cholesterol concentrations. Extravasated erythrocytes and hemosiderin were present in one of these larger lesions. Scarring and some neovascularization were present in the left ventricle of one monkey with a large plaque in the left main coronary artery. None of the monkeys had atherosclerotic lesions in the intracranial arteries. Figure 9 is a photomicrograph of a plaque in a coronary artery of a *C. aethiops*.

The *C. aethiops* fed the atherogenic diet had little cholesterol accumulation in organs or tissues other than in arteries in comparison to other species fed this diet (table IV). Hepatic cholesterol approximately doubled and tendon cholesterol concentration was increased about fourfold in these monkeys compared to monkey chow-fed control animals.

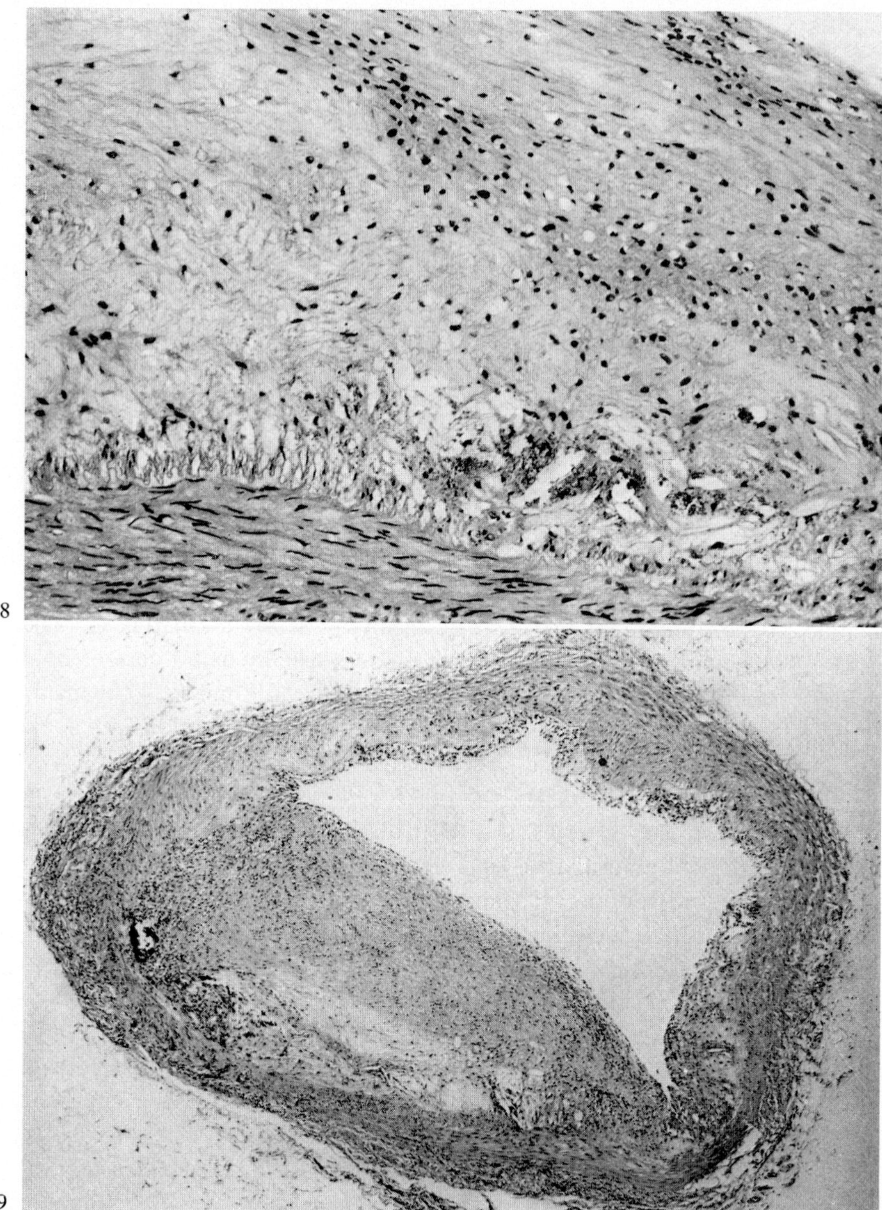

Fig. 8. Photomicrograph of an aortic plaque from a *C. aethiops*. Note lipid clefts and mineralization near the internal elastic lamina, beneath a 'fibromuscular cap'. H.E. × 73.

Fig. 9. Photomicrograph of a coronary artery from a *C. aethiops* fed an atherogenic diet. Note the lumen reduced by a large eccentric plaque. The underlying media is thinned. There are lipid clefts and mineral deposits on the intimal side of the internal elastic lamina. H.E. × 73.

Table IV. Effect of cholesterol containing diet on organ and tissue cholesterol concentration of African green monkeys [1, 2]

	Control diet (N = 4)	Atherogenic diet (N = 4)
Brain	16.61 ±1.44	19.80 ±0.68
Myocardium	1.30 ±0.05	1.16 ±0.11
Lung	3.85 ±0.23	3.52 ±0.23
Liver	4.24 ±0.39	8.17 ±0.76
Spleen	3.87 ±0.41	4.51 ±0.42
Jejunum	2.45 ±0.25	2.55 ±0.28
Ileum	2.36 ±0.13	2.49 ±0.22
Pancreas	1.86 ±0.23	1.71 ±0.13
Kidneys	3.45 ±0.12	3.01 ±0.21
Skin		
Eyelids	2.62 ±0.20	2.71 ±0.23
Ear	2.13 ±0.20	2.50 ±0.14
Thoracic	2.07 ±0.13	2.41 ±0.10
Forearm	2.08 ±0.33	3.06 ±0.62
Tendon	0.53 ±0.07	2.13 ±0.77
Muscle	0.62 ±0.01	0.71 ±0.10
Fascia	0.90 ±0.09	1.65 ±0.51
Adipose	1.54 ±0.10	2.11 ±0.87
Eye	0.72 ±0.21	0.48 ±0.02
Cornea	0.67 ±0.11	1.14 ±0.39

[1] All data are expressed as mg/g of wet weight, mean ±SEM
[2] Adapted from Bullock *et al.* [6].

Control animals in this study consisted of three monkeys fed the same diet without added cholesterol (table I). After the 30th month they were fed liquid diets containing up to 1 mg of cholesterol/Cal for metabolic studies. During the last six months, two of them had serum cholesterol concentrations of approximately 400 mg/dl. Four monkeys that were fed monkey chow were also examined for atherosclerosis. Their lesions consisted principally of fatty streaks. Small plaques, mainly in the aorta, were present in the three monkeys fed the control liquid diets and in one of four monkeys fed monkey chow.

2. *Cercopithecus talapoin*
(*Miopithecus talapoin*, Talapoin Monkey, Mangrove Monkey)
The talapoin is found in the mangrove swamps, swamp forests, and gallery forests in West Central Africa along the Atlantic Coast from Southern

Table V. Individual necropsy data for four talapoins[1] (all values = mean ± SEM)

Monkey No.	Sex	Dietary history	Kidney grade	Serum urea nitrogen	24-hour urine protein	Serum cholesterol concentration	Atherosclerosis, %		
							aorta arch	thoracic aorta	abdominal aorta
				mg/dl	mg	mg/dl			
5494	F	control diet 11 months	end stage	87 ±7 (6)	41 ±10 (4)	175 ±7 (8)	20 FS	10 FS	5 FS
5466	F	test diet 8 month then control diet 2 months	end stage	72 ±12 (6)	128 ±29 (4)	1,296 ±64 (8)	5 FS 75 PL	5 FS 65 PL	40 PL
5520	F	test diet 8 months then control diet 2 months	end stage	53 ±5 (6)	107 ±28 (4)	1,314 ±107 (8)	70 PL	40 PL	10 PL
5635	F	test diet 8 months then control diet 5 months then monkey chow 2 months	end stage	40 ±4 (6)	115 ±23 (4)	1,257 ±211 (8)	10 FS 50 PL	5 FS 37 PL	5 FS 15 PL

Numbers in parentheses represent number of determinations.

FS = Fatty streak; PL = plaque.

[1] Adapted from HAMM *et al.* [11].

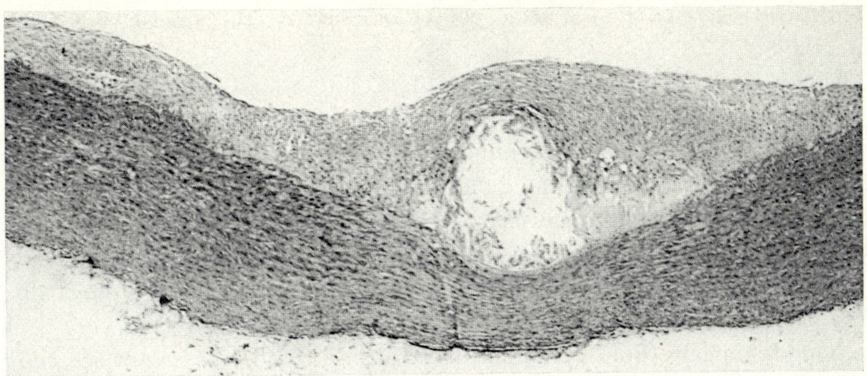

Fig. 10. Photomicrograph of an aortic plaque from a *C. talapoin* fed an atherogenic diet. The space contained mainly lipid before the tissue was processed for paraffin sections. H.E. × 19.

Cameroun to Northern Angola [21]. It is the smallest of the Old World monkeys (females 0.7–0.9 kg and males 1.2–1.4 kg). Female talapoins have a true menstrual cycle and develop a marked swelling of the peroneal skin during the estrogen phase of the menstrual cycle. Because of difficulties in acquiring this species from their country of origin, there are only a few colonies of talapoins in the United States.

Hamm *et al.* [11] fed a diet containing 25% lard and 1 mg cholesterol/Cal (complete diet composition in table I) to ten adult females and five adult males. A control group of three males and four females were fed the same diet without added cholesterol. Monkeys fed the test diet developed extremely high serum cholesterol concentrations which averaged 868 ± 74 mg/dl for the eight-month duration of the study. One female reached a high of 2,044 mg/dl the last month of cholesterol feeding. There was an apparent relationship between the serum cholesterol concentration and indices of renal disease in the monkeys fed the test diet. Test monkeys which had serum urea nitrogen concentrations which averaged less than 30 mg/dl throughout the study had significantly lower serum cholesterol concentrations than did those test animals with serum urea nitrogen levels above 30 mg/dl (694 ± 61 mg/dl versus 1100 ± 88 mg/dl, p <0.01). There was a positive relationship between average serum cholesterol and serum urea nitrogen concentrations in the test monkeys (r = 0.78, p <0.01). Proteinuria was most marked in the test diet group, especially in the females. In the test group there was a positive correlation between serum cholesterol concentration and the average amount of protein excreted in the urine per kg of body weight in 24 h (r = 0.84, p < 0.01).

Since the talapoin is difficult to obtain, no animals were killed for necropsy; however, three test and one control animals have died. These four animals were necropsied and found to have bilateral end stage kidneys.

Table V lists the individual data for these four animals. Experience with large numbers of nonhuman primates of several species ingesting the same test diet suggested to the authors that the three test animals had an unusually large amount of fibrous plaque for their short feeding period. Histologically, the plaques had a well-developed fibromuscular cap covering a core containing foam cells, lipid clefts, and mineral (fig. 10). These lesions are more similar to those of man and the African green monkeys *(C. aethiops)* than lesions from other nonhuman primates.

3. *Cercopithecus ascanius* (Redtailed Monkeys)

NAPIER and NAPIER [21] list five subspecies which live in tropical lowland rain forests of Africa; their food consists almost entirely of vegetables, and they raid crops when available.

STRONG and TAPPEN [28] studied the aortas and coronary arteries from 64 redtailed monkeys [28] shot within 40 miles of Kampala, Uganda. They were divided into three categories: juveniles included any monkey that did not have a complete adult dentition; subadults were monkeys with adult dentition but incomplete epiphyseal union, and adults had complete epiphyseal union and adult dentition. After staining with Sudan IV, 4% of juveniles and subadults and 11% of adults had demonstrable atherosclerotic aortic lesions. All lesions were minute sudanophilic spots involving 1% or less of the intimal surface. There were a moderate number of small musculoelastic intimal thickenings in the main branches of the coronary arteries. There are no other reports about atherosclerosis in these monkeys.

C. Cercocebus

The genus *Cercocebus* contains two groups of mangabeys, the albigena group *(C. albigena, C. atterimus)* and the torquatus group *(C. torquatus, C. atys, C. galeritus)* [3]. Both groups live in Africa. Only *C. albigena* (gray-cheeked mangabey) has been studied to determine the prevalence of naturally occurring atherosclerotic lesions. This species is wholly arboreal preferring mainly swampy areas of forest.

STRONG and TAPPEN [28] studied the aortas and hearts from 49 gray-cheeked mangabeys shot within 40 miles of Kampala, Uganda. The animals were divided into three categories based on identical dentition and skeletal characteristics as those described above for *C. ascanius*. Small yellow-gray intimal patches and small glistening gray-white elevated plaques were present in a few individuals. After staining with Sudan IV, intimal lesions were seen in the aorta of 32% of juveniles and subadults and 85% of adults. There was a similar prevalence of lesions in adult animals of both sexes. The female animals seemed to have slightly more extensive fatty streaking than males. The small elevated plaques were composed of elastic fibers, smooth muscle, and some collagen with only a small amount of fat. There were no coronary lesions producing narrowing, stenosis, or occlusion. There were a moderate number of small musculoelastic intimal thickenings in the main branches of the coronary arteries.

There are no reports of the use of this species in the study of experimental atherosclerosis.

D. Erythrocebus

NAPIER and NAPIER [21] list one species, *E. patas* with four subspecies, in this genus. The patas monkey is found in sub-Saharan Africa. They are ground-living monkeys which inhabit woodland savanna, wooded steppes, open savanna and sometimes sub-desert. They prefer flat, open arid country. Mature males weigh 7.5–12.6 kg and females weigh 4.1–7.1 kg.

PUCAK [15] after finding fatty streaking in the aortas of monkey chow-fed animals, has initiated a study of dietary induced atherosclerosis in 50 patas monkeys. Results of this study have not yet been reported.

References

1 ARMSTRONG, M. L.: Atherosclerosis in rhesus and cynomolgus monkeys; in STRONG Atherosclerosis in primates (this volume).

2 ARMSTRONG, M. L. and MEGAN, M. B.: Arterial fibrous proteins in cynomolgus monkeys after atherogenic and regression regimens. Circulation *48:* Suppl. IV, p. 41 (1973).

3 ARMSTRONG, M. L. and MEGAN, M. B.: Responses of two macaque species to atherogenic diet and its withdrawal; in SCHETTLER and SCHLIERF Atherosclerosis III, pp. 336–338 (Springer, Berlin 1974).

4   Blakely, G. A.; Morrow, A. C., and Morton, W. R.: Intraspecies variation in serum cholesterol levels in imported *Macaca nemestrina*. Lab. Anim. Sci. *23:* 119–121 (1973).

5   Bullock, B. C.; Lehner, N. D. M.; Clarkson, T. B., and Feldner, M. A.: Response of three species of nonhuman primates to an atherogenic diet. Circulation *48:* suppl. II, p. 251 (1973).

6   Bullock, B. C.; Lehner, N. D. M.; Clarkson, T. B.; Feldner, M. A.; Wagner, W. D., and Lofland, H. B.: Comparative primate atherosclerosis. I. Tissue cholesterol concentration and pathologic anatomy (in press).

7   Bullock, B. C.; Paula, R., and Rudel, L. L.: Dietary-induced cirrhosis in the pigtailed macaque. Fed. Proc. Fed. Am. Socs exp. Biol. *33:* 626 (1974).

8   Corey, J. E.; Hayes, K. C.; Dorr, B., and Hegsted, D. M.: Comparative lipid response of four primate species to dietary changes in fat and carbohydrate. Atherosclerosis *19:* 119–134 (1974).

9   Doyle, J. L.; Hollander, W.; Goldman, H. M., and Ruben, M. P.: Experimental atherosclerosis and the periodontium. J. Periodont. *40:* 44/350–48/354 (1969).

10   Gresham, G. A.; Howard, A. N.; McQueen, J., and Bowyer, D. E.: Atherosclerosis in primates. Bıt. J. exp. Path. *79:* 94–103 (1965).

11   Hamm, T. E.; Lehner, N. D. M.; Bullock, B. C., and Clarkson, T. B.: Renal disease, hypercholesterolemia, and atherosclerosis in the talapoin monkey *(Cercopithecus talapoin)* (in press).

12   Howard, C. F.: Spontaneous diabetes in *Macaca nigra*. Diabetes *21:* 1077–1090 (1972).

13   Howard, C. F.: Atherosclerosis in spontaneously diabetic monkeys. Circulation *48:* suppl. IV, p.41, (1973).

14   Jagannathan, S. N.; Madhavan, T. V., and Gopalan, C.: Effect of adrenaline on aortic structure and serum cholesterol in *Macaca radiata*. J. Atheroscler. Res. *4:* 335–345 (1964).

15   Pucak, C. J.: Personal commun. (1974).

16   Kramsch, D. M. and Hollander, W.: Occlusive atherosclerotic disease of the coronary arteries in monkey *(Macaca irus)* induced by diet. Expl molec. Path. *9:* 1–22 (1968).

17   Kramsch, D. M.; Hollander, W., and Renaud, S.: Induction of fibrous plaques versus foam cell lesions in *Macaca fascicularis* by varying the composition of dietary fats. Circulation *48:* Suppl. IV. p. 41 (1973).

18   Kuzuya, H.: Arteriosclerosis in pyridoxine-deficient monkeys. Primates *2:* 99 (1959).

19   Lang, C. M. and Barthel, C. H.: Effects of simple and complex carbohydrates on serum lipids and atherosclerosis in nonhuman primates. Am. J. clin. Nutr. *25:* 470–475 (1972).

20   Malmros, H.; Wigand, G., and Kockum, I.: Experimental hyper-cholesterolemia and hypertriglyceridemis in cynomolgus monkeys fed saturated fat and cholesterol. J. Atheroscler. Res. *5:* 474–482 (1965).

21   Napier, J R and Napier, P H : A handbook of living primates (Academic Press, New York 1967).

22 PICK, B. and GLICK, G.: The influence of hypertension (HT) on diet induced aortic, coronary, and cerebral atherosclerosis in the stumptail macaque. Circulation *46:* Suppl. II, p. 248 (1972).
23 PICK, R. and KATZ, L. N.: Cholesterol-fat induced atherosclerosis in the stumptail macaque *(Macaca speciosa)*. Circulation *40:* Suppl. III, p. 20 (1969).
24 PRATHAP, K.: Spontaneous aortic lesions in wild adult Malaysian long-tailed monkeys *(Macaca irus)*. J. Path. *110:* 135–143 (1973).
25 PRATHAP, K. and LAU, K. S.: Spontaneous and experimental arterial lesion in Malaysian long-tailed monkey; in GOLDSMITH and MOOR-JANKOWSKI Medical primatology 1972, part III, pp. 343–349 (Karger, Basel 1972).
26 STEMERMAN, M. B. and ROSS, R.: Experimental atherosclerosis. I. Fibrous plaque formation in primates an electron microscopic study. J. exp. Med. *136:* 769–789 (1972).
27 STRONG, J. P.: Atherosclerosis in primates. Introduction and overview (this volume).
28 STRONG, J. P. and TAPPEN, N. C.: Naturally occurring arterial lesions in African monkeys. Archs Path. *79:* 199–205 (1965).
29 THORINGTON, R. W.: The identification of primates used in viral research. Lab. Anim. Sci. *21:* 1074–1077 (1971).
30 THORINGTON, R. W. and GROVES, C. P.: An annotated classification of the Cercopithecoidea; in NAPIER and NAPIER Old World monkeys, pp. 629- 647 Academic Press, New York 1970).

THOMAS B. CLARKSON, THOMAS E. HAMM, BILL C. BULLOCK, and NOEL D. M. LEHNER, Department of Comparative Medicine and Arteriosclerosis Research Center, Bowman Gray School of Medicine, *Winston-Salem, NC 27103* (USA)

Prim. Med., vol. 9, pp. 90–144 (Karger, Basel 1976)

Atherosclerosis in New World Monkeys[1]

THOMAS B. CLARKSON, NOEL D. M. LEHNER, BILL C. BULLOCK,
HUGH B. LOFLAND, WILLIAM D. WAGNER

Arteriosclerosis Research Center and Departments of Pathology and Comparative Medicine,
Bowman Gray School of Medicine, Wake Forest University, Winston-Salem, N.C.

Contents

[1] Supported in part by USPHS NIH grants BR00180, RR00236, and HL14164.

I. Introduction

The New World monkeys are groups of nonhuman primates whose natural habitats extend from Central America into South America. Taxonomic groupings are usually based on the complexity of structural and behavioral organization. On this basis the New World monkeys are in a position between the prosimian primates and the Old World monkeys, apes, and man.

The New World monkeys are comprised of two families. The family Callitrichidae consists of five genera of marmosets and tamarins (*Callithrix, Cebuella, Saguinus, Leontideus,* and *Callimico*). The family Cebidae consists of eleven genera (*Aotus, Callicebus, Pithecis, Chiropotes, Cacajao, Cebus, Saimiri, Alouatta, Ateles, Lagothrix,* and *Brachyteles*).

Among Callitrichidae and Cebidae information is available about the usefulness for atherosclerosis research for only the genera *Cebuella, Saguinus, Cebus, Saimiri, Alouatta, Ateles,* and *Lagothrix*. Each of these genera will be discussed separately.

II. Atherosclerosis in the Investigated Genera of New World Monkeys

A. Cebuella

The genus *Cebuella* consists of one species, *C. pigmae* or the pygmy marmoset. These small primates have an adult body weight of less than 200 g and their life-span is said to be six to eight years. The presumed relatively short life, as compared with other nonhuman primates, has suggested their use for studies on the disease of senescence. While no reports can be found on naturally occurring or diet-induced atherosclerosis of this species, one report would suggest that they are unusually susceptible to diet-induced hypercholesterolemia [57]. The authors reported the basal serum cholesterol concentrations to be 167–250 mg/dl. During their experiment pygmy marmosets were fed a liquid diet consisting of condensed milk, 400 ml; corn syrup, 25 ml; the equivalent of four whole eggs, 175 ml; water, 400 ml; plus vitamin and mineral mixtures. Total serum cholesterol concentrations were observed to increase gradually over a four- to eight-month period reaching a maximum of about 800 mg/dl.

B. Saguinus

The genus *Saguinus* consists of four species, only two of which have been investigated as animal models for atherosclerosis, *S. nigricollis* (white-lipped marmoset) and *S. oedipus* (cotton-top marmoset). These small primates (300–500 g) are widely distributed in the rain forest of South America, primarily in the Amazon basin, although they do extend in their distribution up to 1,000 m altitude. They have been widely used in virologic research.

In the summer of 1964, workers in our laboratories, in collaboration with investigators from the Department of Pathology of the Louisiana State University School of Medicine, surveyed the serum lipid concentrations and naturally occurring arterial lesions of 30 adult and 10 juvenile *S. nigricollis*. Free-ranging *S. nigricollis* were found to have a mean total serum cholesterol concentration of 69 ± 3.5 mg/dl SEM, serum triglyceride concentration of 95 ± 7 mg/dl, α-lipoprotein cholesterol of 18 mg/dl $\pm$ 1.8, β- plus pre- β-lipoprotein cholesterol 50 mg/dl $\pm$ 2.9 [32]. Examination of the aortas and the coronary arteries suggested that these animals were quite resistant to naturally occurring atherosclerosis [44]. Of the animals examined, one adult female had a small sudanophilic lesion of the thoracic aorta. No other aortic lesions were seen. No coronary artery lesions were seen among any of the white-lipped marmosets.

While no definitive information is available on the natural occurrence of atherosclerosis among free-ranging *S. oedipus*, they appear to be completely resistant to atherosclerosis when fed a diet devoid of cholesterol. Dreizen *et al.* [14] stated that no gross or microscopic lesions of atherosclerosis were found at postmortem examination of more than 100 *S. oedipus* fed a purified diet containing 25% protein from vitamin-free casein, 20% corn starch as the carbohydrate source, and 10% corn oil.

Two attempts have been made to induce atherosclerosis in this species by the addition of cholesterol to the diet. DREIZEN *et al.* [13] attempted to induce atherosclerosis by feeding a diet high in cholesterol and coconut oil. The marmosets in that experiment developed a malabsorption syndrome characterized by jejunal lipodystrophy, steatorrhea, and osteomalacia which made it necessary to terminate the experiment before it was known if atherosclerosis could be induced. In another experiment the cotton-tops were fed a purified diet consisting of sucrose, 40.6% as the carbohydrate source; vitamin-free casein at 25% as the protein source; 23% lard and 2% corn oil as the fat source; and 5% cholesterol [14]. The total serum cholesterol concentrations increased from a base line range of 80–99 mg/dl to means of 550 mg/dl at 52 weeks and 1,485 mg/dl at 73 weeks. From our own experiences with several nonhuman primate species it is most unusual to see progressive increases in serum cholesterol concentrations over such a long period of time. Usually the concentrations plateau after three to five months of consuming a cholesterol-containing diet. Only in instances where glomerulonephritis has complicated experiments have we seen progressive increases for a year or more. DREIZEN *et al.* [14] necropsied their marmosets at time intervals of a year or less and at 68 and 73 weeks of atherosclerosis induction. Foam cell lesions of the aorta were found after more than a year but not for less than a year on the diet. The lesions were equally prevalent in the thoracic and abdominal aortas; atherosclerosis was particularly common in the pulmonary, renal, pancreatic, splenic, and mesentery arteries. Foam cell lesions occurred in the arteries and arterioles of the tongue and bore a resemblance to those observed in the myocardium. The coronary artery atherosclerosis resembled that found among other New World monkeys in that the proximal main branches were spared and the small intramyocardial arteries were affected.

The marmosets are relatively plentiful, require little cage space, and at least the one species discussed above is slightly susceptible to diet-induced atherosclerosis. Their small size is a disadvantage for studies which require more than a few milliliters of blood or significant amounts of arterial tissue.

C. Cebus

The genus *Cebus* consists of four species; *C. capucinus*, *C. albifrons*, *C. nigrivittatus*, and *C. apella*. Among these, data are available on the natural occurrence of atherosclerosis in *C. albifrons* and *C. apella*. *C. albifrons* has been studied rather extensively in the laboratory. *C. albifrons* is one of the species in which some of the earliest work was done on diet-induced atherosclerosis of nonhuman primates [37]. As a part of a joint study with the Louisiana State University, we examined the occurrence of atherosclerosis among free-ranging *C. albifrons* and *C. apella* [44]. One of 20 adult females, 3 of 17 adult males, and none of the 20 juvenile *C. albifrons* were found to have aortic atherosclerosis. Among the four adult animals with aortic atherosclerosis the lesions were usually very small fatty flecks on the intimal surface. No coronary artery lesions were seen in either adult or juvenile animals. Two adult and 18 juvenile *C. apella* were examined and no atherosclerosis was seen in either the aorta or coronary arteries.

The earlier literature refers to *C. fatuella* which more appropriately should be designated *C. albifrons*. These animals are interesting as animal models of atherosclerosis because rather striking differences are seen in their response to saturated and unsaturated dietary fat, and because there is a large age and sex difference in their susceptibility to diet-induced atherosclerosis.

1. Effect of Dietary Carbohydrate

In an early experiment, PORTMAN *et al.* [49] compared the response of *C. albifrons* to a cholesterol-containing (1 mg/Cal) diet with either sucrose or starch as a source of dietary carbohydrate. After eight weeks the serum cholesterol concentrations of the sucrose-fed animals were higher than those of the starch-fed animals (382 versus 369 mg/dl). The source of the carbohydrate was then reversed and the finding was confirmed (starch, 336 mg/dl, and sucrose, 404 mg/dl). In a more recent experiment, LANG and BARTHEL [23] compared the effect of feeding either dextran or sucrose as a part of a cholesterol-containing diet to *C. albifrons*. No differences in total serum cholesterol concentration was observed, however, the serum triglyceride concentrations were significantly higher among the animals fed sucrose than among those fed the dextran diet. Intimal proliferation was found to be significantly greater in the group fed dextran than in that fed sucrose. No significant difference was found in the aortic cholesterol concentrations.

2. Effect of Type of Dietary Fat

The type of dietary fat fed to *C. albifrons* affects the extent of hypercholesterolemia resulting from a cholesterol-containing (1 mg/Cal) diet. PORTMAN *et al.* [49] found serum cholesterol concentrations of 600–900 mg/dl when the animals were fed lard as compared with 200–300 mg/dl when the dietary fat was corn oil. Similarly, differences were seen in β-lipoproteins with the lard-fed animals having concentrations of about 324 mg/dl as compared to 52 mg/dl among those fed corn oil. In a later experiment, PORTMAN and SINISTERRA [51] investigated the possible mechanisms of this dietary fat effect. They compared the disappearance of intravenously administered [¹⁴C]-cholesterol among monkeys fed cholesterol-containing diets with either lard or corn oil as a source of dietary fat. The biological half-life of the intravenously administered cholesterol was 6.1–7.1 days among the lard-fed animals, and 7.3–9.5 days among the corn oil-fed animals, suggesting either differences in whole body synthesis or in absorption of the dietary cholesterol. To study the relative effectiveness of various fats in the absorption of radio-labeled cholesterol, the absorption values obtained for lard were taken as 1.00 and the other fats were equated to lard. On that basis safflower oil was 0.94, corn oil was 0.56, and coconut oil was 0.44.

The effect of the type of dietary fat on serum cholesterol concentrations, extent of atherosclerosis, and the type of atherosclerosis among *C. albifrons* consuming a cholesterol-containing diet has been studied by WISSLER *et al.* [61]. In that experiment, the monkeys were fed a diet containing about 1 mg cholesterol/Cal, and the source of fat (about 45% of calories) was either butterfat, coconut oil, or corn oil. After 45 weeks, coconut oil-fed monkeys had the highest serum cholesterol concentrations (338 mg/dl) followed by those fed butterfat (297 mg/dl) and corn oil (219 mg/dl). At autopsy all of the coconut oil-fed monkeys and three of four butterfat-fed animals had aortic atherosclerosis. None of the corn oil-fed monkeys had aortic atherosclerosis. Coronary artery atherosclerosis was not seen in any of the animals. The minimal atherosclerosis seen in this experiment may have been due to the young age of the animals as indicated by their weight of about 1,500 g (distinctly juvenile). Also of interest are the reported qualitative differences in the atherosclerotic lesions that developed when the butterfat or coconut oil was fed. The lesions seen in coconut oil-fed animals consisted of small foci of subintimal fat accumulation associated with marked intimal proliferation, while those seen in the butterfat-fed animals contained foam cells with little intimal reaction.

3. Effect of Age and Sex

The effect of age and sex of cebus monkeys has been studied with regard to sterol metabolism by MACNINTCH *et al.* [33] and LOFLAND *et al.* [31]; with regard to arterial metabolism by LOFLAND *et al.* [31]; and finally, with

Table I. The effect of control and cholesterol-containing diets on aortic atherosclerosis of
Cebus albifrons[1]

Group	Sex	Number	Prevalence of lesions (histologic)[2]		Lesion extent[3]	
			thoracic	abdominal	thoracic	abdominal
Adult test	M	4	3/4	3/4	45	34
Adult test	F	6	4/6	4/6	33	23
Young test	M	9	1/9	1/9	25	13
Young test	F	9	3/9	3/9	25	13
Adult control	M	4	1/4	0/4	19	10
Adult control	F	2	0/2	0/2	25	18
Young control	M	3	0/3	2/2	15	11
Young control	F	3	0/3	0/2	20	23

[1] Adapted from BULLOCK *et al.* [2].

[2] Aorta blocks for histologic study were not available in all cases. The data are expressed as the number of animals found on histologic sectioning to have atherosclerotic lesions/ the number of animals examined in that group (rather than the number of animals in the group).

[3] Expressed as the mean percentage of intimal surface with lesions.

regard to the pathologic characteristics of the lesions by BULLOCK *et al.* [2]. Among *C. albifrons* fed cholesterol-containing diets, adult males had the highest serum cholesterol concentrations and young males had lower concentrations than females of either age group. Young females had lower serum cholesterol concentrations than adult females during the first year of consuming a cholesterol-containing diet but higher concentrations during a second year of diet consumption.

Findings on aortic atherosclerosis of *C. albifrons* is summarized in table I. Aortic atherosclerosis was slight and the age and sex differences were not as clear as in atherosclerosis of other arteries. When atherosclerosis did occur, the lesions were most often fatty streaks, however, occasionally raised yellow plaques and slightly raised white plaques were seen, primarily in the thoracic aorta (fig. 1). Histologically, the accumulated lipid was in the intima along the elastic lamina and in smooth muscle cells between layers of the elastic tissue (fig. 2). Cholesterol crystals were seen to accumulate in the inner media (fig. 3, 4); however, the extent of intimal proliferation does not seem proportional to the number of cholesterol clefts in the lesions.

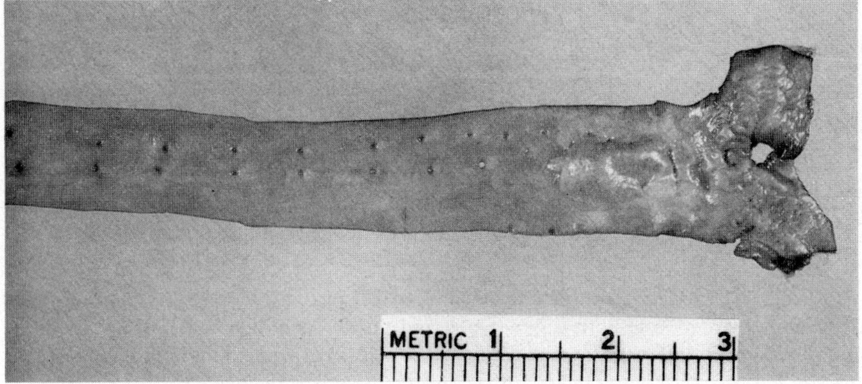

Fig. 1. Large plaques in the aortic arch and upper thoracic aorta of a *C. albifrons* fed an atherogenic diet. The plaques are smaller in the more distal part of the aorta. F. and D.C. Blue No. 1. × 1.8.

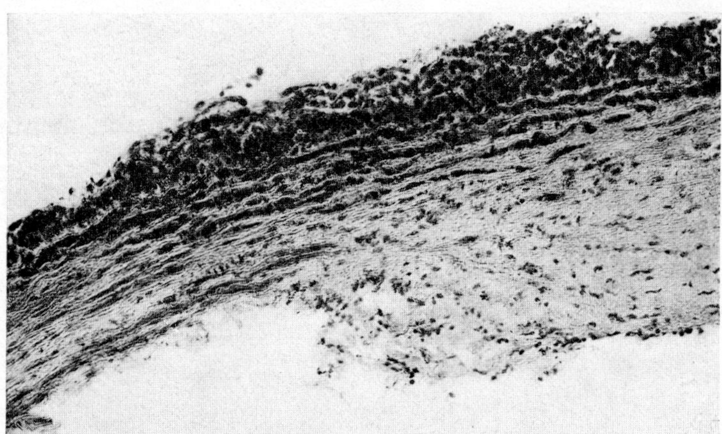

Fig. 2. Section of the thoracic aorta from an adult male *C. albifrons* that had been fed an atherogenic diet. Note abundant sudanophilic material in the intima (top) and media. Frozen section, Sudan IV and hematoxylin. × 90.

The effect of age and sex on the extent and severity of coronary artery atherosclerosis in *C. albifrons* is marked (table II). Among animals fed a cholesterol-containing diet for two years there was a considerable difference in the extent and severity of coronary atherosclerosis between adult and young animals. No difference was seen in comparing the adult males and

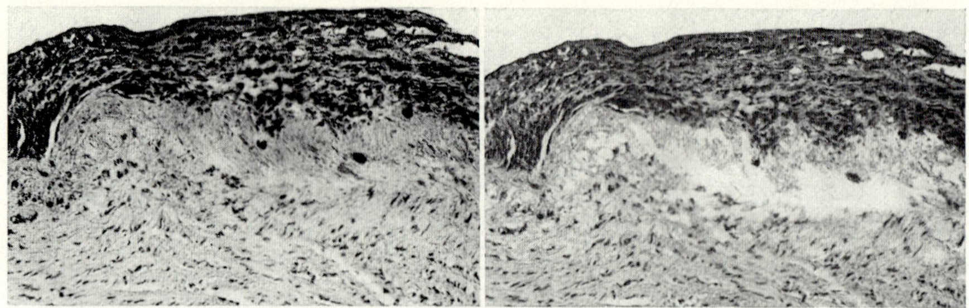

Fig. 3. Frozen section of thoracic aorta from a female *C. albifrons* fed an atherogenic diet for two years. Note both sudanophilic and non-sudanophilic material in the intima and inner media. Photomicrograph on the right was taken using polarized light. Note the marked birefringence in the inner media. This material was removed by lipid solvents [2]. Sudan IV and hematoxylin. × 77.

Fig. 4. Thoracic aorta of adult male *C. albifrons* fed an atherogenic diet. Note lipid clefts betweeen the elastic lamina (left) and foam cells in the intima (right) [2]. Verhoeff-van Gieson. × 90.

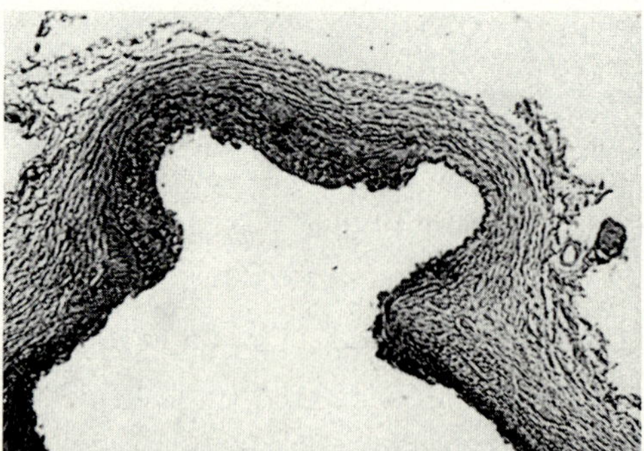

Fig. 5. Sudanophilic intimal thickening of the coronary ostium of a cholesterol fed *C. albifrons* [2]. Frozen section, Sudan IV and hematoxylin. × 16.

Table II. The effect of control and cholesterol-containing diets on coronary artery atherosclerosis of *Cebus albifrons*[1]

Group	Sex	Prevalence[2]	Extent[3]	Severity[4]
Adult test	M	5/5	15	35
Adult test	F	6/6	12	30
Young test	M	5/10	1	14
Young test	F	7/10	5	14
Adult control	M	1/4	0	0
Adult control	F	0/2	0	0
Young control	M	0/3	0	0
Young control	F	1/3	0	0

[1] Adapted from Bullock *et al.* [2].

[2] Expressed as the number of monkeys with coronary artery atherosclerosis, including main and intramyocardial branches/the number of monkeys examined.

[3] Expressed as the percentage of arteries seen with atherosclerotic lesions in 15-step sections of heart.

[4] Expressed as the mean percentage of apparent lumen obliterated in those arteries having atherosclerotic lesions.

---

females. In contrast, however, five times as many lesions were found in young females than in young males fed cholesterol-containing diets. This finding was consistent with increased serum cholesterol concentrations of young females as compared with young males during the second year of atherogenic diet. Microscopically, the features of coronary artery lesions were quite variable. Usually small lesions were seen at the coronary ostia (fig. 5) and arteries of all sizes were affected segmentally. Some lesions were composed almost entirely of foam cells and cholesterol crystals (fig. 6), while others contained minimal lipid finely dispersed in the lesion (fig. 7). A common feature of coronary atherosclerosis in cebus monkeys was the occurrence of plaques with irregular papillary surfaces (fig. 8). Like other New World monkeys, *C. albifrons* also developed lesions in the small intramyocardial arteries (fig. 9).

A consistent finding among *C. albifrons* fed atherogenic diets are large plaques at the carotid artery bifurcation (fig. 10). The prevalence and extent of such plaques is summarized in table III. An age difference in the occurrence of these plaques is not apparent, but the occurrence seems to be affected by the sex of the animal. Cholesterol-fed males of both age groups were more commonly and more severely affected than females, and it is of interest that

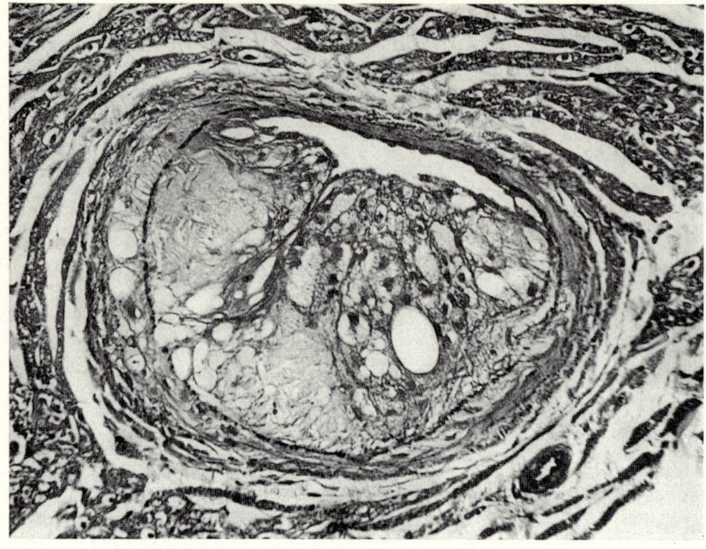

Fig. 6. Coronary artery of a *C. albifrons* fed an atherogenic diet for two years. Note large foamy cells on the right and lipid (cholesterol) clefts above the internal elastic layer at the center and on the left of the photomicrograph [2]. Verhoeff-van Gieson. × 220.

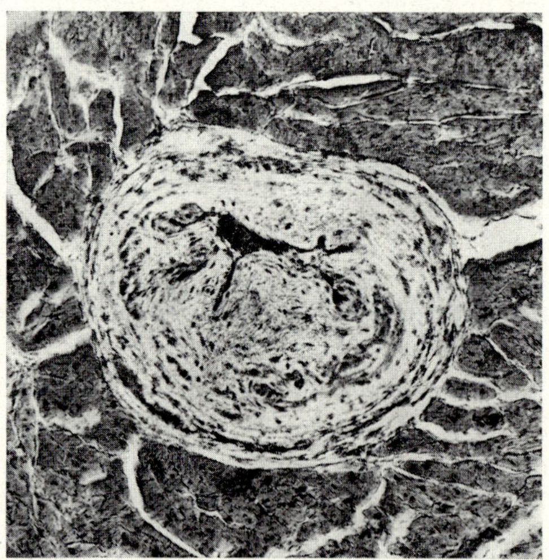

Fig. 7. Intramyocardial artery from an adult female *C. albifrons* fed an atherogenic diet for one year. Note a small amount of diffusely distributed sudanophilic material in a predominantly collagenous lesion [2], Frozen section, Sudan IV and hematoxylin. × 90.

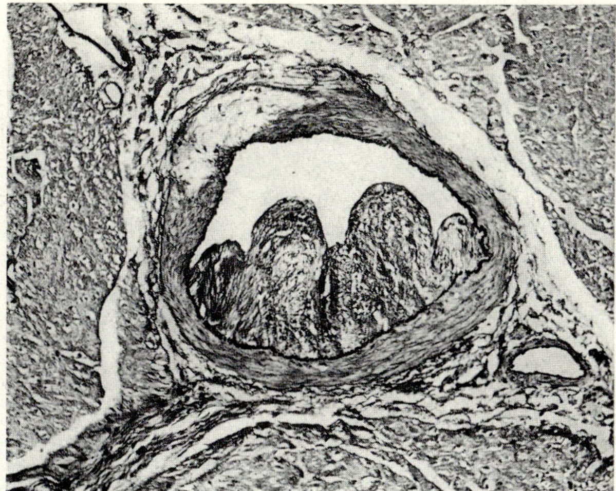

Fig. 8. Coronary artery plaque from a *C. albifrons* female fed an atherogenic diet. The irregular plaque surface is a common finding among cebus monkeys [2]. Verhoeff-van Gieson. × 80.

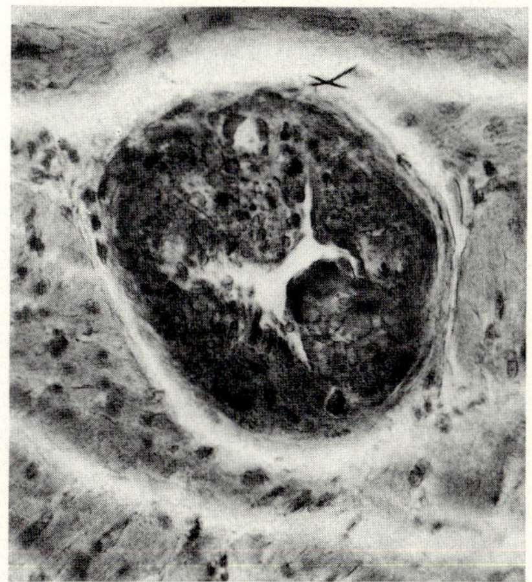

Fig. 9. Small intramyocardial artery narrowed by sudanophilic intimal thickening. This type of lesion is frequently found among cebus monkeys fed an atherogenic diet [2]. Frozen section, Sudan IV and hematoxylin. × 240.

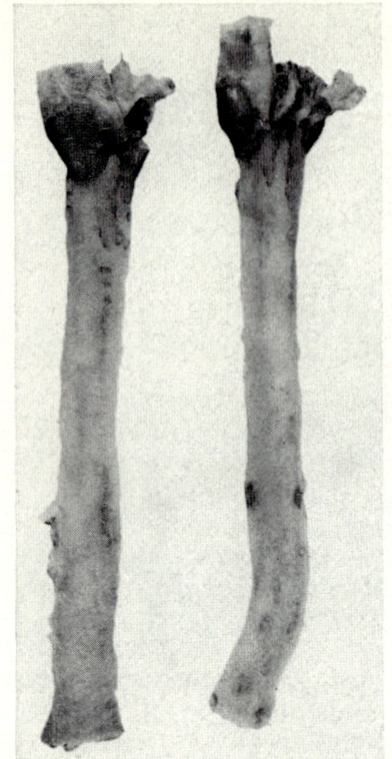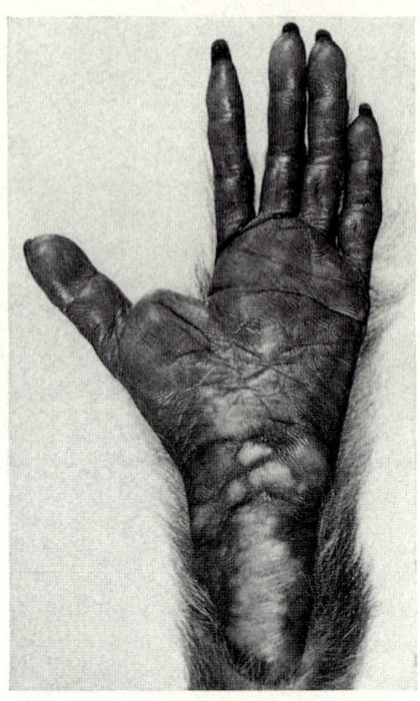

10   11

Fig. 10. Atherosclerotic plaques in the carotid bifurcation of a cebus monkey fed an atherogenic diet [2]. Gross Oil Red O and Sudan IV. × 1,6.

Fig. 11. Cutaneous xanthomas on the plantar surface of an adult male *C. albifrons* [2].

Table III. Atherosclerosis of the carotid artery bifurcation of *Cebus albifrons*[1]

Group	Sex	Number	Carotid artery bifurcation			
			prevalence[2]		severity[3]	
			left	right	left	right
Adult test	M	5	3/5	5/5	1.4	1.8
Adult test	F	7	3/7	3/7	1.8	0.8
Young test	M	11	6/11	8/11	0.9	1.2
Young test	F	10	3/10	4/10	0.8	0.9
Adult control	M	4	1/4	0/4	0.2	0
Adult control	F	2	0/2	0/2	0	0
Young control	M	3	0/3	0/3	0	0
Young control	F	3	0/3	0/3	0	0

[1] Adapted from BULLOCK *et al.* [2].

[2] Expressed as the number of monkeys with atherosclerotic lesions of the artery specified/the number of monkeys examined.

[3] Expressed as the mean for the group on an arbitrary scale of 0–4.

Table IV. Tissue cholesterol concentrations (mg/g of wet tissue) of control and cholesterol-fed *Cebus albifrons*[1]

Tissue	Control[2]	Cholesterol-fed			
		adult		young	
		males	females	males	females
Muscle	0.71	0.77	0.88	0.73	0.83
Liver	4.13	21.67	11.87	7.19	6.51
Skin	1.44	3.41	2.35	1.85	2.11
Spleen	5.5	15.69	8.03	7.76	8.27
Adrenal	61.01	56.88	75.84	45.10	69.91
Testes or ovaries	5.82	6.38	4.82	4.17	2.67
Kidney	3.85	4.38	4.35	3.31	3.45
Heart	1.13	1.32	1.71	0.89	1.43
Lung	4.36	5.10	4.79	3.83	4.40
Jejunum	8.48	3.87	3.28	2.98	3.32
Ileum (prox.)	2.01	3.94	2.21	2.86	3.15
Ileum (dist.)	2.34	3.37	2.56	2.28	2.26
Pancreas	3.36	3.54	3.45	2.96	2.14
Mesenteric fat	3.35	11.92	2.08	2.18	1.18
Thymus	lost	2.50	2.05	1.75	2.34
Foot pad	1.98	7.70	2.60	1.95	2.35

[1] Adapted from LOFLAND *et al.* [28].
[2] Noncholesterol-fed.

this finding among the young monkeys was in contrast to the occurrence of coronary disease in that age group.

Atherosclerotic lesions of *C. albifrons* also are found in the arteries of the tongue, uterus, kidney, liver, as well as, the arteries of the extremities [2]. No lesions of the cerebral arteries have been found in that species.

Cutaneous and tendinous xanthomas appear to be infrequent among cebus monkeys fed cholesterol-containing diets [2]. We have observed one adult male monkey to have raised yellowish nodules on the palmar and plantar surfaces (fig. 11). At necropsy these lesions were found to be composed of large macrophages containing lipid. Similar lesions were found near the insertion of tendons.

4. Tissue Cholesterol Concentrations

In considering nonhuman primate models of atherosclerosis, it is of interest to examine the accumulation of cholesterol in the organs and tissue

Table V. Aortic lipid composition in *Cebus albifrons* fed cholesterol-containing diets with varying fat types[1]

Diet[2]	N	Total lipid	Non-esterified cholesterol	Esterified cholesterol	Total cholesterol	Phospho-lipid	Tri-glyceride	Macro-scopic lesions	Micro-scopic lesions
B	4	16.50	1.61	0.24	1.85	3.60	7.7	1	3
CO	4	15.80	2.02	0.89	2.91	3.70	10.68	1	4
C	4	19.50	1.87	0.23	2.10	3.62	10.47	0	0

[1] Modified from table II of WISSLER *et al.* [61]. All values are means and represent mg/g wet aorta.

[2] All diets are semisynthetic – in addition, B contains 25% butter and 0.435% added cholesterol; CO contains 25% coconut oil, 0.5% cholesterol, and C contains 25% corn oil, 0.5% cholesterol.

of monkeys fed a cholesterol-containing diet. Table IV summarizes the tissue cholesterol concentrations of cholesterol-fed and of control *C. albifrons.* There are increases in the cholesterol concentrations of liver, spleen, skin, and perhaps other tissues, but the increases are not as striking as seen in the macaques and certain other New World monkeys.

5. Arterial Composition

WISSLER *et al.* [61] studied the early accumulation of lipid in the aorta of two- to five-year-old male cebus monkeys fed atherogenic diets containing different types of fats. After 45 weeks, the average serum cholesterol concentrations were 297, 338, 219 mg/dl in the butter, coconut oil, and corn oil groups, respectively. The arterial lipid fraction which showed the most marked differences from diet group to diet group was esterified cholesterol (table V). The concentration was highest in the coconut oil-cholesterol-fed animals.

In the study by MACNINTCH *et al.* [33], male and female young (less than 18 months) and adult (four to five years) cebus were examined one year after they were fed either a control semisynthetic diet containing 25% lard, or an atherogenic diet of similar composition but including 0.5% cholesterol. Grossly, only minimal fatty streaks were present in the aortas of the atherogenic-fed groups. Analysis of the aortic and carotid artery total cholesterol concentrations revealed little accumulation of cholesterol in the aortas after 12 months of atherogenic diet (table VI).

In another study [2] atherogenic diets were fed to groups of young (12–18 months) and adult (four to five years) male and female *C. albifrons* for 24 months. The animals were fed either control semi-synthetic diets

Table VI. Arterial cholesterol concentration of *Cebus albifrons* fed atherogenic (A) and control (C) diets for twelve months[1]

Group	Diet[2]	Thoracic aorta	Abdominal aorta	Right carotid artery
Adult	C	3.1 ± 0.8	2.7 ± 0.3	2.7 ± 0.5
Adult	A	3.7 ± 0.8	2.7 ± 0.3	2.9 ± 1.0
Young	C	2.2 ± 0.5	2.3 ± 0.2	2.1 ± 0.3
Young	A	1.8 ± 0.2	1.8 ± 0.2	1.8 ± 0.2

[1] Adapted from table 7 of MacNintch *et al.* [33]. All values are means $\pm$ SEM and represent mg cholesterol/g wet artery.

[2] Diet C = semisynthetic diet containing 25% lard; A = similar diet as C but containing 0.5% cholesterol.

Table VII. Aortic cholesterol concentration of *Cebus albifrons* fed atherogenic (A) and control (C) diets for 24 months[1]

Group	Diet[2]	Sex	Number	Thoracic aorta		Abdominal aorta		Plasma
				athero-sclerosis[3]	cholesterol concentration	athero-sclerosis	cholesterol concentration	concentration mg/dl
Adult	A	M	4	45	7.47	34	3.39	461
Adult	A	F	6	33	5.53	23	3.29	310
Adult	C	M	4	19	2.81	10	2.39	194
Adult	C	F	2	25	2.84	18	2.34	138
Young	A	M	9	25	2.54	13	2.01	352
Young	A	F	9	25	6.41	13	2.71	455
Young	C	M	3	15	1.93	11	2.00	123
Young	C	F	3	20	2.14	23	1.95	155

[1] Adapted from table IV of Bullock *et al.* [2].

[2] Diet = semisynthetic diet containing 25% lard; A = similar diet as C but containing 0.5% cholesterol.

[3] Expressed as the mean percentage of intimal surface with lesions.

[4] Mean mg cholesterol/g wet aorta.

containing 25% lard or a similar diet with a 0.5% cholesterol additive. At the end of the feeding period, both adult and young animals of either sex showed increased accumulations of cholesterol in the aorta. The increased concentration was seen mainly in the thoracic aorta (table VII). The abdominal aortas of adult *C. albifrons* in the atherogenic group had higher cholesterol concentrations than the control animals, although such differences were not present in the young animals. The increased aortic concentration in the adult male and young female groups fed the atherogenic diet seemed to be related to the higher plasma cholesterol concentration in these groups (table VII).

D. Saimiri

The genus *Saimiri* consists of at least two and possibly three species. Red-backed squirrel monkeys *(S. oersteddi)* are infrequently seen and have not been used in biomedical research. The common squirrel monkey *(S. sciureus)* is one of two phenotypes. One phenotype originates from the Colombian-Brazilian border of the Amazon river and the other from the vicinity of Iquitos, Peru. The phenotypic differences were described and discussed by Cooper [11]. Recently, Jones *et al.* [19] examined the karotypes of squirrel monkeys from different regions. Peruvian and Brazilian squirrel monkeys were found to have the same diploid number of chromosomes (2N = 44) but the number of acrocentric and submetacentric chromosomes varied. These findings suggest that the animals are genetically different and observations on squirrel monkeys should always include descriptive information on their origin.

Of all New World species, squirrel monkeys *(S. sciureus)* have been studied most extensively as animal models of atherosclerosis because of the advantages they offer as animal models for this disease. They are relatively inexpensive, are easily caged and maintained, breed well in captivity, are susceptible to naturally occurring and diet-induced atherosclerosis, and share with man several important aspects of whole body cholesterol metabolism. Their additional advantage in atherosclerosis research is the considerable accumulated knowledge about these primates as laboratory animals. They have been described in a monograph [54]; and in papers on their nutritional requirements [24, 25]; reproduction in the laboratory [6, 22]; anatomy of the coronary arteries [10]; normal clinical laboratory data [27, 38]; normal electrocardiogram [62]; body weights and measurements [42]; diseases that would complicate their use in atherosclerosis research [3, 8].

Table VIII. Aortic atherosclerosis and serum cholesterol concentrations of squirrel monkeys trapped near Leticia, Colombia[1]

| | Males | | Females | | Total serum cholesterol, mg/dl | β- + pre-β- lipoprotein cholesterol, mg/dl |
	juvenile	adult	juvenile	adult		
Normal	19/20	52/90	20/20	33/90	97 ± 2	73 ± 2
Atherosclerotic	1/20	38/90	0/20	57/90	114 ± 3	81 ± 3

[1] Adapted from MIDDLETON *et al.* [43].

1. Naturally Occurring Atherosclerosis

One of the earliest observations on naturally occurring atherosclerosis in squirrel monkeys was that by REWELL [53] on the autopsy findings in a female known to be over 20 years of age. He observed numerous atherosclerotic plaques in the thoracic and abdominal aorta. Later, workers in our laboratories reported on the occurrence, prevalence, and morphology of atherosclerosis in a group of 74 squirrel monkeys maintained in the laboratory for periods of three months to one year [40]. About 85% of these animals had aortic fatty streaks visible after Sudan IV staining, while 7% had raised plaques visible without staining in the abdominal aorta. Coronary artery atherosclerosis affecting the small intramyocardial arteries was present in 30% of the animals. In the same study, measurements were made of blood pressure in 30 of the monkeys. The average blood pressure was 130/80 mm Hg. One of the 30 monkeys was hypertensive with a systolic pressure of 208 mm Hg.

Since it was not possible to eliminate the possibility that confinement and diet had influenced the extent of lesions in the laboratory-maintained group, a study was undertaken on the prevalence and extent of atherosclerosis among free-living squirrel monkeys in the vicinity of Leticia, Colombia [43]. In that experiment, necropsies were done on 120 monkeys (80 adults and 40 juveniles, equally divided between the sexes) in 1964, and another 100 monkeys (equal numbers of adult males and females) in 1965. The findings on aortic atherosclerosis and serum cholesterol concentrations among these free-living squirrel monkeys are summarized in table VIII. It is apparent that aortic fatty streaking was essentially absent among juvenile monkeys and prevalent among adults. Monkeys with fatty streaks or other

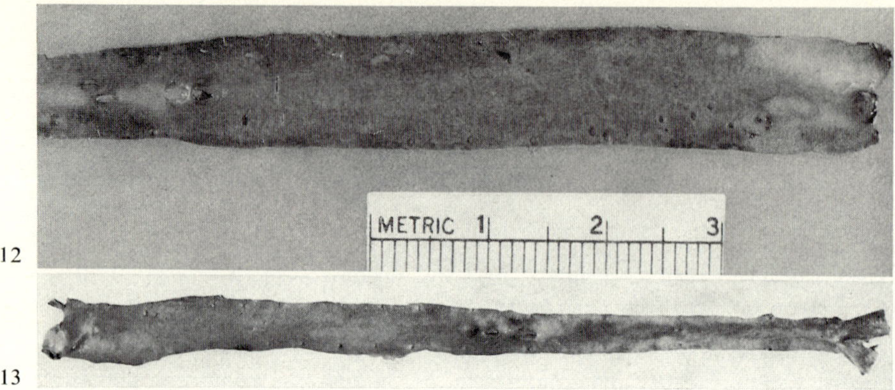

Fig. 12. Thoracic aorta from an adult *S. sciureus* necropsied shortly after capture near Leticia, Colombia. Note fatty streaks (dark in photograph) and a small plaque (right). Sudan IV. × 1,4.

Fig. 13. The thoracic aorta from an adult *S. sciureus* necropsied soon after capture near Leticia, Colombia. Note plaques in the abdominal aorta (right) and extensive fatty streaks. Sudan IV.

atherosclerotic lesions had significantly higher concentrations of total serum cholesterol and β- plus pre-β-lipoprotein cholesterol. Aortic fatty streaks were more frequent among female than male adult monkeys. The lesions that occurred were mainly fatty streaks discernible only after Sudan IV staining (fig. 12) and, occasionally, raised plaques which were usually in the abdominal aorta (fig. 13). Approximately 10% of the adult monkeys had lipid-containing lesions in the small intramyocardial arteries. It is of interest that the monkeys surveyed in their natural habitat had a lower prevalence of both coronary and aortic atherosclerosis than those maintained in the laboratory for three to 12 months and that the mean serum cholesterol concentrations of free-living squirrel monkeys (approximately 100 mg/dl) is about half that of monkeys maintained in our laboratory and fed a low-fat cholesterol-free diet [56].

The ultrastructural morphology of naturally occurring atherosclerosis of squirrel monkeys fed control diets for one year has been described by MCCOMBS *et al.* [34]. It was their conclusion that aortic intimal lesions of squirrel monkeys had numerous similarities in fine structure to those of man and other primates. They found striking morphologic differences in the intimal lesions of the thoracic as compared with the abdominal aorta. In the thoracic aorta, the lesions were three to five cell layers thick and contained densely packed lipid laden cells that resembled blood monocytes. Underlying smooth muscle cells contained lipid, very little extracellular lipid was present, and a fragmented internal elastic lamina was usually associated with the lesions. In contrast, the lesions seen in the abdominal aorta were one to two cell layers thick. They were composed not only of smooth muscle cells containing lipid but also of abundant extracellular lipid. No lipid-laden cells resembling blood monocytes were found.

2. Diet-Induced Atherosclerosis

It has been well-established that the addition of cholesterol to the diets of squirrel monkeys results in hypercholesterolemia and exacerbation of atherosclerosis [41]. The extent to which hypercholesterolemia and athero-sclerosis exacerbation occur is influenced by many factors such as diet consumption, type of fat fed, age, sex, genotype, presence or absence of psychic stress, concomitant exposure to certain coexistent diseases such as diabetes mellitus, hypertension, and hypothyroidism. In this review of diet-induced atherosclerosis of squirrel monkeys, each of these variables will be discussed separately.

a) Composition of Diet

The results of using a variety of diets in our studies on squirrel monkey atherosclerosis have shown that the composition of the diet affects the degree of induced hypercholesterolemia. Moreover, differences resulting from feeding of the various diets must be due to certain interactions not understood at this time. It seems important, therefore, to present here our observations on this matter.

We have used an experimental diet made from natural products (table IX) [26]. The diet provided 26% of the calories as protein, 43% as fat, and contained 1 mg cholesterol/Cal. In 13 monkeys fed this diet for 32 months the serum cholesterol concentrations increased from a base line value of 151 ± 7 (mean $\pm$ SEM) mg/dl to 293 ± 25 mg/dl, and reached a plateau of 228 ± 18 mg/dl.

The diet that we have used most commonly has been a semipurified diet summarized in table X and similar in nutritional composition to the natural products containing about 45% of calories from lard and 1 mg of cholesterol/Cal. When squirrel monkeys are fed this diet, their serum cholesterol concentrations are markedly higher than when the natural products diet is fed. For a random group of squirrel monkeys, the average serum cholesterol response is about 500 mg/dl. The response to the semipurified diet is highly

Table IX. Composition of the natural products diet[1]

Ingredient	Diet, g/100 g
Ground beef steak	10.00
Peanut butter	10.00
Nonfat dry milk solids	25.00
Casein, USP	10.50
Corn meal	19.50
Butter	15.00
Molasses	2.26
Dried egg yolk	1.33
Dried brewer's yeast	3.00
Complete vitamin mix[2]	2.00
Salt mixture[3]	1.00
Cholesterol[4]	0.41

[1] One part of the diet was added to two parts of a hot 2.5% agar solution, thoroughly mixed, and refrigerated until used.

[2] 'Complete Vitamin Fortification Mixture' (Nutritional Biochemicals Corp., Cleveland, Ohio) except vitamin D_3 added 2.5 IU/g diet.

[3] USP XIV Salt Mixture (Nutritional Biochemicals Corp., Cleveland, Ohio).

[4] Crystalline Cholesterol (Nutritional Biochemicals Corp., Cleveland, Ohio).

Table X. Composition of semipurified diet[1]

Ingredient	Diet, g/100 g
Nonfat dry milk solids	30.0
Wheat flour	20.0
Casein, USP	13.0
Lard	25.0
Applesauce	7.3
Complete Vitamin Mix[2]	2.2
USP XIV Salts Mixture	2.0
Cholesterol	0.5

[1] Approximately 0.05 mg cholesterol/Cal lard, and approximately 1 mg cholesterol/Cal.

[2] From Nutritional Biochemicals Corp., Cleveland, Ohio. Vitamin D_3 substituted for D_2 to provide 2.5 IU/g of diet.

Table XI. Composition of purified diet[1]

Ingredient	Diet, g/100 g
Casein, USP	25.0
Lard	25.0
Dextrin	30.0
Complete Vitamin Mix[2]	3.0
Hegsted salt mixture	4.0
Alphacel	12.5
Cholesterol	0.5

[1] Approximately 1 mg cholesterol/Cal.

[2] From Nutritional Biochemicals Corp., Cleveland Ohio. Vitamin D_3 substituted for D_2 to provide 2.5 IU/g of diet.

variable depending upon the genetic background of the monkey [9]. Some monkeys (hyperresponders) have serum cholesterol concentrations of about 1,000 mg/dl after four to five months of diet consumption, while others (hyporesponders) have average serum cholesterol concentrations of about 300 mg/dl.

We have prepared a purified diet (table XI) that resembles as nearly as possible our semipurified diet (table X). The diet provides the same amount of protein from casein, has the same level of lard, and the same of cholesterol on a caloric basis. When this diet was fed to a group of 40 adult male squirrel monkeys, the mean plasma cholesterol concentration of all the animals increased to about half that observed when the same animals consumed the semipurified diet. The hyperresponder monkeys that were fed the semipurified diet (mean serum cholesterol concentration of 1,000 mg/dl) had concentrations of only about 500 mg/dl when fed the purified diet. The average serum cholesterol concentration for the hyporesponder decreased from about 300 to about 200 mg/dl. It is of importance that the relative rank of response did not change significantly. The nonparametric correlation coefficient between the two diet trials was calculated to be 0.86.

Using the semipurified diet described in table X, we have collected information on the effect of the amount of dietary cholesterol on serum cholesterol concentration. The addition of 0.3 mg cholesterol/Cal of diet resulted in an increase in serum cholesterol concentrations from about 172 to about 260 mg/dl. The addition of 0.75 mg cholesterol/Cal increased the serum cholesterol concentration to about 400 mg/dl and a further addition to a

Table XII. Serum and aorta cholesterol concentrations of squirrel monkeys fed various dietary fats

Dietary Group	N	Serum cholesterol mg/dl	Aorta cholesterol, mg/g wet weight[1]		
			total	free	ester
Safflower oil	10	420 ± 32	7.3 ± 0.7	3.7 ± 2.6	3.6 ± 0.4
Lard	10	575 ± 39	7.2 ± 0.8	4.6 ± 0.4	2.6 ± 0.4
Butter	11	671 ± 60	11.6 ± 1.4	6.8 ± 0.6	4.7 ± 1.0
Coconut oil	13	478 ± 37	6.0 ± 0.7	4.1 ± 0.5	2.0 ± 1.3
Controls	30	184 ± 11	5.7 ± 0.9	3.6 ± 0.2	2.1 ± 0.7

[1] Means ± SEM.

level of 1 mg cholesterol/Cal resulted in a small increase to an average of about 500 mg/dl. We have had an occasion to increase the dietary cholesterol levels further to 2 mg/Cal and after a transitory increase the serum cholesterol concentration returns to a mean not different from that obtained with 1 mg/Cal.

For reasons not clear to us, the addition of fat and cholesterol to commercial monkey chow has only a minimal effect on elevating serum cholesterol concentrations of squirrel monkeys.

b) Type of Dietary Fat

The type of dietary fat influences the response of squirrel monkeys to cholesterol-containing diets. We have reported an experiment in which squirrel monkeys were fed for two years on diets containing 1 mg cholesterol/Cal and 45% of calories as either safflower oil, lard, butter, or coconut oil [29].

The effect of the different dietary fats on serum and aorta cholesterol concentrations are summarized in table XII. In this experiment, as well as in others, butter produced the highest serum cholesterol concentration and the most extensive atherosclerosis. Except for coconut oil, safflower oil appeared to be the least atherogenic. The response of squirrel monkeys to coconut oil is unlike that reported for Old World monkeys. Among Old World monkeys coconut oil induced high serum cholesterol concentrations and more extensive atherosclerosis than other dietary fats [60]. Among squirrel monkeys we have found coconut oil to have a very modest effect on serum cholesterol concentration and the extent of atherosclerosis.

Table XIII. Coronary artery atherosclerosis among squirrel monkeys fed various dietary fats

Dietary group	Number examined	Coronary artery atherosclerosis grade[1]			
		none	slight	moderate	severe
Safflower oil	9	3	3	2	1
Lard	10	3	6	0	1
Butter	12	1	5	6	0
Coconut oil	14	8	3	1	0
Controls	6	5	1	0	0

[1] Expressed as the number of monkeys judged to have the grade of coronary disease indicated.

The effect of the various dietary fats on coronary artery atherosclerosis of squirrel monkeys is summarized in table XIII. Again, butter was more atherogenic than the other dietary fats.

c) Effect of Age and Sex

Like cebus monkeys, the age of squirrel monkeys has an influence on the degree of hypercholesterolemia and the extent of atherosclerosis in response to a cholesterol-containing diet. We have conducted an experiment in which we compared the response of young and adult male squirrel monkeys to a diet containing 1 mg cholesterol/Cal and 45% of calories from lard. At the initiation of the experiment the young animals were estimated to be about one year old. The adult animals were estimated to be in excess of four years of age. During the experimental feeding there were no significant differences in the serum cholesterol concentrations with both groups being about 400 mg/dl. In other experiments, however, we have found that young monkeys less than a year of age have less hypercholesterolemia than older monkeys fed a comparable cholesterol-containing diet.

Data on the effect of age on the response of these squirrel monkeys to the cholesterol-containing diet is summarized in table XIV. While the prevalence of aortic atherosclerosis was the same, adult cholesterol-fed monkeys had much more extensive atherosclerosis classified either on the basis of the aorta gross grade or the aorta cholesterol concentration.

The sex of squirrel monkeys affects cholesterol metabolism in a major way for a small segment of the year but has little or no effect on the extent of atherosclerosis. Female squirrel monkeys are anestrus for the majority of

Table XIV. Effect of age on squirrel monkey atherosclerosis

Group	Number of animals	Body Weight[1]	Aortic Atherosclerosis				Aorta cholesterol[4]	
			Thoracic		Abdominal		free	esterified
			pre-valence[2]	extent[3]	pre-valence[2]	extent[3]		
Young control	4	569 ± 30	0	0	0	0	3.95 ± 0.48	0.65 ± 0.14
Adult control	4	915 ± 60	0	0	1/4	1	5.28 ± 1.41	0.85 ± 0.07
Young test	9	819 ± 29	6/9	16 ± 8	5/9	9 ± 3	8.2 ± 1.2	4.9 ± 1.6
Adult test	6	865 ± 38	5/5	43 ± 14	5/5	28 ± 8	22.7 ± 5.7	9.1 ± 1.8

[1] Expressed in grams, means $\pm$ SEM.
[2] Expressed as number of monkeys affected over number observed.
[3] Percentage of intimal surface with atherosclerosis, means $\pm$ SEM.
[4] Expressed as mg/g dry aorta, means $\pm$ SEM.

Table XV. Total serum cholesterol concentrations of cholesterol-fed female squirrel monkeys before and during the breeding season

Group	Number	Non-breeding season[1]			Breeding season[1]			
		December	January	February	March	April	May	June
Safflower oil	5	412 ± 53	372 ± 18	442 ± 46	305 ± 27	244 ± 50	200 ± 21	213 ± 25
Lard	4	376 ± 111	503 ± 175	402 ± 148	295 ± 70	243 ± 24	236 ± 23	250 ± 29
Butter	7	467 ± 37	559 ± 36	471 ± 51	329 ± 13	206 ± 13	182 ± 16	221 ± 24

[1] Values are means $\pm$ SEM and represent mg/dl.

the year and have estrus cycles only during the spring and early summer (March to June). During the breeding season, female monkeys consuming cholesterol have a marked lowering of the serum cholesterol concentration. Table XV summarizes our observations on female squirrel monkeys consuming 1 mg cholesterol/Cal and three types of dietary fat. To our knowledge no information is available on the mechanism of their hypocholesterolemia during the estrus season.

d) Genetic Influence

As we indicated previously, there is considerable individuality in the hypercholesterolemic response of squirrel monkeys to cholesterol-containing

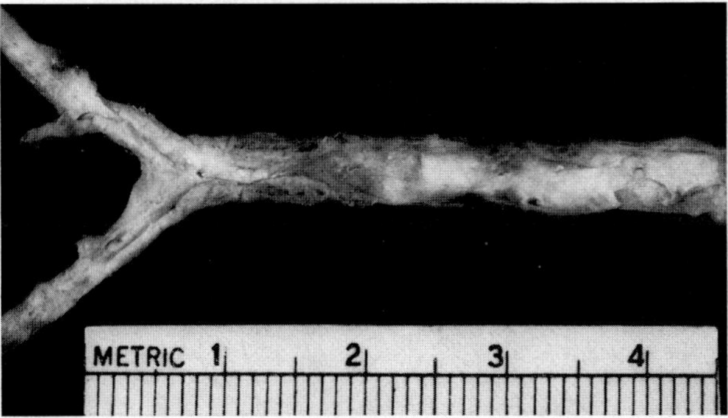

Fig. 14. Intimal surface of the abdominal aorta from a hyperresponding *S. sciureus,* almost completely covered by plaques. Note a small normal area at the lower edge of the aorta at the right of the photograph for comparison [9].

diets. Among squirrel monkeys fed cholesterol, certain individuals (hyper-responders) develop severe hypercholesterolemia while others (hypo-responders) fed the same diet maintain plasma cholesterol concentrations near that of controls. We have conducted an experiment to determine the importance of genetic influence on plasma cholesterol concentrations and, additionally, the extent and severity of the atherosclerotic lesions associated with hypercholesterolemia among the hyperresponder squirrel monkeys [9]. Data derived from this and subsequent studies suggest that about 65% of the variability in plasma cholesterol concentration of cholesterol-fed squirrel monkeys is attributable to genetic factors. Since such a large percent of the variability is genetically determined, it has been possible by selective breeding to develop colonies of hypo- and hyperresponder squirrel monkeys.

Hyporesponder squirrel monkeys fed cholesterol-containing diets, even for long periods, have only minimal arterial lesions. In contrast, the same diets exacerbate markedly the atherosclerosis of hyperresponder monkeys. The lesions resemble qualitatively those described previously for cholesterol-fed monkeys, except that they are more extensive and severe. Lesions frequently associated with hyperlipidemia are seen in organs other than arteries.

An example of the type of aortic lesions seen among hyperresponder squirrel monkeys is illustrated in figure 14. The coronary artery athero-sclerosis of cholesterol-fed hyperresponders is extensive, often calcified, and frequently affects the proximal main branches (fig. 15).

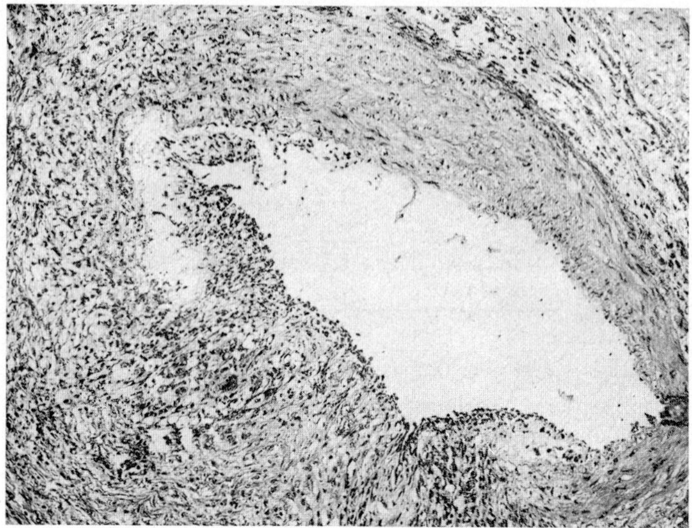

Fig. 15. Section from the heart of a hyperresponding *S. sciureus* fed an atherogenic diet. Lumen of the left main coronary arteries is narrowed by extracellular lipid, foam cells and other cellular elements. HE. × 80.

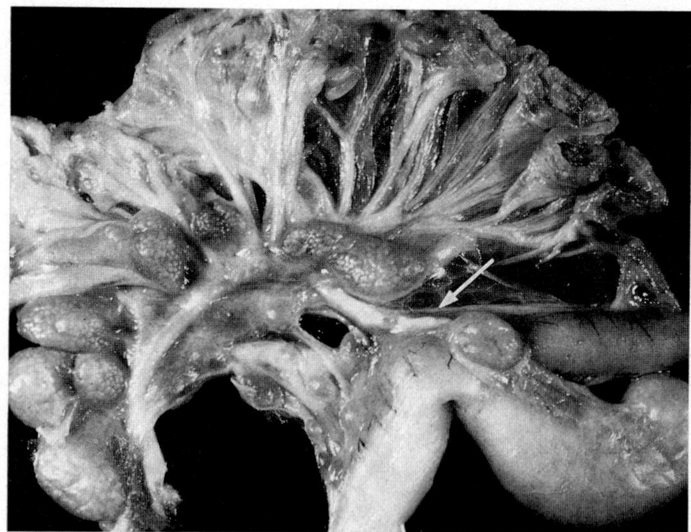

Fig. 16. Tissue from a hyperresponding *S. sciureus* fed an atherogenic diet [9]. Large plaques are visible from the adventitial surface of the mesenteric artery between the lymph node and small intestine.

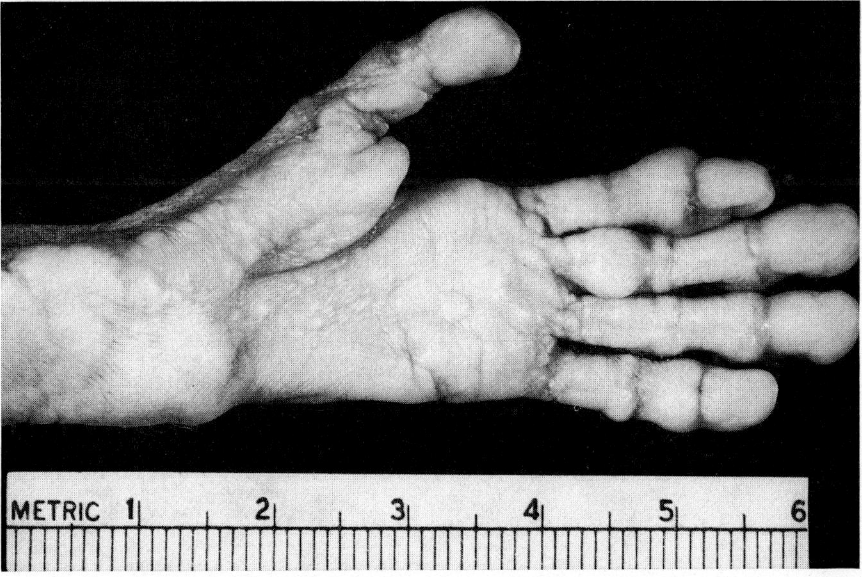

Fig. 17. Cutaneous xanthomas on the palmar surface of a hyperresponding squirrel monkey fed an atherogenic diet [9]. × 1,8.

Arterial lesions in other vascular beds are common among hyperresponders and, as an example, a large mesenteric artery plaque is shown in figure 16. About one-half of the hyperresponders have cutaneous xanthomatosis (fig. 17, 18). Xanthelasma is infrequent. Foam cell lesions are often seen in the aortic valve leaflets (fig. 19) and at the base of the aortic valves. Similar foam cell lesions are seen in the choroid plexus of some of the animals (fig. 20).

e) Psychic Stress

Psychic stress has been shown to influence the response of squirrel monkeys to a cholesterol-containing diet [21]. In that experiment, the monkeys were fed a monkey chow-based atherogenic diet which resulted in a very modest increase in serum cholesterol concentration. The monkeys either remained undisturbed in their cages (group III), were subjected to psychologic stress by the Sidman avoidance procedure (group I), or were placed in the psychologic stressing device but not subjected to the avoidance procedure (group II). Studies on adrenal gland weight and 17-ketosteroids excretion suggested that animals of groups I and II were being subjected to the same amount of stress.

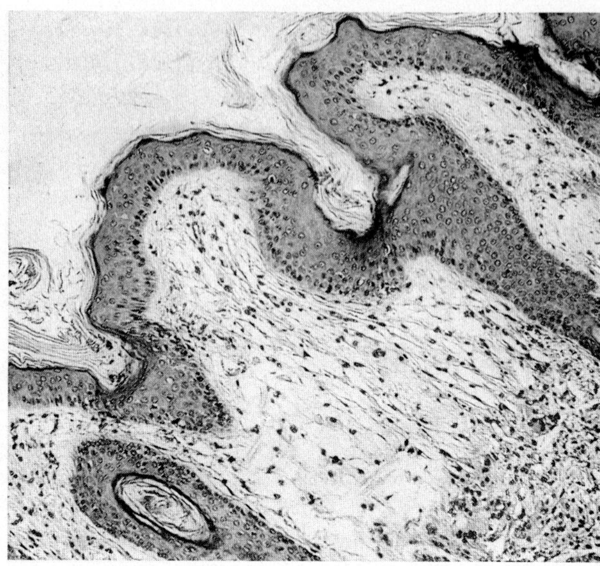

Fig. 18. Cutaneous xanthoma from a hyperresponding squirrel monkey. Note foam cells and multinucleated cells (lower left). HE. × 95.

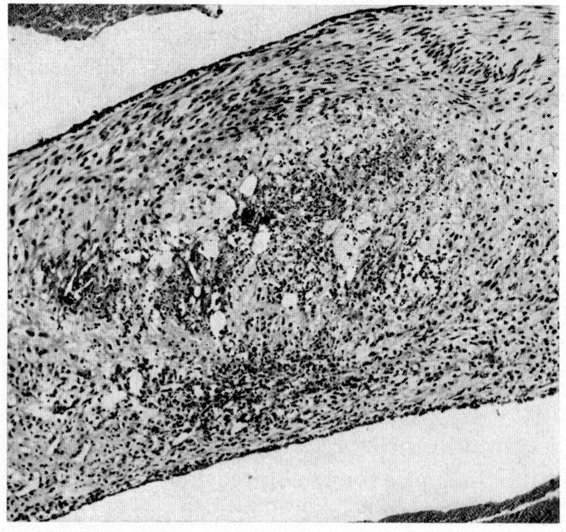

Fig. 19. Left atrioventricular valve from a hyperresponding squirrel monkey markedly thickened by an accumulation of foam cells, lipid clefts (center of valve) and inflammatory cells. The animal was fed an atherogenic diet [9]. HE. × 80.

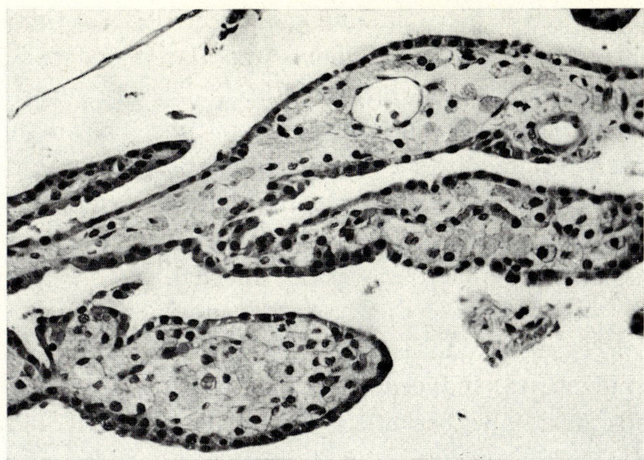

Fig. 20. Foam cells in the choroid plexus of a hyperresponding squirrel monkey fed an atherogenic diet. HE. × 150.

Table XVI. Effect of psychic stress on the response of squirrel monkeys to a cholesterol-containing diet[1]

Group	Serum cholesterol, mg/dl[2]		Aorta	Coronary
	before stress	after stress	cholesterol[3]	athero-sclerosis[4]
I (psychic stress)	180 ± 22	228 ± 27	3.7 ± 5	5/6
II (box control)	181 ± 19	214 ± 22	3.4 ± 3	4/6
III (cage control)	153 ± 11	160 ± 10	2.9 ± 3	0/6

[1] Adapted from LANG [21].
[2] Expressed as means $\pm$ SEM.
[3] Expressed as mean mg cholesterol/g of wet weight $\pm$ SEM.
[4] Expressed as number affected over number examined.

The principal findings in this study are summarized in table XVI. Animals of groups I and II had transient increases in serum cholesterol concentration during the manipulation and this was associated with a higher prevalence of coronary atherosclerosis.

f) Exacerbation by Carbon Monoxide

The exposure of squirrel monkeys to carbon monoxide affects their response to a cholesterol-containing diet [59]. Squirrel monkeys of the Peruvian type were fed a diet containing 1 mg cholesterol/Cal and 45% of calories as lard. One half of the animals was placed in a test chamber for 4 h a day and exposed to air while the other half was exposed in the same way to carbon monoxide for the same length of time. The amount of carbon monoxide given resulted in a carboxyhemoglobin concentration of from 9 to 26% saturation. In this seven-month study the extent and severity of coronary artery atherosclerosis was markedly exacerbated by carbon monoxide while aortic atherosclerosis was not affected.

g) Effect of Hypothyroidism

Hypothyroidism was induced in squirrel monkeys by ablation of their thyroid glands with [131]I. The effect on the development and severity of atherosclerosis was determined after these monkeys and controls had been fed a natural products diet containing cholesterol (1 mg/Cal) for three years. The hypothyroid monkeys had greater serum cholesterol concentrations (414 ± 18 versus 228 ± 18 mg/dl, mean $\pm$ SEM, $p < 0.01$), a greater percentage of serum lipoprotein as β-lipoprotein (59 ± 2 versus 54 ± 1, $p < 0.02$), but similar serum triglyceride concentrations as the euthyroid controls. The hypothyroid monkeys also had abnormal IV glucose tolerance tests and slightly lower systolic blood pressures (table XVII).

Aortic lesions in both groups were limited to fatty streaking. Histologically, these lesions consisted of slight cellular thickenings of the aortic intima and contained small intracellular and extracellular lipid droplets. These lesions were only slightly more prevalent and extensive in the hypothyroid monkeys than in the cholesterol-fed control (euthyroid) monkeys (7/8 versus 8/12, number affected/number in group; mean surface involvement of those affected, 52 versus 31%). Aortic cholesterol concentrations were consistent with the extensiveness of lesions, and were greater in the hypothyroid animals. The greatest difference in the amount of atherosclerotic disease was found in the coronary arteries. Of the total number of cross-sections of coronary arteries observed in 15 sections, the hypothyroid animals had $26.4 \pm 6.1\%$ affected with intimal fat-containing lesions while the controls had $5.5 \pm 1.0\%$ affected, $p < 0.01$ (table XVIII).

h) Effect of Hypertension

The effect of hypertension on atherosclerosis in the squirrel monkey was studied after the induction of renal hypertension. The monkeys developed and maintained systolic blood pressures that were significantly greater than control animals (134 ± 4 versus 111 ± 2 mm Hg, mean $\pm$ SEM, $p < 0.01$).

Table XVII. Average (means ± SEM) serum cholesterol, triglyceride, β-lipoprotein, and systolic blood pressure levels in control, hypothyroid, diabetic, and hypertensive squirrel monkeys fed a natural products diet containing 1 mg of cholesterol/Cal for three years[1]

	Number	Serum cholesterol, mg/dl	Serum triglyceride, mg/dl	β-Lipoprotein[2]	Systolic blood pressure, mm Hg
Control	12	228 ± 18	20 ± 2	1.5 ± 0.1	111 ± 2
Hypertensive	13	278 ± 17	27 ± 4	1.9 ± 0.2	134 ± 4 $p < 0.01$
Hyperthyroid	8	414 ± 18 $p < 0.01$	30 ± 3	2.9 ± 0.2 $p < 0.01$	100 ± 4 $p < 0.05$
Diabetic	7	575 ± 79 $p < 0.01$	27 ± 4	5.2 ± 1.0 $p < 0.01$	124 ± 2 $p < 0.05$

[1] Adapted from LEHNER *et al.* [26].

[2] Serum β-lipoprotein was precipitated with antihuman β-lipoprotein serum. Values are expressed as millimeters of β-lipoprotein precipitate per 60 mm of serum in capillary tube.

Probability values obtained by comparison with control monkeys.

Table XVIII. Aortic cholesterol concentration and coronary artery atherosclerosis of control, diabetic, hypothyroid, and hypertensive squirrel monkeys fed a natural products diet containing 1 mg of cholesterol/Cal for three years[1] (means ± SEM)

	Number	Thoracic aortic cholesterol, mg/g dry wt	Abdominal aortic cholesterol, mg/g dry wt	Coronary arteries with atherosclerotic lesions[2], %
Control	12	10.4 ± 1.1	6.3 ± 0.9	5.5 ± 1.0
Hypertensive	13	15.6 ± 1.7 $p < 0.01$	11.0 ± 0.9 $p < 0.01$	17.7 ± 4.5 $p < 0.05$
Hypothyroid	8	15.9 ± 1.2 $p < 0.01$	10.9 ± 1.3 $p < 0.05$	26.4 ± 6.1 $p < 0.01$
Diabetic	7	29.1 ± 3.5 $p < 0.01$	26.0 ± 7.6 $p < 0.01$	30.7 ± 7.0 $p < 0.01$

[1] Adapted from LEHNER *et al.* [26].

[2] Total number of coronary arteries observed with intimal, fat-containing lesions × 100, divided by the total number of arteries observed in 15 sections of the heart.

Probability values obtained by comparison with control group.

The hypertensive and control monkeys were fed a natural products diet containing 1 mg of cholesterol/Cal for three years. During this time the hypertensive group had an average serum cholesterol concentration of 278 ± 17 mg/dl which was slightly greater but not significantly different than of the controls which averaged 228 ± 18 mg/dl. Serum β-lipoprotein concentrations, serum triglyceride concentrations, and IV glucose tolerance tests were not significantly different between these groups (table XVII).

The prevalence of fatty streak lesions in the aorta was slightly greater in the hypertensive animals (13/13) than in the controls (8/12), but the extent of these lesions in affected animals was about the same (39 versus 31, average percentage of surface involved in affected animals). Microscopically, these lesions consisted of cellular intimal thickenings with fine intracellular and extracellular lipid droplets. The fatty streaks of the hypertensive monkeys appeared to have a greater thickness than the controls but contained less stainable lipid. Additionally, four of 13 hypertensive monkeys had raised lesions which, on the average, affected 13% of their aortic intimal surface. Histologically, these lesions consisted of a pool of lipid-staining material covered by a fibromuscular cap. Aortic cholesterol levels reflected the differences in lesion prevalence and severity, being greater in the hypertensive monkeys. They also had a greater prevalence of lesions in the coronary arteries, 17.7 ± 4.5 versus $5.5 \pm 1.0\%$, of coronary arteries with lesions in 15 sections of the heart (table XVIII).

The higher blood pressure of the hypertensive animals was the only factor identified in this study which could account for the more extensive atherosclerosis in this group. Using the data from the hypertensive and control monkeys, significant positive correlation coefficients were found between systolic blood pressure and aortic cholesterol ($r = 0.51$, $p < 0.01$, $n = 24$) and between systolic blood pressure and percent of coronary arteries with lesions ($r = 0.52$, $p < 0.01$, $n = 25$).

i) Effect of Diabetes mellitus

The effect of diabetes mellitus on the development and severity of atherosclerosis was studied by comparing lesions observed in monkeys with alloxan-induced diabetes and in controls after they had been fed a natural products diet containing 1 mg of cholesterol/Cal for three years. The alloxan diabetic animals had much more extensive and severe atherosclerosis of the aorta and coronary arteries than the controls. The prevalence of fatty streaks in the aorta was seven of seven in the diabetic groups and eight of 12 in the control group, with the affected monkeys in each group averaging about the same percentage of intimal surface involvement, 36.5 versus 31. However, four of the seven diabetic monkeys had extensive raised lesions which affected approximately 30% of their aortic intimal surface. Histologically, these lesions consisted of a thick fibromuscular covering over a core of lipid and amorphous basophilic debris. The base of the lesions often contained sterol clefts and extensive mineralization. The aortic media under these large lesions appeared to be involved, being much thinner than normal, and in three of the four monkeys the media adjacent to the lesions contained abundant lipid and mineral. One of the four monkeys had numerous aneurysmal dilatations 1–4 mm in diameter.

Coronary artery atherosclerosis was also much more extensive in the diabetic group compared to the controls (30.7 $\pm$ 7.0 versus 5.5 $\pm$ 1.0% of the coronary arteries with lesions in 15 sections of the heart, mean $\pm$ SEM). These coronary arterial lesions did not appear to differ in character between the groups, but those of the diabetic monkeys were larger, causing greater luminal stenosis (table XVIII).

In addition to being insulin-deficient, several differences were noted between the diabetic and control monkeys which might account for the differences in atherosclerotic disease. The diabetic monkeys in comparison to the controls had markedly greater serum cholesterol concentrations (575 $\pm$ 79 versus 228 $\pm$ 18 mg/dl), a much greater percentage of serum lipoprotein as β-lipoprotein (68 $\pm$ 4 versus 54 $\pm$ 1), and slightly greater systolic blood pressure (124 $\pm$ 2 versus 111 $\pm$ 2 mm Hg) (table XVII). All of these factors undoubtedly influenced the development and severity of atherosclerotic lesions and for that reason this study did not provide data supporting or negating a direct effect of diabetes on the development of atherosclerosis.

j) Tissue Cholesterol Concentration

As with other primate models of atherosclerosis, it is of interest to consider the accumulation of cholesterol in the tissues and organs of squirrel monkeys fed an atherogenic diet. In table XIX we have summarized the effect on tissue cholesterol concentration of an atherogenic diet (1 mg of cholesterol/Cal) fed to squirrel monkeys for 42 months. Large increases in cholesterol concentration were seen in liver and moderate increases in skin, tendon, and adipose tissue [4].

k) Arterial Composition

PORTMAN *et al.* [48] reported the effect of short and long-term hypercholesterolemia on the composition of intima-inner media of the aortas of

Table XIX. Effect of diet on organ and tissue cholesterol concentration of squirrel monkeys[1, 2]

	Control diet (N = 12)	Atherogenic diet (N = 8)
Brain	15.44 ± 0.23	15.36 ± 0.80
Myocardium	1.34 ± 0.03	1.47 ± 0.07
Lung	3.52 ± 0.31	3.34 ± 0.46
Liver	4.41 ± 0.30	23.71 ± 8.66
Spleen	3.90 ± 0.11	4.85 ± 0.25
Jejunum	2.36 ± 0.09	3.33 ± 0.21
Ileum	2.27 ± 0.06	2.63 ± 0.24
Pancreas	2.19 ± 0.06	2.32 ± 0.14
Kidneys	3.92 ± 0.10	4.34 ± 0.20
Skin		
Eyelids	2.25 ± 0.18	3.30 ± 0.36
Ear	2.68 ± 0.32	3.64 ± 0.83
Thoracic	2.11 ± 0.10	2.60 ± 0.30
Forearm	2.28 ± 0.14	4.26 ± 1.92
Tendon	1.78 ± 0.49	6.13 ± 2.40
Muscle	0.72 ± 0.01	0.75 ± 0.06
Fascia	0.78 ± 0.08	1.70 ± 0.62
Adipose	0.98 ± 0.09	1.51 ± 0.15
Eye	0.77 ± 0.05	0.76 ± 0.06
Cornea	0.78 ± 0.13	0.86 ± 0.20

[1] All data are expressed as mg/g of wet weight, means ± SEM.

[2] Adapted from BULLOCK *et al.* [4].

Table XX. The effect of short-term hyperlipemia (2–5 months) and long-term hyperlipemia (> 6 months) on aortic intima-inner media cholesterol and phospholipid concentrations in the squirrel monkeys[1, 2]

	Number	Cholesterol	Esterified cholesterol	Phospholipid
Control	14	2.79	0.692	7.70
Short-term hyperlipemia	8	5.46	1.04	10.53
Long-term hyperlipemia[3]	12	10.84	7.98	16.34

[1] Modified from table III of PORTMAN *et al.* [48].

[2] All values are means and represent mg/g aorta.

[3] Atherosclerosis was present grossly in the aortas of this group.

Table XXI. Aortic cholesterol in young adult squirrel monkeys fed control and atherogenic diets for 12 months[1]

	Control[2]	Atherogenic[3]
Thoracic aorta		
% atherosclerosis[4]	23.0 ± 12.8	34.0 ± 5.8
Total cholesterol	13.5 ± 3.6[5]	20.6 ± 1.7
% free	72	72
% esterified	28	28
Abdominal aorta		
% atherosclerosis	13.0 ± 4.0	35.0 ± 5
Total cholesterol	10.3 ± 3.8	14.1 ± 2.2
% free	72	65
% esterified	28	35

[1] Modified from table II of St. Clair *et al.* [55].
[2] Semisynthetic diet containing 25% lard by weight.
[3] Semisynthetic diet containing 25% lard and 0.5% cholesterol by weight.
[4] Total surface area involved with gross lesions.
[5] Expressed in mg/g lipid-free dry aorta.

squirrel monkeys. Aortas of animals that were made hyperlipemic for a short time accumulated more lipid in comparison to controls (table XX). The largest relative increases were seen for the nonesterified cholesterol fraction followed by the phospholipid fraction. A small increase was seen in cholesterol ester concentration. However, in long-term hypercholesterolemia which induced an aggravation of atherosclerosis, the aortas showed an accumulation of cholesterol ester eightfold higher than control values while nonesterified cholesterol concentration was approximately fourfold higher and phospholipid concentrations twofold higher than control values.

In another study, St. Clair *et al.* [55] reported esterified cholesterol and nonesterified cholesterol concentrations of thoracic and abdominal aortas of squirrel monkeys fed for 12 months either a diet containing 0.5% cholesterol or a control diet. The concentrations of total cholesterol of both aorta segments of the atherogenic group were higher than the control values (table XXI). In both the control and atherogenic diet groups, the thoracic aorta accumulated more total cholesterol than the abdominal aorta segment. No differences were seen in the aortic free cholesterol to esterified cholesterol ratios, with the exception of the abdominal aorta of the atherogenic group which showed a higher percent of esterified cholesterol.

Table XXII. The fatty acid composition of aortic esterified cholesterol in young adult squirrel monkeys fed control and atherogenic diets for 12 months[1, 2]

Fatty acids[3]	Control		Atherogenic
	1	2	
16:0	18.5	23.7	17.1 ± 1.4
16:1	5.3	6.6	5.4 ± 0.5
18:0	19.3	30.0	15.0 ± 2.7
18:1	29.7	28.8	44.3 ± 2.2
20:4	—	—	2.5 ± 0.7

[1] Modified from table VII of ST. CLAIR *et al.* [55].

[2] Aortic sample represents both thoracic and abdominal segments. Values for the control group represent two individual animals; values for the atherogenic group represent the means $\pm$ SEM for nine animals. All values are expressed as a percent of the total fatty acids in esterified cholesterol.

[3] Number of carbons: number of double bonds.

Table XXIII. Cholesterol concentration in atherosclerotic and nonatherosclerotic areas of squirrel monkeys fed cholesterol-containing diets[1, 2]

	Grossly non-lesioned areas (32)	Fatty streaks (43)	Plaques (17)
Cholesterol	3.60 ± 0.20	4.60 ± 0.30	7.10 ± 0.70
Esterified cholesterol	2.10 ± 0.70	3.00 ± 0.80	11.20 ± 2.80

[1] Modified from table VII of PUCAK *et al.* [52].

[2] All values are means $\pm$ SEM and represent mg/g wet aorta; number of samples analyzed in parentheses.

Table XXII shows the fatty acid composition of aortic cholesterol esters in squirrel monkeys fed for 12 months control diets or diets containing 0.5% cholesterol. The major fatty acid of esterified cholesterol in atherosclerotic squirrel monkeys was oleate which accounted for 44.3% of the total cholesterol ester fatty acids. In the two lesion-free control aortas, cholesteryl oleate constituted 28.8 and 29.7% of total cholesterol esters.

As a portion of a study comparing primate species with regard to lipid composition of normal and atherosclerotic areas, PUCAK *et al.* [52] separated

Table XXIV. Sphingomyelin, lecithin, and cholesterol concentrations in aortic intima-inner media of squirrel monkeys fed control or atherogenic diets for six to ten months[1]

	Control[2]	Atherogenic[3]
Sphingomyelin	2.07 ± 0.28	3.28 ± 0.22
Total phospholipid, %	26.4	26.2
Lecithin	2.72 ± 0.20	5.83 ± 0.32
Total phospholipid, %	36.2	46.4
Total cholesterol	5.06 ± 0.74	13.64 ± 1.20

[1] Modified from table I of PORTMAN *et al.* [48]. All values are means of nine animals per group ± SEM and represent mg/g.

[2] A semipurified diet; 15 % of the calories supplied by corn oil.

[3] A semipurified diet; 45 % of the calories supplied by butter; 1 mg cholesterol/Cal.

Table XXV. Lysolecithin and cholesterol concentrations in aortic intima-inner media of squirrel monkeys[1]

Diet	Lysolecithin	Cholesterol
Control[2]	0.142 ± 0.034	3.51 ± 0.38
Atherogenic[3] (2–9 weeks)	0.449 ± 0.10	5.25 ± 0.50
Atherogenic (8–12 months)	0.800 ± 0.16	17.17 ± 2.01

[1] Adapted from table VI of PORTMAN *et al.* [44]. All values represent mg/g aorta and are presented as means ± SEM.

[2] A semipurified diet; 15 % of the calories supplied by corn oil.

[3] A semipurified diet; 45 % of the calories supplied by butter; 1 mg cholesterol/Cal.

cholesterol-fed squirrel monkey aortas into normal areas, fatty-white streaks, and raised fatty plaques. Relative to concentrations in normal areas, cholesterol and esterified cholesterol of fatty streaks increased to the same extent (table XXIII). In plaques, esterified cholesterol concentrations were higher.

The concentrations of aortic phospholipids of squirrel monkeys have been reported in studies by PORTMAN *et al.* [48]. In several studies these authors observed increases in lecithin, sphingomyelin and lysolecithin con-

Table XXVI. Effect of aortitis on extent of atherosclerosis induction and regression in squirrel monkeys[1]

	Aortitis		No aortitis	
	number	extent[2]	number	extent
Lesion induction	7	20.00 ± 7.06	9	15.11 ± 3.96
Lesion regression	9	20.33 ± 4.71	5	3.80 ± 0.96

[1] Adapted from HAYES *et al.* [17].

[2] Extent of atherosclerosis expressed as mean percentage of intimal surface that was sudanophilic, followed by the standard error of the mean.

centrations in aortic intima-inner media of squirrel monkeys fed cholesterol-containing diets (tables XXIV, XXV). Sphingomyelin concentrations were found to increase in early atherosclerotic lesions before any increase above control values was seen in lecithin concentration. In more severe lesions lecithin was found to be the predominant phospholipid and increased to the greatest extent with increasing severity of atherosclerosis [48].

l) Regression of Diet-Induced Atherosclerosis

Little information is available on the regression of diet-induced athero-sclerosis of squirrel monkeys. MARUFFO and PORTMAN [39] examined the reversibility of atherosclerosis in the coronary arteries of squirrel monkeys. The lesions were induced during a three-month feeding of a diet rich in butter and then were allowed to regress for three to five months during which the animals consumed normal diets. The authors observed some morphologic changes in the lesions, but did not interpret these as regression. Later, HAYES *et al.* [17] suggested on the basis of their data that the early diet-induced lesions of squirrel monkeys regressed if they were not com-plicated with aortitis but did not regress if aortitis accompanied the athero-sclerotic lesion (table XXVI).

3. Problems in the Use of Squirrel Monkeys as
Animal Models for Atherosclerosis Research

Because squirrel monkeys have been used extensively in atherosclerosis research, some of the processes which complicate interpretation of the findings have been identified and will be reviewed here.

Table XXVII. Serum lipid levels of squirrel monkeys with normal and impaired response to glucose tolerance test[1]

Response to glucose Tolerance test	Number	Serum cholesterol[2] mg/dl	Serum-free fatty acids[2], mEq/l
Normal	14	192 ± 10	0.82 ± 0.06
Impaired	16	182 ± 10	0.98 ± 0.12

[1] Adapted from DAVIDSON *et al.* [12].
[2] Means $\pm$ SEM

a) Aortitis

Aortitis characterized by plasma cells and lymphocytes with lesser numbers of eosinophils, mass cells, macrophages, and histiocytes has been found to be present in about 44 % of wild-caught squirrel monkeys [17]. This aortitis affected the extent of atherosclerosis with more extensive disease present among the animals with aortitis. It also strikingly affected the potential for regression of early diet-induced atherosclerosis. Diet-induced lesions would regress in animals free of aortitis, but not in animals suffering from that disease (table XXVI).

b) Glucose Tolerance

LANG [20], working in our laboratories, described an impairment in response to glucose tolerance test in a portion of a population of squirrel monkeys. Because of the interest in possible relationships between impaired glucose tolerance and serum lipid levels, DAVIDSON *et al.* [12] have compared the levels of serum cholesterol and free fatty acids in a group of squirrel monkeys with normal glucose tolerance and in another group whose response was found to be impaired (table XXVII). No differences of lipid levels in the fasting serum of these two groups were observed. Many questions about the possible relationships between this glucose tolerance impairment and lipid metabolism remain unanswered and are deserving of further study.

c) Renal Disease

In squirrel monkeys, renal disease occurs relatively frequently, especially with advancing age, and among animals consuming cholesterol for long periods. Some animals are found to have the nephrotic syndrome with hypercholesterolemia, hypoalbuminemia and sudanophilic globules in the

Table XXVIII. Effect of age on cholelithiasis in squirrel monkeys

Group	Number of animals	Cholelithiasis	
		prevalence[1]	extent[2]
Young control	4	0/4	0
Adult control	4	0/4	0
Young test	9	4/9	78.1 ± 41.2
Adult test	5	5/5	244.9 ± 46.8

[1] Expressed as number of monkeys affected over number observed.
[2] Mean mg of gallstones for group $\pm$ SEM.

urine. Other monkeys that do not have noticeable clinical manifestations of renal disease are often found on histologic examination to have changes consistent with glomerulosclerosis or glomerulonephritis.

An array of kinds of lesions are seen among affected monkeys. Some animals have mostly lesions of the basement membrane while others have only mesangial changes with or without proliferation. Monkeys fed atherogenic diets may have numerous foam cells in glomeruli. By electron microscopy some of these foam cells appear to be endothelial cells.

The effect of diet on squirrel monkey renal disease is not clear. Renal lesions are seen in animals fed commercial monkey chow, although the lesions seems to be more extensive when the animals consume diets containing cholesterol. It is not certain whether the diets induce some of the renal changes or whether a naturally occurring renal lesion is aggravated by the cholesterol-containing diets. The degree to which atherosclerosis is affected by the renal diseases is being investigated currently.

d) Cholelithiasis

It is now well-known that squirrel monkeys consuming cholesterol-containing diets have a high frequency of cholesterol gallstone disease [30, 45]. OSUGA and PORTMAN [45] found that cholelithiasis could be induced by feeding corn oil as 15% of calories added to a semipurified diet, but with no added cholesterol. Similarly, we have found that the type of dietary fat affects the prevalence and extent of cholelithiasis among squirrel monkeys consuming a cholesterol-containing diet [30]. Safflower oil was found to be more lithogenic than either lard or butter. Susceptibility to cholelithiasis is also age-dependent. We have compared the prevalence and extent of chole-

Table XXIX. The extent of naturally occurring atherosclerosis of howler monkeys
(*Alouatta caraya*)[1]

Weight range, kg	Aortic atherosclerotic index[2]			
	number	males	number	females
0–1.9	10	0.25 ± 0.19	7	0
2.0–3.9	33	0.77 ± 0.36	13	1.42 ± 0.72
4.0–6.5	53	4.21 ± 0.66	98	3.26 ± 0.48
6.6–9.7	78	7.49 ± 0.89	0	0

[1] Adapted from MALINOW and MARUFFO [35].
[2] Expressed as mean percentage of aortic intima that was sudanophilic $\pm$ SEM.

lithiasis among young (about two years of age) and adult (over five years of age) squirrel monkeys consuming a semisynthetic diet with lard as the source of fat, and cholesterol added in the amount of 1 mg/Cal. Data are presented in table XXVIII indicating that adult squirrel monkeys are more susceptible to cholelithiasis than are young animals.

E. Alouatta

The genus *Alouatta* consists of five species. Only one of these *A. caraya*, commonly called the black howler, has been investigated for possible usefulness in atherosclerosis research.

MALINOW and MARUFFO [35] have reported an extensive and thorough investigation of naturally occurring atherosclerosis among free-living black howler monkeys (for summary see table XXIX). It can be seen that in this species the extent of aortic atherosclerosis increases with increasing size of the animal. There is also a suggestion of somewhat more extensive lesions in males than in females. The investigators described both typical fatty streaks alone, and focally distributed raised plaques involving from 0.5 to 5% of the aortic intimal surface. In the more advanced plaques the intimal lipid was covered with a highly fibrotic subintimal cap. No calcification, hemorrhage, or ulceration was observed.

Howler monkeys have been found to be very difficult to acquire and to maintain in the laboratory. These difficulties have discouraged their use in atherosclerosis research and to our knowledge they are not being used in any laboratory at this time.

F. Ateles

The genus *Ateles*, the spider monkeys, consist of four species; *A. paniscus*, *A. belzebuth*, *A. fusciceps*, and *A. geoffroy*, the differentiation of which is based on coat color and geographic range. Since differentiation of the species is difficult, most publications indicate *Ateles* sps.

Spider monkeys have long arms, legs and prehensile tails. The adult body weight range is from 5.4 to 6.8 kg. They adapt readily to captivity and present no problems for laboratory study except the requirement of fairly large cages.

1. Naturally Occurring Atherosclerosis

The plasma lipid concentrations and the extent and severity of naturally occurring atherosclerosis of recently trapped spider monkeys have been determined [32]. It is of interest that spider monkeys are one of the two species having the highest total plasma cholesterol concentration of six species of New World monkeys studied, namely 130 $\pm$ 7 mg/dl based on 29 observations. Moreover, they, along with the woolly monkeys, transport more cholesterol in the β-lipoproteins (118 $\pm$ 7 mg/dl) and less in the α-lipoproteins (12 $\pm$ 0.7 mg/dl) than do other New World monkeys. Mostly juvenile animals were examined, about half of which had aortic fatty streaks. All nine adult females examined showed sudanophilic aortic lesions of some degree. In other studies, extensive atherosclerosis has been noted among a small number of spider monkeys maintained in captivity for long periods [5, 7, 16].

2. Diet-Induced Atherosclerosis

Spider monkeys fed an atherogenic diet that usually induces marked hypercholesterolemia in other nonhuman primates became only slightly hypercholesterolemic (increased from control concentrations of about 200 to about 290 mg/dl). Cholesterol feeding for up to 48 months exacerbates the lesions only slightly. The details of these observations follow.

a) Dietary Hypercholesterolemia

EGGEN *et al.* [15] compared the hypercholesterolemia of spider monkeys fed a cholesterol-containing semisynthetic diet with that of baboons and rhesus monkeys fed the same diet. The spider monkeys had a total serum cholesterol concentration of 200 mg/dl, about the same concentration as seen in baboons and about half that of rhesus monkeys. PUCAK *et al.* [52] studied the response of spider monkeys fed two kinds of diets. The first

Table XXX. Effect of cholesterol-containing and control diets fed to spider monkeys[1]

| Group | Number | Total serum cholesterol concentration[2] | Aortic atherosclerosis[3] | | | | Coronary artery atherosclerosis[4] |
| | | | Thoracic | | Abdominal | | |
			fatty streak	plaques	fatty streak	plaques	
Test male	5	334 ± 15	30	–	–	10	1.16 ± 0.37
Test female	4	245 ± 8	40	15	5	10	0.94 ± 0.18
Control male	4	210 ± 4	6	–	–	–	0.47 ± 0.15
Control female	3	190 ± 9	8	–	1	–	0.61 ± 0.27

[1] Adapted from Pucak *et al.* [52].
[2] Mean mg/dl $\pm$ SEM.
[3] Expressed as mean percentage of intimal involvement.
[4] Expressed as mean grade (on a scale 0–5) $\pm$ SEM.

diet, fed for 38 months, contained about 45% of calories derived equally from butter and coconut oil, with cholesterol added in the amount of about 1 mg/Cal, and the remainder of the diet consisting of Purina Monkey Chow (Ralston Purina Company, St. Louis, Missouri). When fed that diet, serum cholesterol concentrations of males increased from 210 ± 4 to 334 ± 15 mg/dl while that of females increased from 190 ± 9 to 245 ± 8 mg/dl. The mean serum cholesterol concentration of the male test monkeys was considerably greater than that of the female test monkeys at every sampling interval.

Since it had been observed that atherogenic diets made with Purina Monkey Chow caused only slight hypercholesterolemia when fed to several other kinds of nonhuman primates, the diet of the animals was changed to a semisynthetic diet containing the same kinds of fat, the same caloric amounts of fat, and the same amount of cholesterol on a caloric basis. This diet was fed for ten months, and the sex difference of serum cholesterol concentrations persisted, but the extent of hypercholesterolemia did not increase. In both dietary situations, about 90% of the plasma cholesterol was transported in the β-lipoproteins.

b) Diet-Induced Lesions

In the study by Eggen *et al.* [15] spider monkeys were found to be relatively resistant to diet-induced atherosclerosis. In the study by Pucak *et al.*

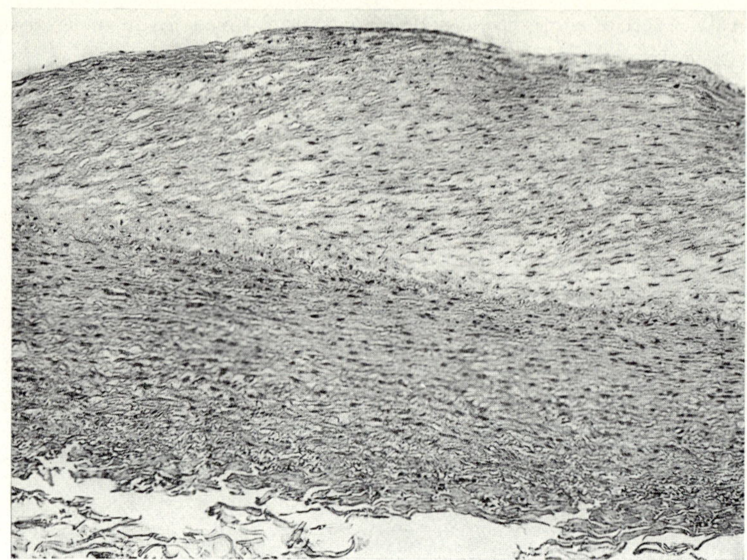

Fig. 21. Plaque in the abdominal aorta of a spider monkey *(Ateles* sp.*)* fed an athero-genic diet for 48 months. Note that the intima, composed mostly of smooth muscle cells with some vacuolization, is about the same thickness as the media [52]. HE. × 90.

[52] again it was found that feeding an atherogenic diet, even for a 48-month period, exacerbated the lesions only slightly. The effect of the diets fed in that experiment are summarized in table XXX.

Most lesions in the aorta were longitudinally oriented, whitish, slightly raised fatty streaks. These lesions were better delineated and more easily seen after the aortas were stained for fat.

The microscopic appearance of the atherosclerotic plaques varied from what appeared to be a smooth muscle thickening with mostly intracellular lipid accumulation to a plaque with a relatively acellular core of lipid capped with smooth muscle cells (fig. 21, 22). Extracellular lipid accumulations were usually near the internal elastic membrane with considerable medial involvement associated with larger lesions.

Fig. 22. Plaque near the origin of a branch of the abdominal aorta from a spider monkey fed an atherogenic diet. Thick fibromuscular cap covers a clear area where lipids were removed during preparation for paraffin embedding [52]. HE. × 90.

Fig. 23. Main left coronary artery of a spider monkey fed an atherogenic diet narrowed by a thickened intima composed of smooth muscle cells and lipid [52]. HE. × 90.

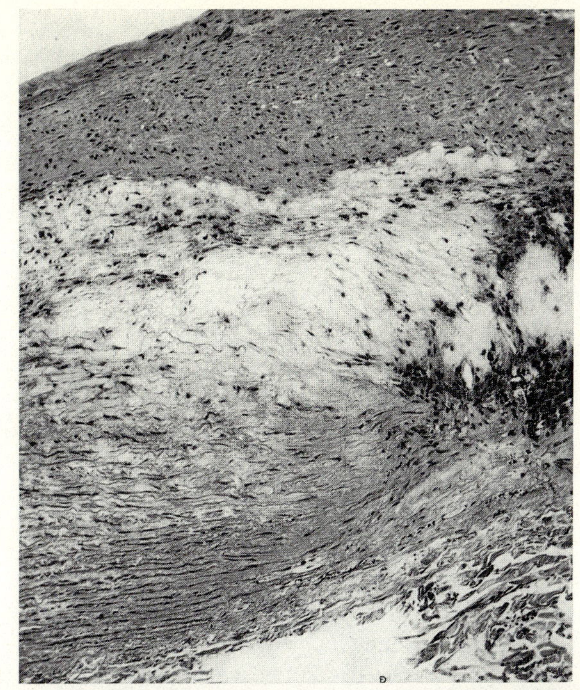

22

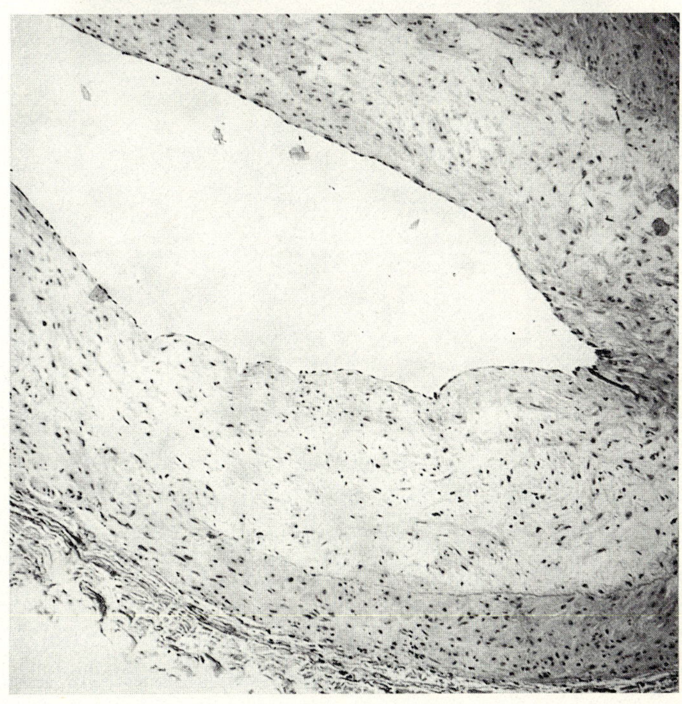

23

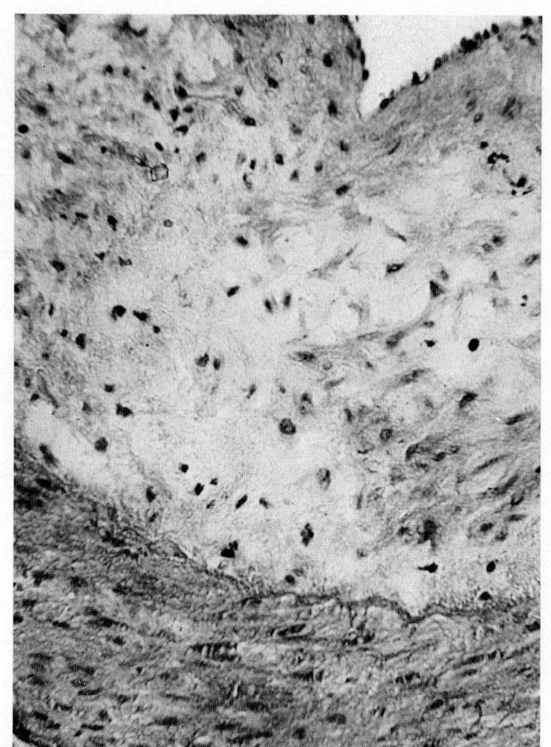

24

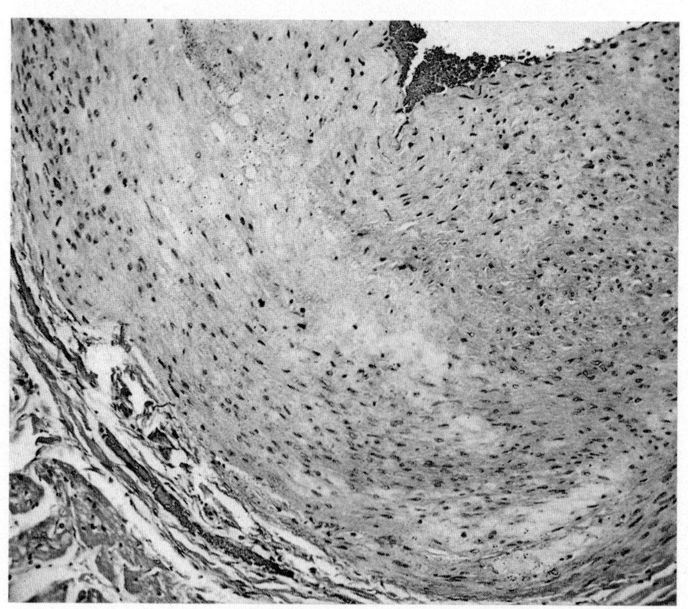

25

Table XXXI. Lipid concentration of aorta without lesion (NA), fatty streak (FS) or plaques (PQ) in control and atherogenic diet-fed spider monkeys[1]

	Control diet[2]		Atherogenic diet[3]		
	NA (5)	FS (4)	NA (5)	FS (3)	PQ (3)
Cholesterol	1.65 ± 0.064	2.51 ± 0.372	1.35 ± 0.111	3.14 ± 1.469	8.26 ± 2.087
Esterified cholesterol	0.29 ± 0.039	0.90 ± 0.258	0.43 ± 0.109	2.67 ± 0.930	3.58 ± 0.488
Triglyceride	0.55 ± 0.081	1.21 ± 0.331	0.52 ± 0.070	1.70 ± 0.624	2.05 ± 0.055
Phospholipid	5.0 ± 0.80	23.3 ± 4.78	5.6 ± 0.49	33.4 ± 0.57	34.1 ± 0.97

[1] Adapted from table VII of Pucak *et al.* [52]. Results given in mg/g wet aorta, means ± SEM; number of observations in parentheses.

[2] 75% Purina Monkey Chow (Ralston Purina Co., Saint Louis, Mo.), 12.5% coconut oil was fed for 38 months followed by a semisynthetic diet containing 12.5% butter and 12.5% coconut oil for the remaining ten months of the study.

[3] Similar dietary regimen as the control animals but included a 0.5% cholesterol additive.

Microscopically, the lesions of the coronary arteries were composed primarily of what appeared to be smooth muscle cells. These lesions appeared to contain relatively more intracellular and extracellular lipid than did aortic lesions (fig. 23–25).

c) Arterial Composition

In the study by Pucak *et al.* [52], aortic tissue was separated into lesion-free areas, fatty streaks, and plaques for analysis of cholesterol, esterified cholesterol, triglyceride, and phospholipid concentrations (table XXXI). Fatty streak tissue from either the control or atherogenic diet-fed animals had higher concentrations of all four lipid fractions studied. The concentrations of phospholipid in fatty streaks were especially elevated and were five times the level found in normal aorta. Atherosclerotic plaques of the atherogenic diet-fed spider monkeys likewise showed a large concentration of phospholipid. The other lipid fractions were elevated above levels in fatty streaks or in lesion-free arteries. Nonesterified cholesterol concentrations were five times higher in plaques in comparison to normal aorta while esterified

Fig. 24. Higher magnification of figure 23. Stellate cells contained or separated by lipids particularly near the wavy internal elastic lamina [52]. HE. × 60.

Fig. 25. Section of the right coronary artery from a cholesterol-fed spider monkey. The thickened intima contained lipid (clear spaces) and a few mineral granules (top, left of center). The media (lower right) also contained lipid [52]. HE. × 90.

cholesterol concentrations were approximately nine times the concentration found in normal aorta. Although mention is made by PUCAK *et al.* [52] that the esterified cholesterol concentration of the aortic plaque in spider monkeys was lower than that of the squirrel monkey plaque (mean concentration 11.20 mg/g), squirrel monkey lesions do not show grossly elevated concentrations of phospholipids [48]. The spider monkey atherosclerotic lesions, both fatty streak and plaque, possess the unique characteristic of grossly elevated phospholipid concentrations not seen in other nonhuman primate species, and might be useful for studies related to this lipid fraction and atherogenesis.

G. Lagothrix

The genus *Lagothrix*, the woolly monkeys, consists of two species, *L. lagothricha* and *L. flavicauda*. These infrequently studied New World monkeys are large (5–10 kg), with short massive limbs and prehensile tails. To our knowledge, only *L. lagothricha* has been studied as a possible model for atherosclerosis. Woolly monkeys are difficult to acquire, and there is usually considerable mortality associated with their adaptation to the laboratory. We have reported on the naturally occurring lesions seen at necropsy in eight woolly monkeys that died in a group of 17 monkeys being acclimated for an experiment [18].

1. Naturally Occurring Atherosclerosis

ANDRUS and PORTMAN [1, 58] examined 15 *Lagothrix* killed in their native habitat and concluded that this species had more naturally occurring microscopic aortic atherosclerosis than did *Cebus*, *Saimiri*, *Callicebus*, or *Alouatta* species.

As a part of our collaborative field study with the Louisiana State University Medical School, we examined the plasma lipid concentrations and the extent and severity of naturally occurring atherosclerosis of recently trapped woolly monkeys *(L. lagothricha)* [32]. Based on the study of 20 animals, woolly monkeys along with spider monkeys have the highest total serum cholesterol concentration (133 $\pm$ 12 mg/dl) and transport more cholesterol in the β-lipoproteins (118 $\pm$ 12 mg/dl) and less in the α-lipoproteins (11 $\pm$ 1 mg/dl) than do other New World monkeys. Only juvenile animals were examined pathologically, and about half of them had aortic fatty streaks. Among the eight adult woolly monkeys necropsied in our diagnostic laboratory, only two animals had sudanophilic lesions in their aorta in excess of 1% of the total animal surface (one animal had 2% involvement in the thoracic aorta and 5% in the abdominal aorta while the other animal had 15% thoracic and 25% abdominal aortic fatty streaks). The monkey with the greatest area of aortic sudanophilic lesion also had focal musculoelastic intimal hyperplasia of the coronary and renal arteries.

2. Diet-Induced Atherosclerosis

PORTMAN and ANDRUS [47] fed a semisynthetic diet containing coconut oil and 0.5% cholesterol to woolly monkeys, and observed serum cholesterol concentrations of about 600 mg/dl. Later, MANN [36] fed cholesterol-containing diets to woolly monkeys and compared their cholesterol metabolism with other nonhuman primates. He reported that woolly monkeys apparently lacked a negative feedback mechanism for controlling endogenous cholesterol synthesis when fed a cholesterol-containing diet.

More recently, we have examined the response of woolly monkeys to cholesterol-containing diets [4]. In this experiment, again the woolly monkeys as a group did poorly. The experiment was initiated with ten adult *L. lagothricha*: seven died during the 42-month study, six of them during the first year. Consequently, our observations on this species are based on the small number of three animals. At the initiation of the experiment, the woolly monkeys had an average total serum cholesterol concentration of about 100 mg/dl. When fed a diet containing 45% of calories as lard and 1 mg of cholesterol/Cal. the serum cholesterol concentration increased only slightly to 150 mg/dl, for comparison, consuming the same diets and at the same time intervals, stumptailed macaques had average serum cholesterol concentrations of 700 mg/dl and African green and squirrel monkeys about 450 mg/dl. We subsequently had occasions to examine woolly monkeys fed 0.5 and 1.0 mg of cholesterol/Cal in liquid form diets, and again only modest increases in serum cholesterol concentrations were seen. After the three woolly monkeys consumed atherogenic diets for 42 months, their arteries were examined. Two of the three animals had fibrofatty plaques in the abdominal aorta. Plaques of one of the monkeys did not contain much lipid but there were foam cells in the deeper part of the plaque. The other animal had a typical 'atheroma' with lipid clefts, a fibrous cap and some mineralization (fig. 26). All the three monkeys had small fatty streaks in the common carotid arteries and all had rather large plaques at the carotid bifurcation. Only one animal had femoral artery lesions. Two of the three monkeys had intimal thickening of the main branches of the coronary arteries. This thickening tended to be fairly uniform around the entire

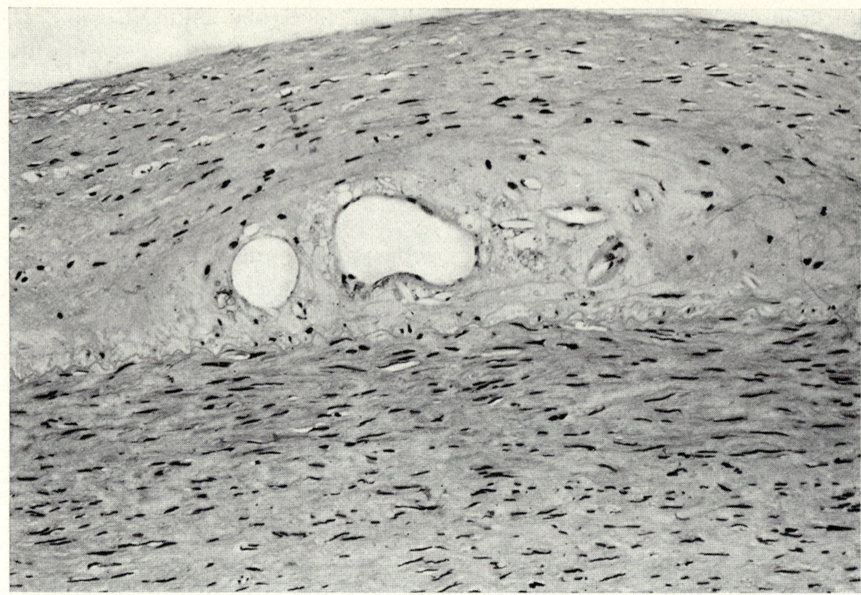

Fig. 26. Plaque from the abdominal aorta of a woolly monkey. To the right of the large endothelialized space note a few granules of mineral. There is a thick fibromuscular cap (top) above the area where lipids had accumulated. The media below the slightly distorted internal elastic membrane is relatively unchanged. HE. × 70.

circumference of the vessels. The mean percentage of the luminal space occupied by these thickenings was about 10%. The third monkey had punctate intimal thickening in five of the sections from the 15 blocks.

While we have emphasized that woolly monkeys are difficult to obtain and maintain, they are of interest for further study. By comparing the arteries of the eight adults studied in our diagnostic laboratory and the arteries of the three animals that consumed atherogenic diet for 42 months, it seems clear that there was marked exacerbation of atherosclerosis in this species with total serum cholesterol concentrations usually only at 150 mg/dl. Further studies of the effect of cholesterol-containing diets on lipoprotein structure and composition may be particularly rewarding.

References

1 ANDRUS, S. B. and PORTMAN, O. W.: Comparative studies of spontaneous and experimental atherosclerosis in primates. Symp. zool. Soc. Lond. *17:* 161–177 (1966).

2   BULLOCK, B. C.; CLARKSON, T. B.; LEHNER, N. D. M.; LOFLAND, H. B., jr., and
    ST. CLAIR, R. W.: Atherosclerosis in *Cebus albifrons* monkeys. III. Clinical and path-
    ologic studies. Expl molec. Path. *10:* 39–62 (1969).
3   BULLOCK, B. C.; LEHNER, N. D. M., and CLARKSON, T. B.: New World monkeys; in
    BEVERIDGE Primates in medicine, vol. 2, pp. 62–74 (Karger, Basel 1969).
4   BULLOCK, B. C.; LEHNER, N. D. M.; CLARKSON, T. B.; FELDNER, M. A.; WAGNER,
    W. D., and LOFLAND, H. B.: Comparative primate atherosclerosis. I. Tissue cho-
    lesterol concentration and pathologic anatomy. Expl molec. Path. *22:* 151–175 (1975).
5   CLARKSON, T. B.; BULLOCK, B. C., and LEHNER, N. D. M.: Pathologic characteristics
    of atherosclerosis in New World monkeys. Prog. biochem. Pharmacol., vol. 4, pp.
    420–428 (Karger, Basel 1968).
6   CLARKSON, T. B.; BULLOCK, B. C.; LEHNER, N. D. M., and FELDNER, M. A.: The
    squirrel monkey as a laboratory animal. National Academy of Sciences, National
    Research Council, Publ. No. 1594, pp. 64–87 (1968).
7   CLARKSON, T. B.; LOFLAND, H. B.; BULLOCK, B. C.; LEHNER, N. D. M.; ST. CLAIR,
    R. W., and PRICHARD, R. W.: Atherosclerosis in some species of New World monkeys
    Ann. N. Y. Acad. Sci. *162:* 103–109 (1969).
8   CLARKSON, T. B.; BULLOCK, B. C.; LEHNER, N. D. M., and MANNING, P. J.: Diseases
    affecting the usefulness of nonhuman primates in nutrition research; in HARRIS
    Feeding and nutrition of nonhuman primates, pp. 233–250 (Academic Press, New
    York 1970).
9   CLARKSON, T. B.; LOFLAND, H. B.; BULLOCK, B. C., and GOODMAN, H. O.: Genetic
    control of plasma cholesterol. Archs Path. *92:* 37–45 (1971).
10 COLBURN, G. L.: The gross morphology of the coronary arteries of the common
    squirrel monkey. Anat. Rec. *155:* 353–367 (1966).
11 COOPER, R. W.: Squirrel monkey taxonomy and supply; in ROSENBLUM and COOPER
    The squirrel monkey, pp. 1–29 (Academic Press, New York 1968).
12 DAVIDSON, I. W. F.; LANG, C. M., and BLACKWELL, W. L.: Impairment of carbo-
    hydrate metabolism of the squirrel monkey. Diabetes *16:* 395–401 (1967).
13 DREIZEN, S.; LEVY, B. M., and BERNICK, S.: Diet-induced jejunal lipodystrophy in the
    cotton-top marmoset *(Saguinus oedipus)*. Proc. Soc. exp. Biol. Med. *138:* 7–11 (1971).
14 DREIZEN, S.; LEVY, B. M., and BERNICK, S.: Diet-induced atherosclerosis in the
    marmoset. Proc. Soc. exp. Biol. Med. *143:* 1218–1223 (1973).
15 EGGEN, D. A.; STRONG, J. P., and NEWMAN, W. P., III.: Experimental atherosclerosis
    in primates: a comparison of selected species. Ann. N. Y. Acad. Sci. *162:* 110–119
    (1969).
16 FINLAYSON, R.: Spontaneous arterial disease in exotic animals. J. Zool., Lond. *147:*
    239–343 (1965).
17 HAYES, K. C.; WESTMORELAND, N. P., and FAHERTY, T. P.: An aortitis in the squirrel
    monkey and its effect on atherosclerosis. Expl molec. Path. *17:* 334–347 (1972).
18 HENDERSON, J. D., jr.; WEBSTER, W. S.; BULLOCK, B. C.; LEHNER, N. D. M., and
    CLARKSON, T. B.: Naturally occurring lesions seen at necropsy in eight woolly monkeys
    (Lagothrix sp.) Lab. Anim. Care *20:* 1087–1097 (1970).
19 JONES, T. C.; THORINGTON, R. W.; HU, M. M.; ADAMS, E., and COOPER, R. W.:
    Karotypes of squirrel monkeys *(Saimiri sciureus)* from different geographic regions.
    Am. J. phys. Anthrop. *38:* 269–278 (1973).

20 LANG, C. M.: Impaired glucose tolerance in the squirrel monkey *(Saimiri sciureus)*. Proc. Soc. exp. Biol. Med. *122:* 84–86 (1966).

21 LANG, C. M.: Effects of psychic stress on atherosclerosis in the squirrel monkey *(Saimiri sciureus)*. Proc. Soc. exp. Biol. Med. *126:* 30–34 (1967).

22 LANG, C. M.: The estrous cycle of the squirrel monkey *(Saimiri sciureus)*. Lab. Anim. Care *17:* 422–451 (1967).

23 LANG, C. M. and BARTHEL, C. H.: Effects of simple and complex carbohydrates on serum lipids and atherosclerosis in nonhuman primates. Am. J. clin. Nutr. *25:* 470–475 (1972).

24 LEHNER, N. D. M.; BULLOCK, B. C., and CLARKSON, T. B.: Ascorbic acid deficiency in the squirrel monkey. Proc. Soc. exp. Biol. Med. *128:* 512–514 (1968).

25 LEHNER, N. D. M.; BULLOCK, B. C.; CLARKSON, T. B., and LOFLAND, H. B.: Biological activities of vitamins D_2 and D_3 for growing squirrel monkeys. Lab. Anim. Care *17:* 483–493 (1967).

26 LEHNER, N. D. M.; CLARKSON, T. B., and LOFLAND, H. B.: The effect of insulin deficiency, hypothyroidism, and hypertension on atherosclerosis in the squirrel monkey. Expl molec. Path. *15:* 230–244 (1971).

27 LEHNER, N. D. M.; MANNING, P. J.; FELDNER, M. A., and BULLOCK, B. C.: Hematologic and blood chemical characteristics of squirrel monkeys *(Saimiri sciureus)*. Technicon Int. Congr. Advances in Automated Analysis, Chicago 1969, pp. 133–138.

28 LOFLAND, H. B.; CLARKSON, T. B.; ST. CLAIR, R. W.; LEHNER, N. D. M., and BULLOCK, B. C.: Atherosclerosis in *Cebus albifrons* monkeys. I. Sterol metabolism. Expl molec. Path. *8:* 302–313 (1968).

29 LOFLAND, H. B.; CLARKSON, T. B., and BULLOCK, B. C.: Whole body sterol metabolism in squirrel monkeys *(Saimiri sciureus)*. Expl molec. Path. *13:* 1–11 (1970).

30 LOFLAND, H. B.; CLARKSON, T. B., and MELCHIOR, G. W.: Cholesterol gallstones in nonhuman primates; in SCHETTLER and SCHLIERF Atherosclerosis III, pp. 392–395 (Springer, Berlin 1974).

31 LOFLAND, H. B., jr.; ST. CLAIR, R. W.; CLARKSON, T. B.; BULLOCK, B. C., and LEHNER, N. D. M.: Atherosclerosis in cebus monkeys. II. Arterial metabolism. Expl molec. Path. *9:* 57–70 (1968).

32 LOFLAND, H. B.; ST. CLAIR, R. W.; MacNINTCH, J. E., and PRICHARD, R. W.: Atherosclerosis in New World primates. Archs Path. *83:* 211–214 (1967).

33 MacNINTCH, J. E.; ST. CLAIR, R. W.; LEHNER, N. D. M.; CLARKSON, T. B., and LOFLAND, H. B.: Cholesterol metabolism and atherosclerosis in cebus monkeys in relation to age. Lab. Invest. *16:* 444–452 (1967).

34 McCOMBS, H. L.; ZOOK, B. C., and McGANDY, R. B.: Fine structure of spontaneous atherosclerosis of the aorta in the squirrel monkey. Am. J. Path. *55:* 235–252 (1969).

35 MALINOW, M. R. and MARUFFO, C. A.: Aortic atherosclerosis in free-ranging howler monkeys *(Alouatta caraya)*. Nature, Lond. *206:* 948–949 (1965).

36 MANN, G. V.: Cystine deficiency and cholesterol metabolism in primates. Circulation Res. *18:* 205–212 (1966).

37 MANN, G. V.; ANDRUS, S. B.; McNALLY, A., and STARE, F. J.: Experimental atherosclerosis in cebus monkeys. J. exp. Med. *98:* 195–218 (1953).

38 Manning, P. J.; Lehner, N. D. M.; Feldner, M. A., and Bullock, B. C.: Selected hematologic serum chemical and arterial blood gas characteristics of squirrel monkeys *(Saimiri sciureus.)* Lab. Anim. Care *19:* 831–837 (1969).

39 Maruffo, C. A. and Portman, O. W.: Nutritional control of coronary artery atherosclerosis in the squirrel monkey. J. Atheroscler. Res. *8:* 237–247 (1968).

40 Middleton, C. C.; Clarkson, T. B.; Lofland, H. B., and Prichard, R. W.: Atherosclerosis in the squirrel monkey. Archs Path. *78:* 16–23 (1964).

41 Middleton, C. C.; Clarkson, T. B.; Lofland, H. B., and Prichard, R. W.: Diet and atherosclerosis of squirrel monkeys. Archs Path. *83:* 145–153 (1967).

42 Middleton, C. C. and Rosal, J.: Weights and measurements of normal squirrel monkeys *(Saimiri sciureus).* Lab. Anim. Sci *22:* 583–586 (1972).

43 Middleton, C. C.; Rosal, J.; Clarkson, T. B.; Newman, W. P., and McGill, H. C., jr.: Arterial lesions in squirrel monkeys. Archs Path. *88:* 352–358 (1967).

44 Newman, W. P.; Middleton, C. C.; Clarkson, T. B.; Rosal, J., and Strong, J. P.: Naturally occurring arterial lesions in five species of New World primates. Archs Path. *98:* 173–176 (1974).

45 Osuga, T. and Portman, O. W.: Experimental formation of gallstones in the squirrel monkey. Proc. Soc. exp. Biol. Med. *136:* 722–726 (1971).

46 Portman, O. W. and Alexander, M.: Changes in arterial subfractions with aging and atherosclerosis. Biochim. biophys. Acta *260:* 460–474 (1972).

47 Portman, O. W. and Andrus, S. B.: Comparative evaluation of three species of New World monkeys for studies of dietary factors, tissue lipids and atherogenesis. J. Nutr. *87:* 429–438 (1965).

48 Portman, O. W.; Alexander, M., and Maruffo, C. A.: Nutritional control of arterial lipid composition in squirrel monkeys: major ester classes and types of phospholipids. J. Nutr. *91:* 35–46 (1967).

49 Portman, O. W.; Hegsted, D. M.; Stare, F. J.; Bruno, D.; Murphy, R., and Sinisterra, L.: Effect of the level and type of dietary fat on the metabolism of cholesterol and beta lipoproteins in the cebus monkey. J. exp. Med. *104:* 817–828 (1956).

50 Portman, O. W.; Osuga, T., and Alexander, M.: Nutritional influences on gallstones, bile lipids, and arterial lipids; in Schettler and Schlierf Atherosclerosis III, pp. 389–392 (Springer, Berlin 1974).

51 Portman, O. W. and Sinisterra, L.: Dietary fat and hypercholesteremia in the cebus monkey. II. Esterification and disappearance of cholesterol-4-C^{14}. J. exp. Med. *106:* 727–742 (1957).

52 Pucak, G. J.; Lehner, N. D. M.; Clarkson, T. B.; Bullock, B. C., and Lofland, H. B.: Spider monkeys *(Ateles* sp.) as animal models for atherosclerosis research. Expl molec. Path. *18:* 32–49 (1973).

53 Rewell, R. E.: Uterine fibromyomas and bilateral ovarian granulosa cell tumors in senile squirrel monkey, *Saimiri sciurea.* J. Path. Bact. *68:* 291–293 (1954).

54 Rosenblum, L. E. and Cooper, R. W.: The squirrel monkey (Academic Press, New York 1968).

55 St. Clair, R. W.; Lofland, H. B., and Clarkson, T. B.: Influence of atherosclerosis on the composition, synthesis, and esterification of lipids in aortas of squirrel monkeys *(Saimiri sciureus.)* J. Atheroscler. Res. *10:* 193–206 (1969).

56 ST. CLAIR, R. W.; MACNINTCH, J. E.; MIDDLETON, C. C.; CLARKSON, T. B., and LOFLAND, H. B.: Changes in serum cholesterol levels of squirrel monkeys during importation and acclimation. Lab. Invest. *16:* 828–832 (1967).

57 SOBEL, H.; MONDON, C. E., and MEANS, C. V.: Pigmy marmoset as an experimental animal. Science *132:* 415–416 (1960).

58 STARE, F. J.; ANDRUS, S. B., and PORTMAN, O. W.: Primates in medical research with special reference to New World monkeys; in PICKERING Proc. Conf. Research with Primates, pp. 59–66 (Tektronix Foundation, Oregon 1963).

59 WEBSTER, W. S.; CLARKSON, T. B., and LOFLAND, H. B.: Carbon monoxide aggravated atherosclerosis in the squirrel monkey. Expl molec. Path. *13:* 36–50 (1968).

60 WISSLER, R. W.; VESSELINOVITCH, D.; GETZ, G. S., and HUGHES, R. H.: Aortic lesions and blood lipids in rhesus monkeys fed three food fats. Fed. Proc. Fed. Am. Socs exp. Biol. *26:* (1960).

61 WISSLER, R. W.; FRAZIER, L. E.; HUGHES, R. H., and RASMUSSEN, R. A.: Atherogenesis in the cebus monkey. I. A comparison of three food fats under controlled dietary conditions. Archs Path. *74:* 312–322 (1962).

62 WOLF, R. H.; LEHNER, N. D. M.; MILLER, E. C., and CLARKSON, T. B.: The electrocardiogram of the squirrel monkeys. *Saimiri sciureus.* Am. J. appl. Physiol. *26:* 346–351 (1969).

THOMAS B. CLARKSON; NOEL D. M. LEHNER; BILL C. BULLOCK; HUGH B. LOFLAND, and WILLIAM D. WAGNER, Department of Comparative Medicine and Arteriosclerosis Research Center, Bowman Gray School of Medicine, *Winston-Salem, NC 27103* (USA)

Prim. Med., vol. 9, pp. 145–223 (Karger, Basel 1976)

Arterial Metabolism in Primates[1]

OSCAR W. PORTMAN and D. ROGER ILLINGWORTH

Departments of Nutrition and Metabolic Diseases, Oregon Regional Primate Research Center, Beaverton, and Department of Biochemistry, University of Oregon Medical School, Portland, Oreg.

Contents

[1] Supported in part by USPHS, NIH grants RR 00163 and HL 09744.

I. Introduction

To understand the development of atherosclerosis, one must combine the knowledge of morphology and of metabolic potential of the arterial wall with that of the changes which occur in those tissues under various conditions. Despite numerous studies of arterial composition and metabolism, the picture is far from complete. Such studies in man have been hampered by the difficulty of obtaining fresh arterial tissue as well as by such technical difficulties as identifying, dissecting, and analyzing different areas of arteries and kinds of lesions. These difficulties are partially circumvented by the use of experimental animals which makes it possible to obtain fresh tissue and to control some experimental conditions. However, susceptibility to atherosclerosis varies widely among different species, and there are doubts whether the observations on animals are applicable to man. Nonhuman primates have been used by a number of investigators to make their observations more relevant to man, but substantial differences in susceptibility to arterial disease occur in various species of nonhuman primates. In addition, the expense of experimenting with nonhuman primates and the scarcity of healthy animals of known age and genetic makeup have restricted the use of primate animals for arterial studies.

In this review we have emphasized experiments on nonhuman primates but have tried to relate the observations to man and to nonprimate species. In a few cases, we included areas in which little work has been done with nonhuman primates but which seemed important to the understanding of arterial biochemistry and pathology. A review of some aspects of arterial metabolism appeared in 1969 [206]; we have, therefore, emphasized work done since that date.

In any histological and chemical study of the arterial wall, three major variables in the experimental animals seem particularly important: age, species, and the existence of 'spontaneous' or experimentally induced atherosclerosis. Whether the changes that occur with aging (the most obvious of which is intimal thickening) but without overt atherosclerosis are related to the atherosclerotic lesion is difficult to establish. Age clearly plays a role in the susceptibility of experimental animals to 'spontaneous' or uninduced lesions as well as to nutritionally induced atherosclerosis. There are marked differences in the smooth muscle cells, elastic fibers, and other histological and cytological features of the arterial wall of young and old animals even when the increases in lipids are not histologically demonstrable. The relationship between the early lipid-rich arterial lesions induced in experimental animals and the older, more complicated ones may be similar to the relationship between the fatty streak and the more complex lesions in man. The lesions most often studied in experimental animals probably resemble most closely the fatty streak of human arteries.

Some animals are very resistant to nutritionally induced atherosclerosis, for example, several carnivores, including domestic cats and dogs. The herbiverous rabbit, on the other hand, is extremely susceptible. Many of these differences in susceptibility are the result of differences in susceptibility to hyperlipemia. Even among primates, species differ widely in their susceptibility to atherosclerosis. These differences are partly, but not completely, due to differences in susceptibility to increased plasma lipids or specific classes of lipoproteins.

Although the composition of plasma and the composition or morphological appearance of the arterial wall have been determined in a number of nonhuman primates, only a few comparative studies in a single laboratory have been reported. The cholesterol and β-lipoprotein concentrations of three species of New World monkeys (*Cebus albifrons*, *Saimiri sciureus*, and *Lagothrix lagotricha*) and chimpanzees *(Pan troglodytes)* were shown to be affected by diet, but there were no species differences on any of the seven diets tested [3, 4, 215]. After 6 and 12 months on the experimental diets, all four species also showed a high positive correlation between the percentage of aortic surface covered with sudanophilic and fibrotic lesions and the mean plasma cholesterol concentration. However, all four species differed widely in the slopes of the lines relating mean plasma cholesterol levels and the percentage of aortic surface with lesions (fig. 1). Although age was an important factor in determining the susceptibility of cebus monkeys to atherosclerosis [159], this species regardless of age, seemed more resistant [42, 43, 162, 341] than the highly susceptible squirrel monkey [42, 43, 166]. Individual squirrel monkeys varied greatly in susceptibility to nutritionally induced hypercholesterolemia and atherosclerosis [151]. Our experience with this variability is illustrated in figure 2. Striking differences in the occurrence of sudanophilic lesions in two species of free-ranging Central African monkeys

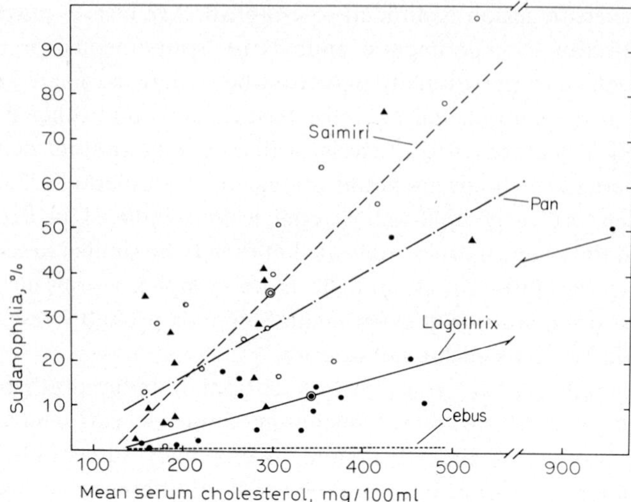

Fig. 1. The relationship between the percentages of the intimal surface of the aorta that were covered with sudanophilic and fibrotic lesions and the mean plasma cholesterol level of cebus *(C. albifrons)*, squirrel *(S. sciureus)* (○), and woolly *(L. lagotricha)* (●) monkeys, and chimpanzees *(P. satyrus)* (▲). The plasma cholesterol levels for primates on four diets were not affected by differences of species, but the degrees of sudanophilia were. There was a difference in arterial susceptibility to hyperlipemia. Correlation coefficients (r) of aortic sudanophilia versus mean plasma cholesterol values were 0.84, 0.93, and 0.70 for woolly and squirrel monkeys and chimpanzees, respectively. From PORTMAN and ANDRUS [215] and ANDRUS *et al.* [4].

(Cercocebus albigena and *Cercopithecus ascanius)* were observed [317]. Numerous studies documented the susceptibility of rhesus monkeys *(Macaca mulatta)* to nutritionally induced hyperlipemia and atherosclerosis [6, 161, 320, 321, 344]. The closely related cynomolgus monkey (*Macaca irus* or *fascicularis*) seems to be particularly susceptible to coronary atherosclerosis [140]. *Macaca nigra* sometimes develop spontaneous diabetes, hyperlipemia (including hypertriglyceridemia), and aortic sudanophilia [115, 116]. Although the free-ranging baboons develop some lesions [100, 156], they are somewhat resistant to nutritionally induced hyperlipemia and atherosclerosis [88]. The strong indications that arterial lesions cannot be explained entirely by plasma lipoprotein patterns suggest important arterial wall differences among different primate species. The fact that various nonprimate species differ in their susceptibility to atherosclerosis, which cannot be explained entirely by differences in plasma lipids, prompted us to examine the reactions of the arterial wall *per se.* The special problems of sampling different layers

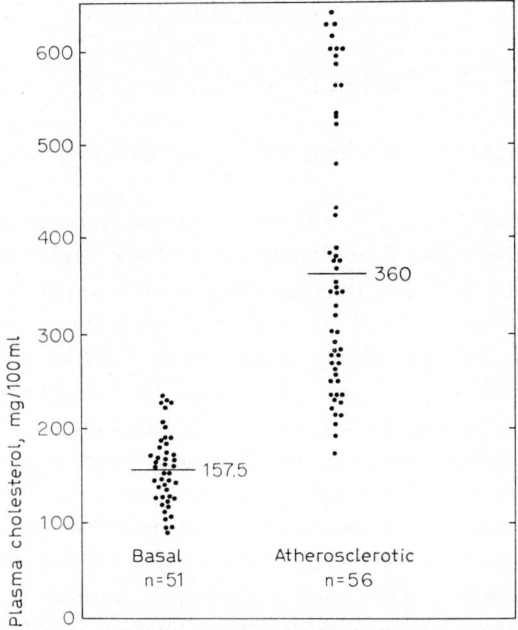

Fig. 2. The mean plasma cholesterol levels of two series of squirrel monkeys fed a basal semipurified diet or one producing hyperlipemia. The basal diet provided 15% of calories as corn oil, the diet producing hypercholesterolemia 45% of calories as butter plus cholesterol at 0.1 g/100 kcal.

and diseased versus nondiseased arteries add to the difficulties inherent in any metabolic study. On the one hand, *in vivo* studies are difficult to interpret; on the other, tissues seldom perform *in vitro* as they do in the intact animal. Therefore, any conclusions about changes that precede or result from incipient atherogenesis must be based on a very cautious interpretation of numerous *in vivo* and *in vitro* studies.

II. Lipid Composition and Metabolism of the Arterial Wall

Most lipid in the normal arterial wall and practically all of it in the plasma exist in specific lipoproteins or in membranous structures. Before attempting an intensive evaluation of the significance of their macromolecular organization, we will first consider the concentrations and reactions of lipids and then deal with their role in tissue structures and lipoproteins.

A. Free Cholesterol

All vertebrate cells, including those of the aorta, contain cholesterol. The molar ratio of cholesterol to phospholipid is approximately 1.0 in cellular plasma membranes and less than 1.0 in subcellular organelles. Therefore, the ratio of cholesterol to phospholipid is less than 1.0 in normal arterial tissue of all species [207, 213]. The mean concentration of ester cholesterol in the aortic intima plus inner media of newborn rhesus monkeys is 0.06 mg/g (wet weight), or 5% of the total cholesterol, compared with 0.28 mg/g, or 15%, for adult monkeys. When squirrel monkeys were fed a semipurified diet which elevated their plasma lipids, both the free and the esterified cholesterol concentrations increased in the aortic intima plus inner media. At first, the increase was concentrated in the free fraction (fig. 3), but later, as arterial cholesterol increased, the increase concentrated in the esterified fraction. Despite diffuse and severe atherosclerosis, the concentration of esterified cholesterol seldom exceeded that of free cholesterol. There is considerable variation in the percentage of arterial cholesterol in its free form when atherosclerosis is induced in experimental animals [47]. Most of the cholesterol in the discrete arterial lesions of man is esterified, regardless of whether intracellular or extracellular lipid predominates. During the course

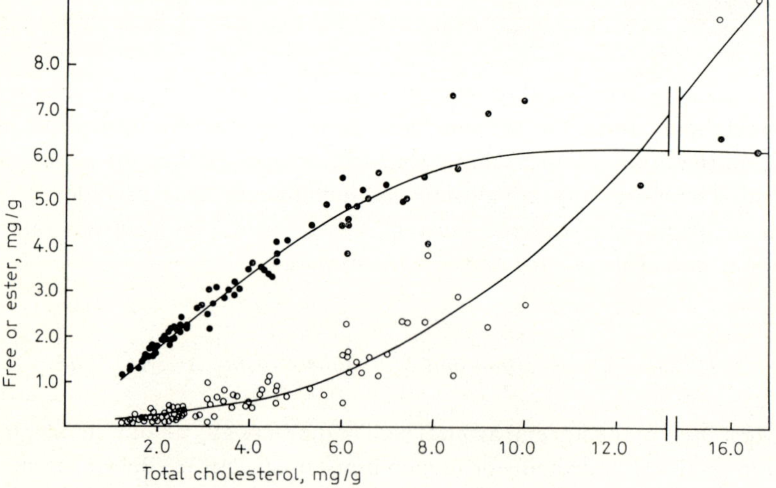

Fig. 3. Relationships of the free and of the esterified cholesterol concentrations to the total cholesterol in the aortic intima plus inner media of squirrel monkeys. ● = Free cholesterol; ○ = esterified cholesterol. From Portman *et al.* [213].

of atherogenesis, the molar ratio of free cholesterol to phospholipid in aortic intima plus inner media of squirrel monkeys does not exceed 1.0 until the lesions become severe. The molar ratio of free cholesterol to phospholipid in atherosclerotic rabbit and human arterial tissue greatly exceeded 1.0, which may be the maximum value for normal plasma membrane. (The plasma membrane is the subcellular fraction with the highest cholesterol/ phospholipid ratio.)

The equilibration of free cholesterol is rapid between plasma lipoproteins and most tissue membranes, including those of the aorta. On the other hand, the rate of cholesterol synthesis from intermediates such as acetate is exceedingly low in most nonhepatic tissues, particularly when compared with the rates of synthesis in the liver. Arterial cholesterol synthesis is probably somewhat more active in young animals than in mature ones. There is also considerable evidence that acetate is incorporated into nonsaponifiable constituents other than cholesterol by the arterial tissue of some species. It is extremely difficult to determine or even to define the meaning of net turnover of cholesterol in arterial tissue when there is a fairly rapid exchange of arterial free cholesterol with the miscible pool of cholesterol; the quantity of free cholesterol for membrane proliferation and renewal in arterial cells is undoubtedly derived largely from plasma by cholesterol exchange reactions. Apparently there is direct or indirect exchange of plasma cholesterol with the cells and subcellular structures of the arterial wall. Therefore, there is no *a priori* requirement for cholesterol to be synthesized *in situ*. In view of the rapid rate of *in vivo* and *in vitro* cholesterol exchange between plasma lipoproteins and the aorta wall, and the very low rate of cholesterol synthesis by the arterial wall *in vitro* (which cannot be studied easily *in vivo* because of the plasma-tissue exchange), we assume that cholesterol synthesis by the arterial wall is not important, particularly in mature animals. The failure of free cholesterol specific activity in arteries to equal that in the plasma as observed in cockerels [67] or man [81] after administration of [4-^{14}C]-cholesterol could indicate arterial synthesis of cholesterol. Perhaps the cockerels were growing and did have synthesis *in situ*, and the human arteries contained lesions or other areas which exchanged cholesterol very slowly with plasma. Several investigators have shown by direct radioassay and by radio-autography that in animals or human patients treated with radio-cholesterol the specific activity of plaque free cholesterol is much less than that of normal arterial wall, the periphery of plaques, or plasma [98]. Other reports indicate low rates of arterial synthesis of cholesterol [179, 312].

Opposed to the proposition that cholesterol synthesis in the arterial wall is not important are the findings that different preparations incorporate labeled acetate or mevalonate into squalene and into cholesterol and other sterols [9, 61, 80, 272, 282, 334]. The main points of difference between investigations which showed arterial synthesis of cholesterol and those which did not are: (1) age of the experimental animals [68]; (2) species [9]; (3) identification of the nonsaponifiable products other than cholesterol [53, 149, 150, 260, 334], and (4) the type of tissue preparation used. Squalene and sterol (largely noncholesterol) are synthesized by intact pigeon tissue but not by tissue homogenates [260]. Rigorous identification criteria demonstrated that cholesterol and cholestanol are synthesized by human aorta. Specific intermediate steps in cholesterol synthesis have been established in subfractions of homogenates of hog [285] and bovine aortas [322, 327].

Current views on the structure of plasma membranes and lipoproteins are consistent with the evidence that the uptake (exchange) by normal aorta of free cholesterol is much more active than that of the esterified form; however, JENSEN [129] has emphasized the activity of normal rabbit aorta in simultaneously taking up and hydrolyzing plasma cholesteryl esters. In studies of atherosclerotic rabbits *in vitro* [183] and of normal rat aorta [69], free cholesterol was shown to transfer into and out of aortas when lipoproteins were present. The transfer is considered a physicochemical exchange since it is not prevented by heating of the aortas.

B. Cholesteryl Esters, Including Cholesterol Esterification and Cholesteryl Ester Hydrolysis

Cholesteryl esters play a prominent role in the genesis of atherosclerosis. They are the most prevalent class of lipids in the active atherosclerotic lesions, and on a percentage basis, their increase during atherogenesis is highest among the major constituents of lesions. Even if their accumulation is only the result of other initiating factors, the sequence of increase and distribution in different kinds of lesions may reveal much about the way in which lesions develop. Since cholesteryl esters are a heterogeneous series with different fatty acyl groups, their source may be revealed by the distribution of kinds of esters in lesions.

In most tissues except plasma, cholesteryl esters are much less prevalent than free cholesterol. Since cholesteryl esters occupy the core position in plasma lipoproteins and presumably aggregate in masses in those tissues

Table I. The mean level of free and esterified cholesterol in the intima + inner media and in the outer media + adventitia of squirrel monkeys[1]

	Free cholesterol, mg/g	Esterified cholesterol, mg/g
Intima + inner media		
Control	2.74	0.435
Hyperlipemic	13.76	8.21
Outer media + adventitia		
Control	1.91	0.361
Hyperlipemic	2.44	1.026

[1] Nine animals were maintained for one year on a diet which resulted in hypercholesterolemia (mean = 300 mg%) and 12 were maintained on a control diet (mean plasma cholesterol = 155 mg%).

where they are found, they would not be expected to exchange readily, and they do not. Evidence of the origin of cholesteryl esters has been based on (1) the rates of accumulation of cholesteryl esters and their histological distribution, (2) the patterns of fatty acyl moieties in plasma and the arterial wall, and (3) studies of cholesterol ester uptake, cholesterol esterification, and cholesteryl ester hydrolysis by arterial tissue *in vitro*.

During the development of atherosclerosis in rabbits [181, 182], the rate of ester and free cholesterol accumulation is logarithmic with time, but the slope of the ester curve is steeper. As stated previously (sects. II. A), we [213] observed a similar increase in squirrel monkey aortas, but differentiated two phases of cholesterol accumulation (fig. 3). Slight but diffuse intimal thickening became apparent when total cholesterol reached about 6 mg/g wet weight of intima plus inner media, and it was at about that concentration that the rapid increase of cholesteryl esters with increasing total cholesterol began. The line of cleavage in the dissection of our squirrel monkey aortas was the middle of the media. During the course of nutritionally induced atherogenesis, much less cholesteryl ester accumulated in the outer media plus adventitia than in the intima plus inner media (3-fold versus 19-fold – table I). In cebus monkeys on atherogenic diets for one year, the total cholesterol in the full thickness of aorta was much less than that for squirrel monkeys, and despite some lesions, only about 20% of arterial cholesterol was esterified [153].

C. Patterns of Cholesteryl Ester Fatty Acids in Plasma and the Arterial Wall

The fatty acid composition of human plasma cholesterol esters is altered to some extent by diet; however, cholesteryl linoleate is the predominant plasma cholesteryl ester unless the diet contains very little linoleate. This is true for many species of nonhuman primates that we have tested (cebus, woolly, squirrel and rhesus monkeys, and chimpanzees) as well as for the rabbit. In rats, cholesteryl arachidonate (arachidonic acid is the product of the elongation and desaturation of linoleic acid) predominates. The pattern of plasma cholesterol esters is nearly identical with that produced *in vitro* by the plasma-esterifying enzyme, lecithin: cholesterol acyltransferase (LCAT) [222, 318, 319]. Although nascent cholesteryl esters appear predominantly in the high density lipoproteins (HDL) [319], those of the low density lipoproteins (LDL) also seem to be derived indirectly from the LCAT reaction. Cholesteryl linoleate is prominent in human LDL, the fraction that carries most of the cholesterol esters, particularly in man and animals which are susceptible to atherosclerosis. The cholesteryl esters formed by the liver and the intestinal wall under the influence of pancreatic cholesteryl ester synthetase are unusually rich in cholesteryl oleate. Some human arterial lesions are also rich in cholesteryl oleate. Cholesteryl oleate predominates in all rabbit atherosclerotic lesions and in most lesions of the squirrel monkey.

Even in the rhesus monkey fetus at term there is a detectable concentration of cholesteryl esters in the intima plus inner media [207]. In spite of the fact that free linoleic acid is transferred across the primate placenta more efficiently than free oleic or palmitic acids [216], the linoleate concentration (percentage of total fatty acids in the fraction) of plasma cholesteryl esters in the fetus is less than half that of the maternal fraction (14.4 and 37.2%). The difference in the concentration of cholesteryl linoleate in the aortic intima plus inner media is even more striking: 1.9% for the fetus, 33.3% for the adult. The very low concentration in the fetus is difficult to explain since in normal mature rhesus or squirrel monkeys the percentage of arterial cholesteryl linoleate is only slightly lower than that of the plasma, and the cholesteryl oleate is only slightly higher.

Figure 4 shows the ratio of cholesteryl oleate to cholesteryl linoleate in plasma and aortic intima plus inner media, and in plasma of squirrel monkeys fed two different diets. Although plasma cholesteryl ester composition shifted toward oleate with introduction of the atherosclerosis-inducing diet (which contained low levels of linoleate), the shift was even more dramatic

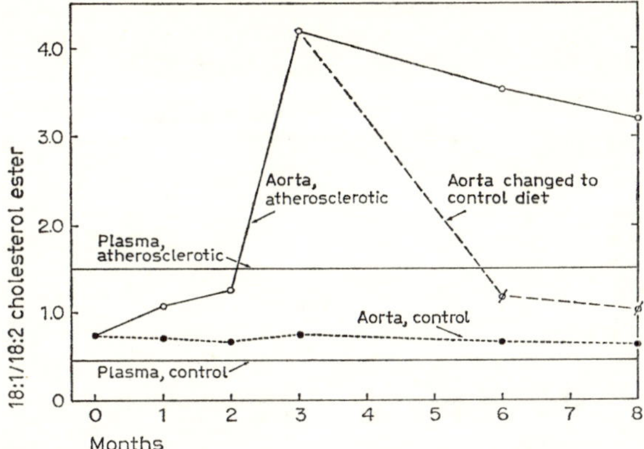

Fig. 4. A comparison of the cholesterol ester composition (the ratio of oleic to linoleic acids) of aortic intima plus inner media and plasma from a group of normocholesterolemic and a group of hypercholesterolemic squirrel monkeys. Experimental diets were initiated at time zero. After three months, a part of the hypercholesterolemic group was transferred to the control diet. From PORTMAN [206].

in the arteries. Similar changes have been observed in rabbit atherosclerosis. Cholesterol oleate was predominant in isolated foam cells from atherosclerotic rabbit aorta [203].

The composition of human arteries is presented in the excellent studies by SMITH [288, 289] and SMITH and SLATER [292, 293, 295, 296]. The quantity of cholesteryl esters increases with age in progressively thickened but otherwise lesion-free aorta, and the fatty acid spectrum resembles that of plasma cholesteryl esters. Fatty streaks and other lesions in which intracellular lipid predominates are rich in cholesteryl oleate and differ greatly from the plasma pattern. On the other hand, 'gelatinous lesions' and other older lesions containing mainly extracellular lipid have cholesteryl esters like those of plasma, i.e. with cholesteryl linoleate predominating. In human aortic lesions, moreover, cholesteryl oleate predominates intracellularly whereas cholesteryl linoleate is the main extracellular ester [143]. We have also occasionally observed a preponderance of cholesteryl linoleate in the intima plus inner media of very severely atherosclerotic squirrel monkey aorta. In general, however, experimentally induced arterial lesions of primates and nonprimates [47] have contained predominantly cholesteryl oleate. The cholesteryl oleate of the fatty streak and of lesions rich in intracellular lipid obviously results from some selectivity in the uptake of different cholesterol esters or in cholesterol esterification or hydrolysis.

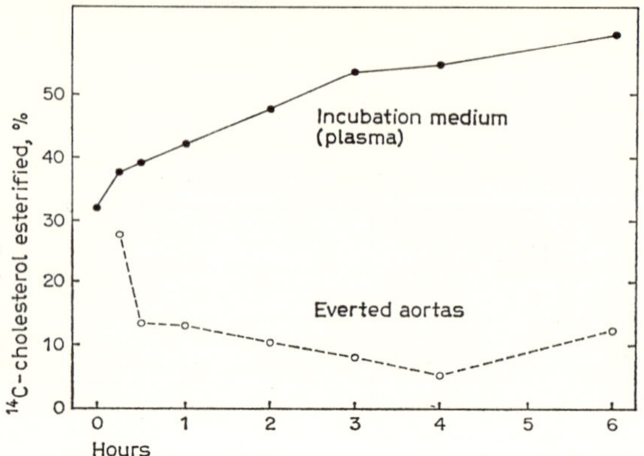

Fig. 5. The relative efficiency of uptake of free and ester cholesterol by everted rat aortas *in vitro*. The percentages of [14C]-cholesterol esterified in the incubation media and in the aorta are shown as a function of time after the unlabeled aortas were placed in a medium containing free and esterified radiocholesterol in plasma lipoproteins. From PORTMAN[206].

Cholesteryl esters with trienoic fatty acids increase in human and rabbit atherosclerotic lesions [33, 65, 90, 91, 143]. The trienoic fatty acid is derived from linoleate and has been identified as eicosatrienoic acid (20:3 — 8, 11, 14) [91].

Esterified cholesterol is not as readily transferred from plasma lipoproteins to the artery wall as is the free form (fig. 5). *In vivo* studies of cholesteryl ester uptake by experimental animals are invariably complicated by the possibility of esterification and hydrolytic reactions after cholesterol has entered the arterial wall. Thus, the ultimate equilibration of plasma and aortic free and ester cholesterol specific activities at a substantial interval after the administration of radiocholesterol [153] does not indicate the form in which the cholesteryl esters entered the wall. That some selectivity by the arterial wall must be involved is indicated by the finding of high concentrations of cholesteryl oleate, even in severe lesions which are rapidly accumulating cholesteryl esters.

We have, however, found even in short-term *in vitro* experiments with everted squirrel monkey carotid arteries or rat aortas that the apparent uptake of cholesteryl esters is as rapid as that of free cholesterol when the intimal surface is damaged mechanically. A similar effect of mechanical injury to the endothelial surface *in vivo* was also shown by studies of different cholesteryl esters and of labeled cholesteryl esters in different parts of rabbit aortic lesions [26].

The demonstration of cholesterol esterification in the aorta wall *in vitro* has long been hampered by difficulties in preparing a free cholesterol substrate. *In vitro* experiments with fatty acids or fatty acid precursors have been more successful; many of the early efforts, however, showed very low cholesteryl ester synthesis, particularly with short incubations, using normal aorta, or with fatty acyl substrates other than oleate or a precursor of oleate. Intact tissue, isolated foam cells, and cell-free preparations were all active under certain conditions. In agreement with studies on rabbits [22, 104, 224, 225] and pigeons [259] were the cholesterol ester synthetases in cell-free preparations of rhesus monkey aortas [36]. An acid synthetase, which was independent of ATP and CoA, was concentrated in the 105,000 g supernatant, whereas the alkaline ATP-CoA-dependent enzyme was concentrated in microsomes. The activities of both enzymes, which were not reversible, were elevated 5- to 100-fold in rabbits fed cholesterol and in rhesus monkeys fed cholesterol and cholic acid. We found very low cholesterol esterification activity in cell-free preparations of control or atherosclerotic squirrel monkey intima plus inner media with linoleic, oleic, or palmitic acid plus ATP and CoA or the corresponding acyl CoAs [209].

It has been extremely difficult to obtain acceptable enzyme kinetics with cholesteryl ester hydrolase preparations, largely because of the physical properties of the cholesteryl esters. Whether these esters are a part of plasma lipoproteins or are within solid tissues, they are probably always aggregated in a 'core' position. Thus, it is difficult to establish what is the substrate and the extent to which labeled cholesteryl esters are diluted by the endogenous substrate. In spite of these limitations, cholesteryl ester hydrolase has been demonstrated in several nonprimates and in squirrel and rhesus monkeys, and in baboons. HOWARD and PORTMAN [119] concentrated their studies on the 100,000 g supernatant fraction cell-free preparations of aortic intima plus inner media. Recently, activity in both microsomes and high-speed supernatant from rat and rhesus monkey aortas was determined [35]. The enzymes had broad alkaline pH optima and were not affected by addition of bile salts. Cholesteryl linoleate or oleate was hydrolyzed more rapidly than cholesteryl palmitate or stearate. Some investigators have failed to find an effect of atherosclerosis on ester hydrolase activity; others showed a role of that reaction in determining the composition of arterial lesions [34].

Considering the uncertainties that exist regarding cholesteryl ester hydrolase activity in atherosclerosis and the lack of any other evidence for selective retention of specific cholesteryl esters, we are left with the activity of cholesterol ester synthetase as the most feasible explanation for the increase of esters, particularly cholesteryl oleate, during atherogenesis in experimental animals. This enzyme appears to be activated during the early course of atherogenesis when the free cholesterol component and the free cholesterol-to-phospholipid ratio of the intima plus inner media are increasing. The transfer of excess free cholesterol from the plasma membrane to the intracellular compartment of smooth muscle cells, where it can be esterified, would serve to keep the composition of the plasma membrane constant.

D. Total Phospholipids

The role of phospholipids as a constituent of normal cellular and subcellular structures has received more attention than that of cholesterol. It is somewhat surprising that the priorities for studies of the role of these lipids in atherogenesis have been reversed. Because of their greater diversity of composition, the changes in arterial phospholipids with aging and atherosclerosis should be as informative as those for cholesterol and cholesteryl esters. The factors that are common to the phospholipids, namely the fatty acyl groups and phosphate-containing polar head groups, give these compounds the hydrophilic and lipophilic characteristics that are responsible for their interfacial properties and explain their suitability for membrane constituents. All biological membranes contain phospholipids, but some do not contain cholesterol. Nevertheless, all mammalian membranes contain some cholesterol. Studies of the properties of mammalian membranes and of artificial membranes made of phospholipids and cholesterol show that both groups of chemicals play an important role. Not only do different sterols have unique effects on artificial membrane systems, but also the nature of the phospholipids exerts a major influence on permeability. The variables in the phospholipids are the polar head groups and the hydrocarbon chains, particularly the fatty acyl groups. The proportions of sterols and the kinds of phospholipids are probably genetically determined: a highly specific composition is observed for membranes of different species, organs, and subcellular constituents.

Concentrations of the three choline-containing phospholipids – lecithin, sphingomyelin, and lysolecithin – are highly variable in the arterial wall. The former two compounds usually make up 70% or more of the total phospholipid. Most of the change in the phospholipid composition of the inner portion of the artery wall that occurs with the development of atherosclerosis or aging *per se* is in the choline-containing compounds. JACKSON and GOTTO [126] have recently reviewed some aspects of phospholipids in medicine, including their role in atherosclerosis.

There is general agreement that total phospholipids in the innermost layers of the arterial wall increase with atherosclerosis in man [32, 41, 54, 125, 287, 288, 333]. Aging alone resulted in an increase of approximately 45% in total phospholipid of the human aortic intima between 10 and 60 years of age [288]. A 55-percent increase was observed during the same period in normal intima plus most of the media [79]. We [207, 213] found that between birth and adulthood of rhesus monkeys, the phospholipid content of normal aortic intima plus inner media on a wet weight basis remained almost constant, although the types of phospholipids and the fatty acids in the different phospholipids did vary. Since the total DNA content of the aortic intima plus inner media declined with age, total phospholipid per unit of DNA rose. The same pattern of change was found in aging rat aortas, whereas phospholipid increased with age in rabbit aortas, regardless of whether the results were expressed on the basis of wet weight or DNA content [78].

The development of atherosclerosis in rabbits is accompanied by a sizable increase in the total phospholipid content of the aortic inner layers [155]. However, as in man [125, 288], cholesterol concentrations increase more than phospholipid concentrations; and the molar ratios of cholesterol to phospholipid and even of free cholesterol to phospholipid sometimes exceed one. A threefold increase in the total phospholipid of the intima plus media was observed in rabbits fed an atherogenic diet for 23 weeks [78]. Squirrel monkeys, which had developed severe atherosclerosis on an atherogenic diet for six months or more, had up to threefold increases in the aortic intima plus inner third of the media but minimal increases in the outer two thirds of the media plus adventitia [213]. Phospholipid concentration increased about 50% in the full thickness of the aorta wall of the rhesus monkey in the areas which had atherosclerotic plaques [226]. A sixfold increase was observed in the fatty streaks and plaques of spider monkeys fed diets containing cholesterol; the molar ratio of free cholesterol to phospholipid was about 0.2, and for total cholesterol to phospholipid was less than 0.4. PUCAK et al. [226] believe that these exceedingly low ratios account for the higher resistance of the spider than of the rhesus monkey to atherosclerotic plaques [161, 184, 321, 343]. The phospholipid composition of the coronary arteries (full thickness) of rhesus monkeys with severe nutritionally induced atherosclerosis was increased by only 35%, whereas the concentrations of esterified and free cholesterol were elevated by 21 and 2.7

times, respectively, above normal values [5]. Molar ratios of total cholesterol/ phospholipid increased from 0.50 to 2.8 and of free cholesterol/phospholipid from 0.38 to 0.74.

It appears that the changes in the arterial concentrations of individual kinds of phospholipids with aging and atherosclerosis are more complex than the changes of total phospholipids, and that the differences between species are substantial.

E. Sphingomyelin

1. Composition

Ether-insoluble phospholipid (sphingomyelin) levels increase in arteries with aging in man [1, 32, 41, 287, 300, 333]. The percentage of phospholipid as sphingomyelin in normal intima increases from 30% at 25 years to 46% by 37 years [287]. The concentration of sphingomyelin is higher in the normal media than in the intima [288]. Values for lecithin/sphingomyelin plus lyso-lecithin in the intima decrease from 0.97 to 0.86 between 15 and 53 years, whereas the values for the media decrease from 0.69 to 0.37. Thus, the changes in arterial sphingomyelin are prominent in the media and are probably a phenomenon of smooth muscle cells. In rhesus monkeys, we [207, 212] observed a nearly threefold increase in the sphingomyelin content of intima plus inner media (at least 85% of the tissue was media) between birth and maturity (table II). On the basis of (1) the high sphingomyelin concentration in plasma membranes of other cells, (2) the subcellular fractionation of aortic homogenates, and (3) the histological appearance of fetal and adult medial cells, we suggested that the increase in sphingomyelin content was related to a change in the plasma membrane [205–207, 212]. Subsequent experiments [211, 214] in rhesus monkeys showed that the age-

Table II. Sphingomyelin concentrations in the intima plus inner media of aortas from term fetal, one year, and adult (> 3 years) rhesus monkeys[1]

	Total phospholipid, %	Tissue, mg/g	DNA, mg/mg
Term fetus	8.92	0.47	0.234
One year	11.12	0.58	0.483
Adult	25.9	1.38	1.68

[1] Modified from PORTMAN and ALEXANDER [211].

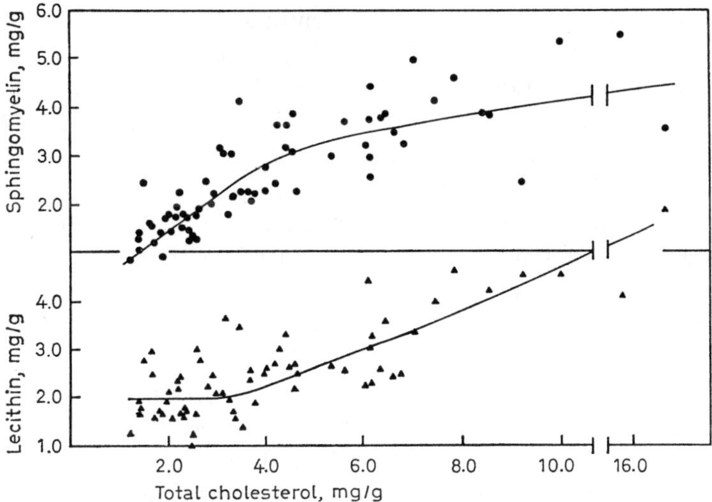

Fig. 6. Relationships of the lecithin and of the sphingomyelin concentrations to the total cholesterol in the intima plus inner media of squirrel monkey aortas. From PORTMAN *et al.* [213].

related increase in arterial sphingomyelin could be due to an increase in the plasma membrane component of medial cells. Sphingomyelin accounts for most of the age-related increase in the phospholipid of normal intima plus media of human aortas [79]. Sphingomyelin concentrations in aorta increase in proportion to the logarithm of age in man [250]. The sphingomyelin content (per unit of DNA) in the normal aortic intima plus media of 24-month-old rats or rabbits is three times higher than the corresponding values for one-month-old animals [78, 302]. The values for lecithin in adult rat and rabbit aortas were increased by about 50 and 100%, respectively, compared with one-month-old animals.

Atherosclerotic lesions also have a higher content of sphingomyelin than normal arterial tissues, particularly in man [32, 288]. There is a progressive increase in sphingomyelin concentration from the normal intima to fatty streaks, fatty nodules, fibrous plaques, and finally calcified plaques. It should be emphasized, however, that the most marked effect on human arterial sphingomyelin is its increase in the media with aging.

The sphingomyelin content of the intima plus inner media of the squirrel monkey increases during the course of atherogenesis [213]. Unlike the situation in man, the change in monkeys occurs early in the course of feeding a diet which induces hyperlipemia. The shape of the curve (fig. 6) relating total cholesterol and sphingomyelin concentrations in the aortic intima plus inner media is similar to the curve of total versus free cholesterol.

Atherosclerosis in rabbits was also associated with about a fourfold increase in the concentration of arterial sphingomyelin [78, 155, 179, 349]. The percentage increase in sphingomyelin was about equal that for lecithin but the absolute increase was less.

BÖTTCHER and VAN GENT [32] have described the effect of atherosclerosis on fatty acids of arterial sphingolipids, and PANGANAMALA *et al.* [193] have reported on the spectrum of long chain bases in sphingomyelins in human atherosclerotic aorta. Cerebrosides have also been measured in human aortic tissue [1, 30, 31, 85, 105, 223, 269, 270].

Sphingomyelin is unevenly distributed within subcellular constituents and among different classes of lipoproteins. Therefore, in explaining the increasing concentrations of arterial sphingomyelin with age in rhesus monkeys or during the development of atherosclerosis in squirrel monkeys, we have concentrated on the distribution of sphingomyelin in different subfractions of arterial tissue. Much of the work on the separation and analysis of these subfractions has equal relevance to other chemical constituents of the arterial wall, particularly to free cholesterol.

2. Lipid Composition of Different Subcellular Structures of the Aorta with Emphasis on Sphingomyelin and Free Cholesterol

Red blood cells and myelin, as well as the plasma membrane of all mammalian cells, are disproportionately rich in sphingomyelin and free cholesterol [7, 73, 214, 284] which are present in very low concentrations in mitochondria. The concentrations of these compounds in microsomes are intermediate between those for plasma membranes and mitochondria, and are highly variable in different tissues [164]. Microsome fractions originate from several cytological structures, smooth and rough endoplasmic reticulum, the plasma membrane, Golgi membranes, and often lysosomes. The differences in the ratios of sphingomyelin and free cholesterol to total phospholipid in microsome preparations of different tissues are probably to a large extent the result of the proportions of membrane derived from different subcellular sites. The differences in microsomes according to origins are, in turn, dependent on the morphology of the cells in question. Arterial tissue contains a great deal of elastin which is difficult to disrupt without either enzymic action [200, 201, 237] or grinding of very small and thin preparations of intima and inner media in an all-glass homogenizer. With the latter type of tissue disruption, it has been possible to obtain mitochon-

Table III. The molar ratios of free cholesterol (Chol) and of sphingomyelin (Sph) to total phospholipid (PL) in subfractions of the aortic intima plus inner media and liver of rhesus monkeys[1]

	Aorta		Liver	
	Chol/PL	Sph/PL	Chol/PL	Sph/PL
Total homogenate	0.052	0.23	0.16	0.050
8.10^3 g·min pellet	0.49	0.19	0.13	0.046
10^5 g·min pellet	0.49	0.17	0.09	0.025
Plasma membrane	0.97	0.35	0.99	0.126
Endoplasmic reticulum	0.39	0.19	0.19	0.036
3.10^6 g·min supernatant	0.38	0.14	0.10	0.064

[1] From PORTMAN and ALEXANDER [211].

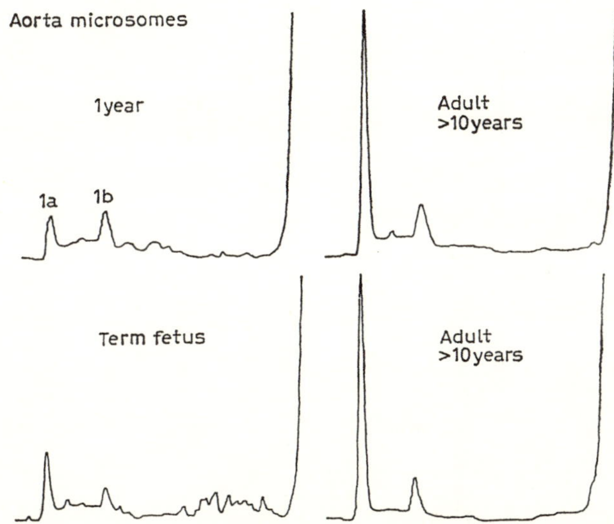

Fig. 7. A comparison of the quantities of plasma membrane (fractions 1 a + 1 b) from identical weights of aortic intima plus inner media of four rhesus monkeys: a fetus at term, a one-year-old, and two monkeys over ten years old. These curves are light scattering patterns of Ficoll gradients. The plasma membrane fractions were rich in sphingomyelin, free cholesterol, and 5′-nucleotidase and were composed of vesicles with the bilayer configuration. From PORTMAN and ALEXANDER [211].

dria, one of the most fragile of subcellular constituents, which were morphologically and enzymically intact [112, 336]. In order to prevent postmortem
effects on composition, to eliminate possible effects of elastase on the plasma
membrane, and to recover quantitatively the tissue subfractions, we chose
[211, 212, 214] to use the nonenzymic disruption of arterial tissue. The homo-
genization of arterial tissue completely disrupted the plasma membranes and
converted them to vesicles which behaved like microsomes on differential
subfraction in sucrose. With a modification of the procedure of KAMAT and
WALLACH [131], the microsomes were subfractionated into endoplasmic reticulum and plasma membranes. The plasma membrane fraction from adult
rhesus monkey aortic intima plus inner media, as well as from liver, had
higher ratios of sphingomyelin and free cholesterol to phospholipid and

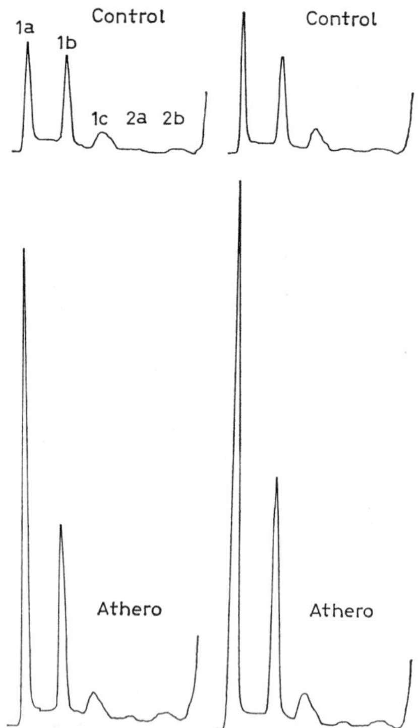

Fig. 8. A comparison of the quantities of plasma membrane (fractions 1a + 1b) from
identical weights of aortic intima plus inner media of control (normal plasma cholesterol)
squirrel monkeys and monkeys with gross atherosclerosis which had been hyperlipemic for
11 months. These curves are light scattering patterns of microsomes on Ficoll gradients.
From PORTMAN and ALEXANDER [211].

higher concentrations of 5'-nucleotidase than any other fractions (table III). The quantity of the plasma membrane fraction from light-scattering measurements per unit weight or DNA increased markedly between the birth and maturity of the rhesus monkey. The change from birth to one year of age was much less than that from one to three years of age (fig. 7).

Nutritionally induced hyperlipemia and atherosclerosis in squirrel monkeys resulted in a complex pattern of increased lipids in the different subcellular fractions of homogenates of the aortic intima plus inner media [211]. There was a pronounced increase in the plasma membrane fraction with short-term (two to five months) or long-term hyperlipemia (fig. 8) and minimal but diffuse intimal thickening, but the major increase in total cholesterol and phospholipid in severe atherosclerosis was in the 3.10^6 g. min supernatant (fig. 9). Presumably the latter originates from the cytosol fraction of cells and from extracellular material. The increase with severe atherosclerosis in phospholipid concentration in this fraction, which was also rich in cholesteryl esters, corresponds to the increase of lecithin in intact intima plus inner media with severe atherosclerosis. Cholesteryl esters also increase in the cellular debris fraction which is rich in elastin.

The increase (with aging in rhesus or with atherosclerosis in squirrel monkeys) in the total quantity of plasma membrane either per unit weight or per quantity of DNA in aortic intima plus inner media was also indicated by a marked increase of 5'-nucleotidase (a plasma membrane marker) activity in total homogenates of intima plus inner media but not in the outer media plus adventitia.

Others [63, 73, 75, 203, 225, 303, 310] have studied the composition of different fractions of homogenates of rabbit aorta during the development of atherosclerosis. With increasing severity of atherosclerosis, the quantities of lipid increased in all fractions (including those from cellular membranes), but most pronouncedly in those lipoproteins rich in plasma or their derivatives (spherical particles without limiting membranes on negative staining). LANG and INSULL [145] also studied a similar cholesteryl ester-rich fraction from human aortic lesions.

3. Ultrastructure of Arterial Smooth Muscle Cells and Changes in Sphingomyelin and Other Lipids with Aging and Atherosclerosis

The morphology of arterial smooth muscle cells and some of the factors that affect them have been reviewed by RHODIN [231]. PARKER [194] and GEER et al. [89], who early detailed the ultrastructure of arterial smooth muscle cells, emphasized the changes in the different cellular membranous

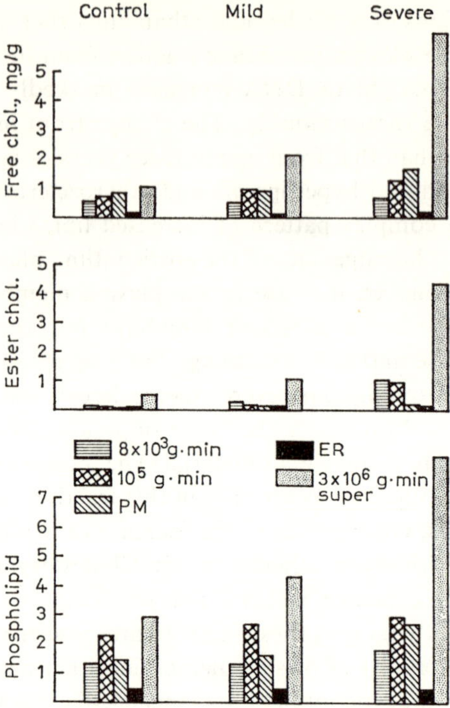

Fig. 9. The effects of mild and severe atherosclerosis in the squirrel monkey on the quantity of free and esterified cholesterol and phospholipid in five subfractions of homogenates of the aortic intima plus inner media. The fractions isolated by differential and density gradient centrifugation were: (1) a pellet formed on centrifugation at 8×10^3 g·min (elastin, myofilaments, collagen, and a few unbroken cells); (2) a pellet formed at 10^5 g·min (connective tissue elements + mitochondria); (3) PM (plasma membrane); (4) ER (endoplasmic reticulum), and (5) the 3×10^6 g·min supernatant (cytosol + extracellular lipid). Adapted from PORTMAN and ALEXANDER [211].

components (which normally contain most of the lipid) with atherosclerosis. One of the principal features of smooth muscle cells, as well as of endothelial cells, is an extensive array of invaginations from the plasma membrane and of vesicles near the plasma membrane. The cytoplasm of adult smooth muscle cells is filled mostly with myofilaments in addition to a prominent Golgi apparatus, some mitochondria, and rough endoplasmic reticulum at each apex of an elongate nucleus, whereas the medial smooth muscle cells of rabbits and rats up to one month of age are filled with intracellular organelles. The outer surface of cells from older animals [302, 302a, 310] is markedly distorted by numerous cellular projections rich in plasmalemmal vesicles. This trend toward an increase in sphingomyelin-rich plasmalemma derivatives and a decrease in intracellular organelles (which have much less sphingomyelin) has also been observed in rats [56], mice [136], and in rhesus monkeys [205].

A similar increase in sphingomyelin early in the course of the development of atherosclerosis in squirrel monkeys [213] has not been well elucidated. Plasmalemmal vesicles of smooth muscle cells increase with advanced atherosclerosis accompanied by a disproportionately greater increase of intracellular structures [195]. The endoplasmic reticulum/plasma membrane ratio increases from 0.18 to 1.5 [196]. In severe atherosclerosis in rabbits or squirrel monkeys, lecithin is the most elevated phospholipid. As indicated previously, this increase in lecithin is partly extracellular. The increase of plasmalemmal vesicles with aging or with relatively short-term hyperlipemia indicates the possibility of increased lipoprotein transport within the vesicular lumen and into smooth muscle cells. The possible transport of lipoproteins across endothelial cells in vesicles has been reviewed [307, 310].

If the increase in arterial sphingomyelin with aging and early atherosclerosis is associated with plasma membranogenesis, and if the relative composition of the new membranes remains constant, then those constituents not derived intact from the plasma would have to be synthesized at an increased rate. Arterial lecithin synthesis increases with the induction of atherosclerosis, but protein content relative to weight or DNA of squirrel monkey intima plus inner media is not affected by atherogenesis. Early atherosclerosis does not affect the arterial synthesis of protein by the human aorta [55].

4. Uptake of Sphingomyelin from Plasma

Although the increase of sphingomyelin in normal aortic intima plus inner media of rhesus monkeys with age, and of squirrel monkeys during atherogenesis, is apparently related to an increased quantity of plasma membrane, the quantity of sphingomyelin relative to total phospholipid never reaches levels reported for human arteries. This difference can be explained by the greater age of the human patients available for study. Furthermore, control values for sphingomyelin in the plasma of rhesus [207, 212] and squirrel monkeys [210, 213] as well as in the LDL of baboons [199], chimpanzees [109, 110], and squirrel monkeys, were lower than for man [25, 283]. The increases in sphingomyelin and sphingomyelin/lecithin ratios of β-lipoproteins with hyperlipemia were also lower than for man.

Most sphingomyelin exchanges between LDL and HDL and between tissue culture cells and lipoproteins. The exchange is somewhat slower than for lecithin, apparently because of a sphingomyelin component in biological structures which is bound very tightly. This stronger binding is illustrated by patterns of phospholipid release from liver microsomes treated with increasing concentrations of deoxycholic or cholic acids [PORTMAN, unpublished]. Sphingomyelin labeled with [1-^{14}C]-palmitate is transferred from plasma to the arterial wall, apparently mostly by exchange, and this process is more rapid with atherosclerotic than with normal everted carotid arteries from squirrel monkeys [210]. In rabbits, a significant increment of aortic sphingomyelin comes from the plasma and the rate of entry is increased more than 60 times by atherosclerosis [180].

5. Synthesis and Hydrolysis of Sphingomyelin by the Arterial Wall

In rats, rabbits, and man [78, 79, 229, 310] the phospholipase activity against lecithin, lysolecithin, and phosphatidylethanolamine increases with age. However, the activity responsible for hydrolyzing sphingomyelin to ceramide plus phosphorylcholine [107, 132] does not rise in rats and rabbits and falls at least 50% between birth and 60 years of age in man [79]. The sphingomyelinase activity in homogenates of aorta of rabbits fed cholesterol for up to 23 weeks changed relatively little. We [210] observed a statistically significant rise in the sphingomyelinase activity of cell-free preparations of intima plus inner media from squirrel monkeys as a result of nutritionally induced atherosclerosis.

Since CHERNICK et al. [52] first demonstrated phospholipid synthesis by the arterial wall, a number of investigators have shown (in several species, under different experimental conditions, and with various substrates) that nutritionally induced atherosclerosis is associated with increased phospholipid synthesis [54, 152, 169, 178, 196, 278, 338, 348]. The incorporation of choline into sphingomyelin, as well as into lecithin, increases with age in rats, rabbits [78], and rhesus monkeys, and with atherosclerosis in squirrel monkeys [PORTMAN, unpublished].

Cell-free homogenates of squirrel monkey aorta incorporated very low levels of palmitate from palmitoyl CoA into sphingomyelin, but substantial synthesis of ceramide occurred [210]. Sphingosinephosphorylcholine acted as an acceptor of palmitate in aortic cell-free preparations. Atherosclerosis stimulated the rate of ceramide and sphingomyelin synthesis in preparations that contained sphingosinephosphorylcholine. The importance of the sphingosinephosphorylcholine pathway is unclear [210, 221].

F. Lecithin

The predominant phospholipid in the inner layers of the normal arterial wall of most experimental animals is lecithin. Lecithin exceeds sphingomyelin by twofold in the normal aortic intima plus inner media of young sexually mature squirrel monkeys [213], rabbits, and rats [78, 155, 169, 310, 349]. It rises only slightly with age. In the normal aortic intima plus inner media of mature rhesus monkeys, lecithin is approximately equal to sphingomyelin and does not change much with age [207, 212]. The lecithin content of normal human aortic intima without visible lesions of the media, is also quite constant with age [79, 288, 310].

Nutritionally induced atherosclerosis in rabbits is associated with about the same percentage increases in arterial lecithin and sphingomyelin concentrations [78, 155, 169, 310]. The increase in lecithin concentrations in the aortic intima plus inner media of squirrel monkeys [213] is minimal during the early course of lipid accumulation, but as accumulation becomes more pronounced almost the total increase in phospholipid is composed of lecithin. The lecithin content in human atherosclerotic lesions of all grades of severity is higher than in control intima but does not increase with severity of lesions [288].

G. Lysolecithin

One of the most dramatic changes in the composition of the aortic intima plus inner media of squirrel monkeys with the development of atherosclerosis is a large increase in the concentration of lysolecithin [204–206, 209, 217, 219, 220]. Few investigators have made a serious attempt to measure lysolecithin in the arterial wall. This deficiency is the result of two factors: (1) the difficulty of measuring lysolecithin when other classes of phospholipids are much more prevalent in the sample, and (2) the belief that postmortem changes (especially in human arteries which are not generally available immediately after death) will result in the appearance of lysolecithin as an artifact. On the contrary, lysolecithinase activity is greater than lecithinase A activity, and the longer the arterial tissue stands before being extracted into chloroform-methanol, the lower the concentration of lysolecithin. We have solved the analytical problem by means of a thin-layer procedure which completely separates lysolecithin from sphingomyelin, and by use of a very sensitive system for detection (fig. 10).

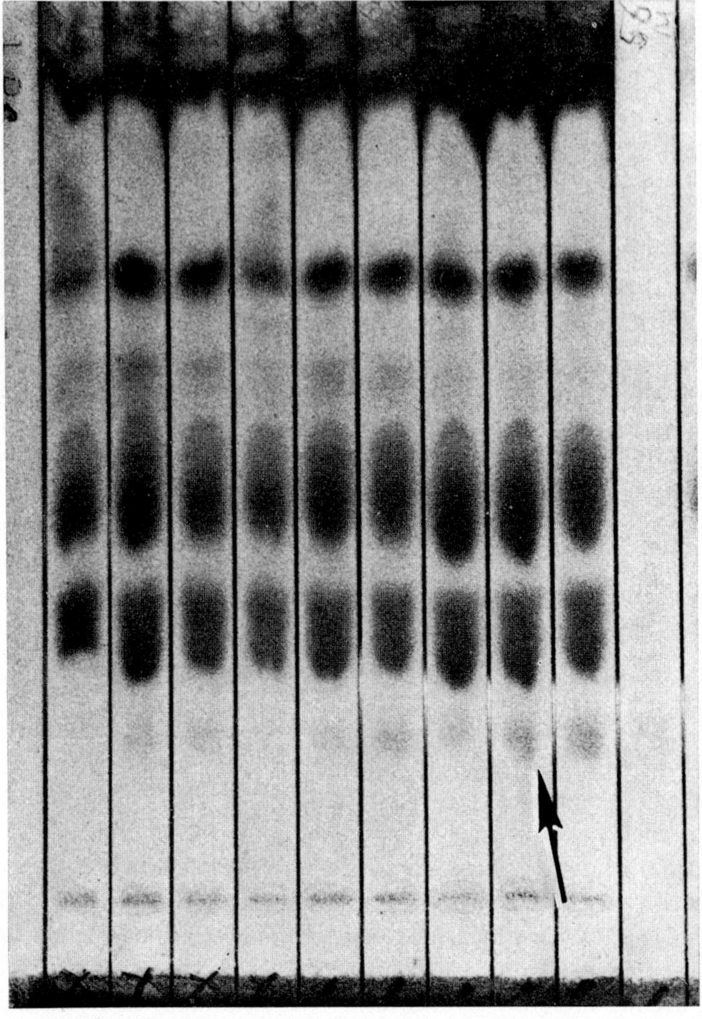

Fig. 10. Separation by thin-layer chromatography of lysolecithin from other arterial tissues. By overloading these plates (4–6 µg phospholipid phosphorus) lysolecithin can be determined even though it is usually a minor constituent; 1 µg lysolecithin (0.06 µg phosphorus) can be measured accurately. Solvent system is chloroform-methanol-H_2O (62:38:7).

Table IV. The concentration of lysolecithin (mg/g wet tissue) in the intima + inner media and in the outer media of aortas from squirrel monkeys *(S. sciureus)* and rabbits with and without nutritionally induced atherosclerosis[1]

	Control	Atherosclerotic
Monkey		
Intima + inner media	0.035	0.605
Outer media	0.030	0.045
Rabbit		
Intima + inner media	0.054	1.88
Outer media	0.025	0.135

[1] From PORTMAN and ILLINGWORTH [219].

Table IV shows the mean values for lysolecithin concentrations in the intima plus inner media and in the outer media plus adventitia of control squirrel monkeys and rabbits and of animals with nutritionally induced atherosclerosis. As atherosclerosis developed, the concentration of lysolecithin increased in the inner but not in the outer layers of the arteries. EISENBERG *et al.* [78] also showed a dramatic increase in lysolecithin concentration in the aorta of rabbits fed cholesterol.

H. Control of Lecithin and Lysolecithin Concentrations in the Arterial Wall

Several kinds of information may help to determine the origin of arterial lecithin. Most evidence indicates a difference between the fatty acyl groups of plasma lecithin on the one hand and lecithin from the aortic wall on the other [206]. Although arterial lecithin is not derived exclusively from the plasma, it exchanges between different classes of plasma lipoproteins, between plasma lipoproteins and cells, and between subcellular organelles. The histological distribution of arterial lecithin, outlined in a preceding section, is mostly in cellular membranes; however, as experimental atherosclerosis develops, the extracellular part of lecithin increases and behaves like lipoprotein, perhaps of plasma origin.

Direct evidence for the synthesis of lecithin by the arterial wall was produced by various *in vitro* and *in vivo* studies with labeled phosphate, choline, glucose, acetate, malonate, fatty acids, and lysolecithin containing labeled fatty acids and choline. Because several reactions are involved in the formation, breakdown, and replacement of substituents of the glycerol phosphatides without eliminating the complete molecule, it is difficult to translate the rate of incorporation of any intermediate into *de novo* synthesis of lecithin. Soon after CHERNICK *et al.* [52] had demonstrated that ^{32}P-phosphate is incorporated into arterial phospholipids, arterial phospholipid synthesis was shown to be increased in rabbits with experimentally induced atherosclerotic lesions [278, 348, 349]. Fatty acids derived from ^{14}C-acetate *in vivo* are incorporated into phospholipids by the atherosclerotic aorta of the rabbit [179]. Effects of atherosclerosis on the incorporation of endogenous fatty acids into phospholipids of the aorta of pigeons and rabbits were also investigated [152, 338]. ^{14}C-Fatty acids were incorporated into phospholipids (including lecithin) by various arterial preparations. Slices of rabbit aorta incorporated more linoleic acid into lecithin than arterial tissue from rat, dog, and baboon, in descending order [308]. The *in vitro* incorporation of fatty acids into aorta phospholipids was stimulated by atherosclerosis in rabbits [196, 197], squirrel monkeys [204, 209], and man [54]. Fatty acyl CoAs were also utilized more actively for lecithin synthesis by cell-free homogenates of atherosclerotic intima plus inner media from squirrel monkeys [209]. Linoleoyl CoA was used more effectively than palmitoyl CoA. Utilization of palmitoyl CoA for lecithin synthesis by cell-free homogenates of rabbit aorta increased dramatically with atherosclerosis [104].

Study of the incorporation of different precursors into phospholipids of rabbit aorta showed choline > ethanolamine > glycerol > acetate [169]. In the squirrel monkey, we have also found that choline was incorporated into aorta phospholipids much more efficiently than was glycerol. Lecithin synthesis from both precursors by the aortic intima plus inner media of squirrel monkeys and rabbits was accelerated by nutritionally induced atherosclerosis. Choline and phosphate were used with equal efficiency for lecithin synthesis by the rabbit aorta [178]. The aorta of normal rabbits and rhesus monkeys incorporates fatty acids and phosphate into phosphatidylinositol better than into lecithin [28, 169, 176, 178]. Considerable difference was found in the histological and cytological distribution of lecithin formed from labeled choline or oleic acid in the aortas of rats and rabbits [304].

Glucose, which is an important substrate for the glycerol portion of lecithin, was also incorporated into aortic phospholipids [324] at an increased rate with atherosclerosis [22, 113, 114, 118, 142, 152, 196, 316]. Most radioactivity was incorporated into the glycerol moiety of phospholipid or triglyceride when radioglucose was incubated with aorta preparations. The studies of HOWARD and co-workers [14, 113, 114, 118] and VOST [325] suggest the possibility that the glycerol from glucose first appears in arterial triglyceride and then in phospholipid; such a pattern could explain the high incorporation of labeled substrates into arterial triglycerides and the extremely low mass of triglyceride in the inner arterial layers, even with atherosclerosis.

The stimulation by glucose or glycerol-3-PO_4 of fatty acyl incorporation into lecithin by cell-free preparations of aorta suggests that some lecithin synthesis occurs *de novo* [312]. In fact, some enzymes, which are characteristic of the total synthesis scheme of KENNEDY, have been demonstrated in the rhesus monkey aorta [92, 93]. CDP-choline diglyceride phosphorylcholine transferase was reduced in the microsomal fraction of the aortas after only 90 days of feeding the monkeys with peanut oil, a particularly atherogenic factor in that species.

In cell-free preparations [204, 205, 208, 209] of intima plus inner media from squirrel monkey aortas, fatty acyl activation is probably a rate-limiting step in lecithin synthesis. The second limiting step is presumably the availability of acceptor molecules such as glycerol-3-phosphate. Lysolecithin appears to be quantitatively the most important fatty acyl acceptor in the aorta and in many other tissues [144]. This is difficult to establish with certainty since neither glycerol-3-phosphate nor lysolecithin accumulates under normal circumstances and both turn over rapidly. STEIN and co-workers [76, 309, 311, 312] were the first to emphasize the role of the aorta in taking up lysolecithin from the medium *in vitro* and in catalyzing the esterification and hydrolysis of lysolecithin. The rate of incorporation of linoleic acid from linoleoyl CoA into lecithin by cell-free preparations of squirrel monkey aortic intima plus inner media was about equal to that of lysolecithin labeled with [1-^{14}C]-palmitate. The utilization of linoleic acid by atherosclerotic homogenates was markedly greater than that by controls; this was due to the greatly increased concentration of lysolecithin, a fatty acyl acceptor in the atherosclerotic tissue.

Arterial lysolecithin concentrations appear to be controlled by at least four activities: (1) utilization of lysolecithin for lecithin synthesis; (2) formation by lecithinase A_1 and A_2 activity in the artery wall; (3) lysolecithinase activity resulting in the formation of glycerylphosphorylcholine and free fatty acids, and (4) the transfer of lysolecithin between plasma and the arterial wall. The plasma lysolecithin is the product of the enzyme LCAT plus lysolecithin transferred from tissues.

Lysolecithin, lecithin, and phosphatidylethanolamine are hydrolyzed at alkaline pH by cell-free preparations of the aortic intima plus inner media of squirrel monkeys [205, 208]. The latter two hydrolases are Ca++-dependent. Regardless of the substrate used, one of the products formed is always a free fatty acid; thus, the activities are not those of a transferase. The order of rates of hydrolysis of three glycerylphosphatides is lysolecithin > phosphatidylethanolamine > lecithin. Preparations of atherosclerotic or control aortic intima plus inner media did not differ significantly in their hydrolytic activities. The rates of phosphatide acyl hydrolases observed by EISENBERG *et al.* [77] in human and rat aortas were rather similar to those we observed in squirrel monkeys. They [78] did not observe much effect of atherosclerosis in rabbits but did find an increase with age in all the hydrolases against glycerylphosphatides in rabbits, rats, and man [79, 310]. PATELSKI *et al.* [198], using different methods of assay, found higher phospholipase activities in rat and rabbit aortas than EISENBERG *et al.* [77, 78].

Phospholipid acid hydrolases are probably present in the aorta since they have been found in other tissues [86, 165, 314]. They are probably in the lysosomes, which are prominent in arterial tissue [200, 201].

Isolated umbilical arteries of man and the carotid arteries of dogs take up lysolecithin from a medium containing albumin and hydrolyze or acylate the lysolecithin to lecithin [76]. Paradoxically, lysolecithinase is much greater than lysolecithin acylation activity, as we confirmed with everted monkey carotid arteries and rat aortas [206]. *In vivo*, the activities of lysolecithinase and fatty acyl CoA: lysolecithin fatty acyl transferase are similar [219, 220]. Minimal atherosclerosis does not affect the uptake by everted carotid arteries but results in a marked accentuation of uptake by monkey or rabbit aortas [219, 220]. This effect was confined to the intima plus inner media.

The subcellular localization of the various hydrolytic and synthetic activities for the glycerylphosphatides has been reviewed by VAN DEN BOSCH *et al.* [29]. The plasma membrane of liver seems to be as active as the microsomes in the acylation of lysolecithin [214, 313]. The microsome fraction of aortic intima plus inner media had most of the fatty acyl CoA: lysolecithin fatty acyl transferase [204, 209]. This microsome fraction is, however, derived largely from the plasma membrane. When intact aorta was incubated with labeled lysolecithin, choline, glycerol, or linoleate and the subcellular fractions from the aortic intima plus inner media were isolated in the presence of inhibitors which greatly retarded lecithin exchange between subcellular components [122], the lecithin of the endoplasmic reticulum had the highest specific activity. This observation is consistent with the conclusions about lecithin synthesis from choline, glycerol, and linoleate by liver *in vivo* [303a]. Preliminary observations on subcellular fractions indicate that lysolecithinase

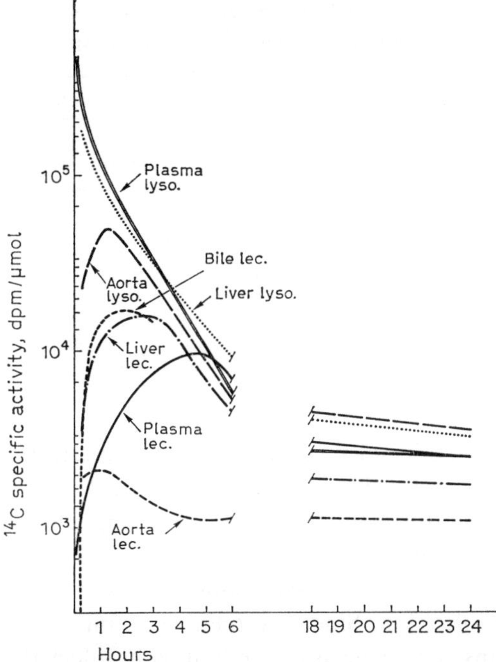

Fig. 11. The metabolism of lysolecithin, a normal constituent of plasma, in the squirrel monkey. Specific activities of lysolecithin in plasma, liver, and aortic intima plus inner media and of lecithin in the same tissues plus common duct bile are shown as a function of time after injection of labeled lysolecithin bound to plasma proteins. The use of doubly labeled lysolecithin indicated that lecithin was formed mainly by direct acylation. Adapted from PORTMAN *et al.* [217] and PORTMAN and ILLINGWORTH [220].

is most active in the endoplasmic reticulum of rabbit aorta. Incubations with intact aorta, however, suggest that this activity is most prominent in the plasma membrane [219, 220].

Lysolecithin in the plasma equilibrates very rapidly with lysolecithin in the other tissues (fig. 11) of the squirrel monkey [219, 220], including the brain [123] and normal aortic intima plus inner media [206] but relatively slowly in the atherosclerotic intima plus inner media. Thus, a pool of body lysolecithin is distributed in plasma and tissues which is formed by tissue phospholipases and the plasma enzyme LCAT [96]. The LCAT enzyme is widely distributed in mammalian species, including cebus, rhesus, and squirrel monkeys [205, 222]. With hyperlipemia and atherosclerosis, there are increases in LCAT activity, total body lysolecithin production rate, and in the concentration of plasma lysolecithin [206].

Table V. Binding capacities (V) and binding affinities (K_m) of different human plasma proteins for lysolecithin from reverse dialysis experiments[1]

	Albumin	Lipoproteins		
		low density	high density	very low density
V μmoles lysolecithin/mg protein	0.147	1.59	2.04	1.22
Moles lysolecithin/moles protein	10.3	795	204	–
K_m (M $\times 10^5$)	1.85	0.4	0.69	0.36

[1] From Portman and Illingworth [218].

The concentration of lysolecithin in hyperlipemic squirrel monkey plasma is increased by only about 50%, but there is a much higher increase in the proportion of lysolecithin bound to LDL [206, 218]. The binding capacities and affinities of LDL and HDL for lysolecithin are much greater than those for albumin (table V). Of the tissue subfractions, the plasma membrane binds the most lysolecithin. Since the relative affinities of different plasma and tissue fractions for lysolecithin undoubtedly affect the distribution of lysolecithin, the high concentrations of arterial lysolecithin with atherosclerosis could reflect the changes in the plasma membranes of smooth muscle cells, or in the LDL in the arterial wall.

Atherosclerosis has a pronounced effect on the uptake of ^{14}C-palmitate- and ^{3}H-choline-labeled lysolecithin by the intima plus inner media of squirrel monkeys *in vivo* and of monkeys and rabbits *in vitro* [217, 219, 220]. With atherosclerosis, there is a greater total flux of lysolecithin between the plasma and arterial wall.

I. Fatty Acid Synthesis by the Arterial Wall

Although various intermediates are utilized in the synthesis of fatty acids by the arterial wall, it has been relatively little investigated. For example, the only intensive study of fatty acid synthesis in the arteries of nonhuman primates is that of Howard [112] on squirrel monkeys. Whereat [337] reviewed the literature on the mechanisms of arterial fatty acid synthesis up to July, 1969. Howard [112] examined the subcellular fractions of squirrel monkey aorta for their ability to use labeled acetate, acetyl CoA, and malonyl CoA for fatty acid synthesis. Like WHEREAT, he found that acetate was most actively utilized by the microsomes. Most of the fatty acyl synthetase activity with malonyl CoA was about equally divided between the microsomes and cytosol. With malonyl CoA, most of the synthesis with cytosol was *de novo* whereas with microsomes, chain elongation predominated. With acetyl CoA, the mitochondria and microsomes produced only chain elongation. Esterification of the labeled fatty acids occurred only when mitochondria or microsomes were present. These findings, together with our work using preformed fatty acids [204, 209], indicate that the subcellular localization of lipid metabolic activities is similar to that of other tissues. They also indicate that ultracentrifugal fractions from aorta are similar to those of other tissues. The ratio of malonyl CoA to acetyl CoA utilization for fatty acid synthesis by the aortic cytosol fraction suggested that the acetyl CoA carboxylase reaction was rate-limiting; however, direct measurement of malonyl CoA formation from acetyl CoA and bicarbonate did not support this suggestion [112].

Several investigators have compared fatty acid synthesis by atherosclerotic and normal arterial tissue *in vitro* (it is impossible to make meaningful comparisons *in vivo*); none, however, have done so with primates. Most of these studies made on the rabbit and pigeon showed increased fatty acid synthesis in the atherosclerotic aorta [66, 149, 152, 257, 335, 336, 338].

J. Triglyceride Levels and Triglyceride Synthesis

Triglyceride concentrations of the inner layers of the normal arterial wall are very low and do not increase much with atherosclerosis. This is a paradox since the incorporation of various substrates into arterial triglycerides is high and increases with atherosclerosis; in addition, clinical and epidemiological studies indicate that very low density lipoproteins which are rich in triglyceride are atherogenic in man. We have shown that neither atherosclerosis in squirrel monkeys [213] nor increasing age in rhesus monkeys [207] significantly increased the very low levels of triglyceride in the aortic intima plus inner media. On the other hand, normal and atherosclerotic outer media plus adventitia had high levels of triglyceride even when the outer segment was carefully cleaned of adhering fat. In our opinion, it is impossible to prepare an arterial sample free of the outer media and adventitia by dissecting these layers from the opened artery; samples of the intima plus inner media must be dissected from the opened artery. Normal or atherosclerotic arterial samples that contain more than 0.2 mg of triglyceride/g wet weight are probably heavily contaminated with the outer layers of the arterial wall which are not involved in the process of atherosclerosis. SMITH [288] found that the concentration of triglyceride in the normal intima increased slightly with age. When compared with normal intima, the fatty streak and the raised fatty nodule had small absolute increases in triglyceride but very sharp drops in triglyceride as a percentage of total lipid.

In accord with the general findings in other species, immediately after the injection of labeled free fatty acids into rhesus monkeys [216] and of labeled glucose into squirrel monkeys [117], the radioactivity of triglycerides in plasma, liver, and other tissues is high. The specific activity of lecithin peaks later than that for triglyceride. Thus, active triglyceride lipolytic activity, coupled with a relative slow lecithin turnover, is probably responsible for the sequence of peaks of radioactivity. Moreover, the conditions for incubating aortas during *in vitro* studies may fail to simulate the *in vivo* condition in some way that is crucial to the distribution of precursors in triglycerides and phospholipids. For example, the degree of oxygenation [113, 114, 118, 142] affected the distribution of products from labeled glucose and acetate. The failure of triglyceride levels to rise in the aortic intima plus inner media during the process of atherogenesis, even though triglyceride-rich very low density lipoproteins (VLDL) in the plasma probably accentuate the process, may be explained by the hypothesis recently proposed by ZILVERSMIT [346]: since lipoprotein lipase is rich in the endothelium, a conversion of VLDL to LDL on the endothelial surface may result in the formation of an atherogenic moiety that enters the arterial wall.

III. The Role of Lipoproteins in Arterial Metabolism and Atherogenesis

From the preceding discussion it is clear that, with the possible exception of cholesterol, the arterial smooth muscle cell contains enzymes necessary to synthesize and metabolize most if not all of the lipids found in the arterial wall. The predominant involvement of the intima and inner media in the development of atherosclerotic lesions suggests, however, that interactions between plasma constituents, in particular the circulating lipoproteins and the arterial wall, serve not only to modulate cellular metabolism, but that uptake of lipid or intact lipoproteins provides an exogenous source of lipid for the artery. In this section, we will focus on the relationships between the lipids in the arterial wall and those of the plasma lipoproteins. In particular, we will emphasize possible mechanisms whereby phospholipids and cholesterol may be incorporated into the artery.

A. Structural Aspects of Plasma Lipoproteins

Since plasma lipoproteins have been thoroughly reviewed in a previous chapter [252] and in other recent reports [264, 265, 267], this section is limited to those aspects of lipoprotein structure which are pertinent to lipoprotein-arterial interactions. The lipoproteins of nonhuman primates display many similarities to those of man [252]. Although the precise structure of the individual lipoproteins remains to be elucidated, abundant evidence [147, 264, 265, 267] supports the concept of a central core of apolar cholesterol esters and triglycerides surrounded by a surface layer composed primarily of phospholipids, free cholesterol, and protein. Free cholesterol is envisaged as being interposed between adjacent phospholipid molecules. One important consequence of this arrangement is that the maximum amount of cholesterol that can be accommodated in a stable configuration is one mole per mole of phospholipid [99, 146]. Variations such as multisubunit aggregates, the penetration of protein into the hydrophobic milieu, or even a bilayer arrangement like that recently proposed for LDL [163] are not inconsistent with a general lipid core model. Such an arrangement of the lipid constituents is also predicted from physicochemical studies which demonstrate only a slight solubility of free cholesterol or phospholipid in triglyceride cholesterol ester mixtures at physiological temperatures [268, 286]. Both triglycerides and cholesterol esters display some solubility in lamellar phospholipid-cholesterol liquid crystals. Triglycerides, however, are more surface-active and would be expected to preferentially occupy any regions at the periphery of a lipoprotein not covered by phospholipid, protein, or free cholesterol. Despite wide variations in size and composition, the plasma lipoproteins all constitute a series of molecular aggregates held together primarily by hydrophobic London-van der Waal forces between apolar regions of lipids and other lipids or proteins as well as by hydrogen bonds between the phosphate head groups of phospholipids and the 3-hydroxyl of cholesterol. In their normal state lipoproteins, like cell membranes, are hydrated and are all surrounded by a layer of organized water about 20 Å in thickness.

B. The Exchange of Lipids
between Lipoproteins and Cell Membranes

Considerable experimental evidence supports the view that the lipids at the surface of lipoproteins are in a dynamic state and that they readily exchange with those of other lipoproteins and cell membranes. Thus, exchange of labeled free cholesterol [239], lecithin, and sphingomyelin [121, 122, 251] between the LDL and HDL has been demonstrated in a number of species including the squirrel monkey [121]. The complete isotopic equilibration observed between the lipoproteins also indicates that the pools of these three lipids are probably homogeneous and that all of the free cholesterol and phospholipid is available for exchange. The rapid *in vivo* exchange of phosphatidylcholine between rabbit LDL and HDL labeled *in vivo* with ³H-labeled leucine and Me-¹⁴C-choline is illustrated in figure 12. The considerably faster turnover of labeled phosphatidylcholine and sphingomyelin

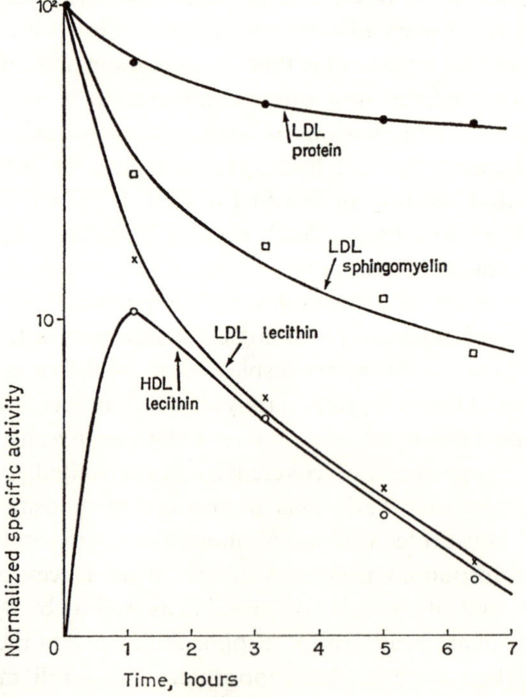

Fig. 12. Time-course of the *in vivo* metabolism of the phospholipid and protein moieties of doubly labeled LDL in rabbits. From ILLINGWORTH and PORTMAN [122].

than of protein is also consistent with exchange with a large pool of unlabeled lipids in cellular membranes. The core lipids, triglycerides and
cholesterol esters, appear not to undergo significant exchange [97, 239] although a net transfer of triglycerides from VLDL to HDL concomitant with
reciprocal movement of cholesterol esters from HDL to VLDL has been
demonstrated [185, 227].

Cell membranes, like the periphery of lipoproteins, are composed primarily
of phospholipids, glycolipids, protein, and variable amounts of free cholesterol. Various models for the structure of cell membranes have been
proposed [16, 37, 126, 281]; the most recent ones are consistent with a
bilayer of phospholipids in the liquid crystalline state, interspersed with
various structural and enzymatic proteins. Up to one mole of free cholesterol
per mole of phospholipid can pack between adjacent phospholipids of the
bilayer. Introduction of cholesterol reduces the area occupied by each phospholipid molecule and results in a more condensed membrane [146]. Thus,
the surface of plasma lipoproteins and the exterior half of the bilayer of
cell membranes appear to share many structural and compositional similarities. Because of their abundance, ease of isolation, and lack of intracellular
organelles, erythrocytes have been widely used as a model in which to
study the interactions of lipids between lipoproteins and cell membranes.
Since erythrocytes, endothelial cells and, to a variable and as yet undefined
extent, the arterial smooth muscle cells are in direct contact with plasma or
its constituents, a consideration of lipid-lipid interactions between lipoproteins and red cells is clearly of value in understanding how the lipid content
of the arterial smooth muscle cell may be increased. Unlike lipoproteins, not
all of the lecithin, sphingomyelin [230, 277], or free cholesterol [8, 17, 72,
173] in the red cell membrane is capable of exchange. On the other hand,
exchange of the two other principal membrane phospholipids – phosphatidylserine and phosphatidylethanolamine – with extracellular lipids is very
low or nonexistent. These observations are consistent with an asymmetric
arrangement of lipids in the membranes whereby most of the lecithin and
sphingomyelin is located in the exterior half of the bilayer and the charged
phosphatidylserine and phosphatidylethanolamine in the interior [37]. Although the lipids of both artificial [71] and biological [263] membranes are
capable of a rapid lateral movement, the transfer of phospholipids from
one side of the bilayer to the other, termed 'flip-flop' by KORNBERG and
MCCONNELL [138] is slow. The speed at which unesterified cholesterol flips
from the exterior to the interior of the membrane is not known. Therefore, the
rapidly exchangeable pools of membrane lipid would seem to correspond to

those in the exterior half of the bilayer, whereas the slow and nonexchange-able pools would correspond to lipids located in the interior part of the membrane. Since cholesterol and phospholipid appear to exchange readily between lipoproteins and erythrocytes, the question arises whether increases in the lipid content of individual lipoproteins or in the absolute concentrations of lipoproteins with a normal lipid composition can induce changes in membrane lipid composition. According to currently available data, pathological conditions characterized by an abnormal composition of the plasma lipoproteins, notably increased proportions of free cholesterol and phospholipid, are associated with changes in erythrocyte lipids. Thus, plasma from patients with LCAT deficiency, hepatic liver disease, or cholestasis, as well as from guinea pigs fed cholesterol, all contain abnormal cholesterol-rich lipoproteins in which the free cholesterol/phospholipid ratio exceeds unity. In all instances, the erythrocytes are atypically large and misshapen and contain increased concentrations of free cholesterol [96]. That these erythrocyte changes can be induced *in vitro* is shown by the fact that incubating normal erythrocytes together with these cholesterol-rich lipoproteins results in a net transfer of cholesterol to the erythrocytes [58, 187, 262]. Similarly, when cholesterol-rich erythrocytes were incubated with normal plasma, cholesterol was transferred to lipoproteins. This net transfer of cholesterol between lipoproteins and erythrocytes is not, however, limited to conditions in which the lipoproteins are abnormally lipid-rich but takes place when normal plasma and erythrocytes are incubated together for 24 h. In this case, because of LCAT activity, the free cholesterol content of the plasma slowly decreases and a net transfer of cholesterol from the erythrocytes to the cholesterol-depleted plasma occurs concurrently [72, 172]. Changes in the phospholipid content and composition of erythrocytes related to an abnormal phospholipid content of lipoproteins have also been reported [190, 191, 332] but in general perturbations in erythrocyte phospholipids are not as pronounced as those observed for cholesterol. In contrast to these observations, increases in the concentration of lipoproteins whose lipid composition is normal do not elicit similar changes in erythrocytes either *in vivo* [10, 59, 177] or *in vitro* [228, 262]. We may therefore conclude that the composition and cholesterol content of membrane lipids, specifically those of the erythrocyte plasma membrane, can be modified by abnormally lipid-rich or lipid-poor lipoproteins in the surrounding media. The basic mechanisms of lipid exchange and of mass transfer from the lipid-rich donor to a lipid-poor recipient, which is invariably superimposed on simultaneous lipid exchange processes, are probably the same. In both cases, the concept of a transient collision complex [101] seems most plausible. It postulates that the rate of exchange or transfer is limited by the diffusion of cholesterol or phospholipid in the complex and by the binding forces holding these lipids in their parent lipoprotein or cell membrane. This proposed mechanism is probably valid only for the transfer and exchange of lipids between extracellular lipoproteins and cellular plasma membranes; at the intracellular level, specific carrier proteins are probably involved [103, 188]. On this basis, the faster rates of exchange and greater mass transfer of cholesterol than of phospholipids, observed between lipoproteins and erythrocytes, would be attributable either to lower hydrophobic binding forces, or to a more rapid diffusion of cholesterol in the collision complex, or to a combination of both of these factors.

In the preceding discussion, we have provided evidence that the lipid composition of the erythrocyte plasma membrane is not static and may change rapidly in response to variations in plasma lipoprotein composition. Although the erythrocyte provides an excellent general model in which to evaluate the interaction of lipids between lipoproteins and cell plasma membranes, the lack of intracellular organelles precludes any studies on how changes in plasma membrane composition affect cellular metabolism or the accretion of lipids by intracellular constituents. It is especially important to know whether variations in plasma membrane lipids, specifically an increase in the content of cholesterol or phospholipid, can induce corresponding changes in the intracellular content of either the lipids themselves or their metabolic products, e.g. cholesterol esters. An insight into these questions, with particular reference to the smooth muscle cells of the artery, can be gained by looking at these and other diploid cells grown under controlled conditions in tissue culture.

C. Lipid Transfer in Tissue Culture Cells

The popularity of tissue culture as a means of studying cellular growth and metabolism has increased rapidly in recent years. Although the explants or subcultures are normally grown in serum-supplemented medium, the extent of intracellular lipid deposition in different lines of cultured cells grown in the presence of serum from normal and hyperlipemic animals varies widely [248, 254]. In contrast, the ability of various phospholipids [124, 202] and free cholesterol [11, 44, 249] to exchange between lipoproteins and a variety of cells grown in tissue culture appears to be general and is probably structurally inherent in the lipoprotein complexes contained in both plasma lipoproteins and cellular plasma membranes. Therefore, the plasma membranes of cells cultured in the presence of abnormally lipid-rich lipoproteins would be expected to acquire cholesterol or phospholipids in a manner similar to the erythrocyte. In diploid cells, however, the subsequent exchange and transfer of these lipids between the plasma membrane and the intracellular organelles could also occur. Intracellular incorporation of plasma membrane lipids can occur by pinocytosis in which small segments of the membrane 'bud off' and are incorporated into the cell as vesicular structures. In addition, excess cholesterol and phospholipid molecules originally incorporated into the external half of the plasma membrane would be expected to slowly redistribute by a process of 'flip-flop' into the inner half of the bilayer. Subsequent transfer to intracellular organelles may be mediated by 'exchange proteins' similar to those involved in the intracellular transport of fatty acids [188], the exchange of various phospholipids between intracellular organelles [103, 158, 340], or by the soluble sterol carrier protein recently implicated in cholesterol biosynthesis [70]. In addition to stimulating a two-way exchange of phospholipids between membranes, the phospholipid exchange proteins also appear to promote a net transfer of phospholipids between membranes of different lipid composition [130, 339]. The extent of lipid exchange and transfer is clearly dependent upon both the number of collisions the carrier proteins make with the membranes, and the number of lipid molecules displaced during each collision. Transfer of phospholipids and cholesterol from the plasma membrane to intracellular organelles or lipoproteins would be favored by increases in the surface area of the plasma membrane and by decreases in the surface area per molecule. Both of these criteria would result from an increase in the content of cholesterol and phospholipid; equilibration of their optimal concentrations between the plasma membrane and the intracellular membranes would be expected to take place rapidly in the absence of intracellular metabolism. Removal of intracellular lipid, whether by hydrolysis or esterification, would disturb the equilibrium and favor net intracellular transport of lipid from the plasma membrane. Thus, the cellular content of cholesterol and of phospholipid would be partly dependent on exchange equilibriums between extracellular lipids and those of the cellular plasma membrane, as well as between those of the plasma membrane and intracellular membranes or lipoproteins. Differences in the metabolic capabilities of cells from various tissues may explain their apparent susceptibility or resistance to accumulate intracellular lipids. Thus, cells highly capable of esterifying cholesterol would be expected

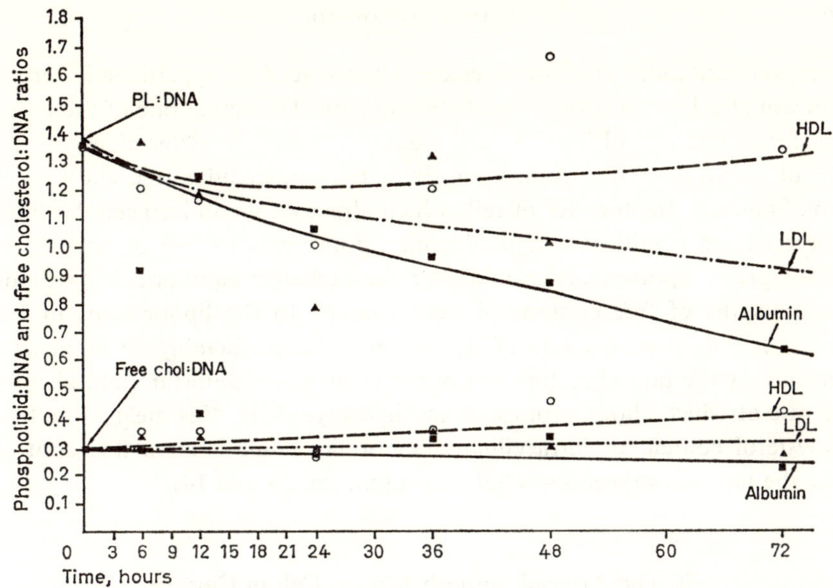

Fig. 13. The influence of variations in the lipid content of the growth medium on the phospholipid/DNA and free cholesterol/DNA ratios of cultured human prostatic cells. From ILLINGWORTH *et al* [124].

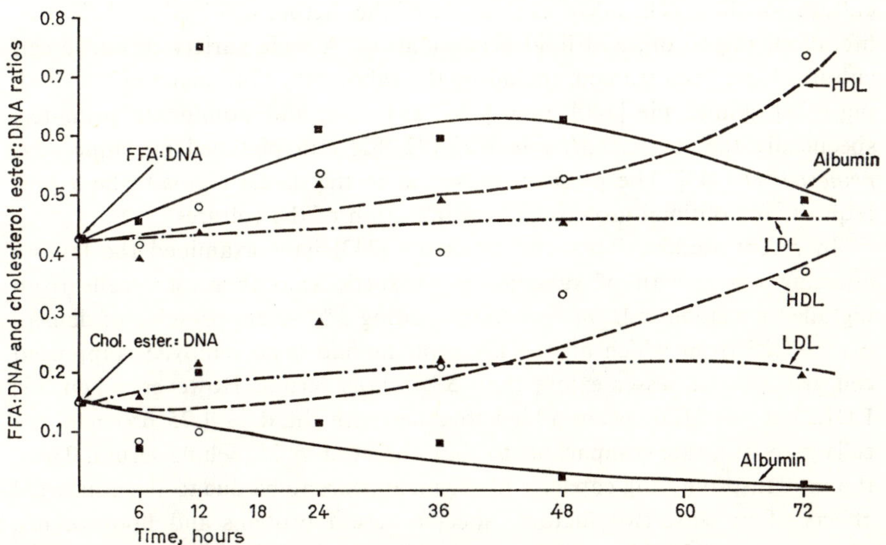

Fig. 14. The influence of variations in the lipid content of the growth medium on the free fatty acid/DNA and esterified cholesterol/DNA ratios of cultured human prostatic cells. From ILLINGWORTH *et al.* [124].

to accumulate more cholesterol esters when placed in hyperlipemic serum than would cells with a lower synthetic capacity. The actual rate of exchange transfer of cholesterol into the cell would, however, be determined by the rate of removal of free cholesterol from the intracellular free cholesterol pool. Similarly, the transfer of cells which already contain intracellular lipid droplets from a milieu of hyperlipemic lipoproteins to one of normal or lipid-depleted lipoproteins should alter the exchange equilibria in favor of net transport of free cholesterol from the cell to the lipoproteins in the media. In this case, the rate of regression could predictably be influenced not only by the rate of restoration of the exchange equilibrium, but also by the rate at which cholesterol esters are hydrolyzed. By this means, the free cholesterol content of intracellular membranes could exert a fine control over the fate of intracellular cholesterol (c.f. fig. 13 and 14).

D. The Arterial Smooth Muscle Cell in Culture

Smooth muscle cells constitute the principal arterial cells and their numbers increase in atherosclerosis. They are probably also the source of extracellular collagen and elastic fibres. In view of this central role, tissue cultures of these cells allow evaluation of the factors involved in their proliferation, migration, and lipid accumulation. A wide variety of aortic cell cultures have been studied, including the rabbit [49, 174], pigeon [256, 298], pig [128], guinea pig [240], man [175, 234, 235] and nonhuman primates, specifically the rhesus *(M. mulatta)* [82, 83] and pigtailed macaque *(M. nemestrina)* [243]. The presence of serum in the media seems to be a prerequisite for optimal growth and proliferation of the cultures.

In recent studies, ROSS and GLOMSET [243] have examined the factors affecting the growth of subcultures of aortic smooth muscle cells from pigtailed macaques *(M. nemestrina)*. Adding 5% serum proteins of density (d) > 1.25 from which all the lipoproteins had been removed, supported cell growth to a lesser extent than 5% whole serum. However, when 5% LDL, but not HDL, were added together with the d > 1.25 proteins, the cells grew at a rate comparable to that observed in 5% whole serum. Thus, the growth-promoting capacity of serum appears to be due to the concerted effects of at least two factors: specific serum proteins and lipoproteins, particularly LDL. The addition of insulin at physiological concentrations also promoted the growth of *M. nemestrina* cells [315]. Some of the growth-promoting factors present in the d > 1.25 proteins from monkey serum are

Table VI. The effects of lipoproteins on the growth and mitosis of primary cultures of aortic smooth muscle cells from rhesus monkeys[1]

Addition to culture medium	% increase in surface area after 10 days growth	% of cells labeled with ^{3}H-thymidine
5% NS	25	8.7
5% HS	112	39.8
VLDL from 5% HS	31	6.5
LDL from 5% HS	83	35.0
HDL from 5% HS	2	4.9
Protein residue from 5% HS	11	3.8
LDL from 5% NS	23	7.3
LDL from HS diluted 1:4.5	78	29.4
LDL from NS concentrated 2-fold	9	6.6

[1] From FISHER-DZOGA *et al.* [82]. NS = Normal serum; HS = serum from hypercholesterolemic monkeys.

not present in plasma and appear to be released from platelets during the clotting process [244]. This finding suggests that adherence of platelets to injured areas of the epithelium and the concomitant release of stimulatory proteins are the key factors in the proliferation of medial smooth muscle cells *in vivo*.

Experiments indicate that the replacement of 'normal' serum with that from hyperlipemic animals elicits a pronounced increase in the growth and lipid content of cultured aortic cells [62, 82]. An increase in the synthesis of DNA, together with an increase in the surface area occupied by medial explants from rhesus monkey aorta, was observed [82] after the substitution of 5% hyperlipemic rhesus monkey serum and 5% fetal calf serum for the 10% fetal calf serum normally used (table VI). When the hyperlipemic serum was replaced by 5% of VLDL, LDL, or HDL from hyperlipemic rhesus monkeys, both DNA synthesis and cell growth in the presence of LDL were stimulated five times more than control samples grown in normal serum. LDL from hyperlipemic monkeys promoted five times more thymidine incorporation into DNA than equivalent concentrations of LDL from normal monkeys. These findings clearly demonstrate that the stimulatory properties of hyperlipemic LDL cannot be attributed to increases in the total lipid content of the medium *per se*. Along with its stimulatory effects on the growth of aortic smooth muscle cells, serum from hyperlipemic animals also promotes an accumulation of lipid droplets within the cellular cytoplasm [50, 82]. In this respect the cells resemble those isolated from 'fatty streaks' [295, 296] which contain numerous amorphous lipid droplets rich in cholesterol esters.

Factors responsible for the intracellular lipidosis frequently seen in tissue cultures have been studied in a variety of cell lines, many of which show characteristic patterns of lipid accumulation similar to the arterial smooth muscle cells [264]. A direct correlation exists between the nonesterified fatty acid content of serum and the extent of sudanophilic intracellular droplets present in cultured human embryonic skin and muscle fibroblasts [48, 253]. Analysis of the intracytoplasmic droplets produced in a variety of cultures from nonarterial tissue by high extracellular free fatty acid concentrations has demonstrated that more than 90 % of the lipid is triglyceride with only 3–5 % as cholesterol esters [94, 157, 171]. This contrasts with the increased lipids of both human MAF fibroblasts [15] and rabbit [49, 82] or human aortic smooth muscle cells [236] grown in the presence of hyperlipemic serum in which the increase in esterified cholesterol exceeds that of triglyceride. In absolute terms, however, the concentration of both of these apolar lipids was increased.

An increase in the concentration of free cholesterol relative to other lipids, particularly phospholipids, in the serum LDL has also been implicated as a stimulatory factor in cellular proliferation and lipidosis [255]. Intracellular cholesterol increases rapidly in mouse fibroblasts incubated in a medium supplemented with serum and a dispersion of cholesterol in Tween-80 [12]. Labeling of the dispersion with ^{14}C-cholesterol demonstrated its preferential uptake by the cells. A similar observation was made after the addition of cholesterol bound to albumin [154], but microcrystalline suspensions of cholesterol elicited virtually no intracellular inclusions. The importance of altering the cholesterol/phospholipid ratio of lipids and lipoproteins has also been emphasized in studies on mouse lymphocytes [247]; the increases in the free cholesterol/phospholipid ratio promoted accretion of cholesterol by the cells, whereas decreases in the ratio stimulated egression of cholesterol from the cells into the growth medium. The extracellular presence of phospholipid dispersions containing lecithin or sphingomyelin was found to stimulate the transfer of labeled free cholesterol from Landschütz ascites cells to the medium [306]. A similar stimulation was also observed in the presence of delipidated human HDL, but not with albumin. Lecithin and sphingomyelin, together with apo HDL, acted synergistically and in their combined presence, cholesterol transfer into the medium was some 10–15 times greater than with albumin.

Table VII. The effects of hypercholesterolemia on the free cholesterol/phospholipid ratios
of plasma lipoproteins

		Free cholesterol/phospholipid molar ratio					
		man	baboon	chimpan- zee	rhesus monkey	rabbit	guinea pig
		(a)	(b)	(b)	(a)	(c)	(d)
LDL	N	0.84	0.47	0.61	0.68	0.81	0.55
	HC	0.97	0.43	0.95	0.78	1.24	2.00
HDL	N	0.32	0.28	0.35	0.37	0.53	0.73
	HC	0.36	0.35	0.36	–	1.09	1.79

From: (a) Morris and Lee [170] and Scanu *et al.* [266]; (b) Howard *et al.* [109];
(c) Camejo *et al* [46], Chen [49], and Garlick and Courtice [87]; (d) Sardet *et al.* [261].
N = Normal; HC = hypercholesterolemic.

Thus, in addition to specific serum proteins, increases in the concentrations of plasma free fatty acids and proportional increases in the relative concentrations of free cholesterol on the surface of LDL, combined with absolute increases in this cholesterol-rich lipoprotein, are probably involved in both the intracellular lipidoses and cell proliferation characteristically observed in the presence of hyperlipemic serum. Table VII shows the effects of hypercholesterolemia on the free cholesterol-to-phospholipid molar ratios from the lipoproteins of a variety of species. Considerable species variations are apparent, but the ratio is generally higher in LDL than HDL and is more affected by hypercholesterolemia. In the LDL, these values undoubtedly represent the average for the heterogeneous population of lipoproteins, and the cholesterol/phospholipid ratio in 'young' cholesterol-rich remnant particles is probably higher than those of the average. On the basis of artificial bilayer studies [146], free cholesterol/phospholipid ratios should be associated with an unstable excess of surface free cholesterol, which would readily be transferred to other lipoproteins or cell membranes, and would therefore promote accretion of cellular cholesterol. In cases where the flux of both free fatty acids and cholesterol is in favor of cellular accumulation, the two lipids may act synergistically to promote the intracellular formation of cholesterol esters. On the other hand, conditions favoring cellular accretion of only free cholesterol would promote *de novo* fatty acid synthesis and esterification with endogenously produced acids. This theory is supported by recent studies on the effects of hyperlipemic serum on ^{14}C-acetate incorporation into the lipids of cultured rabbit aortic smooth muscle cells [51]. Synthesis of free fatty acids, phospholipids, and triglycerides was higher in the presence of hyperlipemic serum whereas incorporation of labeled acetate into cholesterol was markedly suppressed. Similar observations have been reported in human MAF fibroblasts [15]. In both of these studies, *de novo* synthesis accounted for an insignificant proportion of the cholesterol which accumulated within the cells, most of which was derived from the growth medium. Growth of rabbit aortic smooth muscle cells in hyperlipemic serum resulted in a ninefold increase in cholesterol esterification [49]. This pronounced increase is similar to that observed in aortic tissue from pigeons [256, 259], rabbits [64], and squirrel monkeys [258].

E. Mechanisms of Lipid Uptake by Cells in Tissue Culture

At least two theories, which may be termed molecular and particulate, have been formulated to explain the mechanism of accretion of cellular lipids in tissue culture. The molecular theory assumes that the surface lipids of lipoproteins and cellular membranes are in a dynamic state of exchange equilibrium and that changes in the lipid content of one are mutually reflected in the other. Since a thorough treatment of this theory and its relation to intracellular lipidoses has already been developed in this review, we will not discuss it further. Direct cellular uptake of whole lipoprotein molecules provides the basis for the particulate theory of lipid uptake. The intracellular demonstration of an intracellular LDL antigen in cultures of rhesus monkey aortic smooth muscle cells [82, 83, 134] incubated in homologous sera lends strong support to this theory. Cultures exposed to hyperlipemic serum stained positively with horseradish peroxidase-labeled antibody to LDL but not with antibody to HDL [82, 83]; these findings may, however, reflect a much lower concentration of HDL, than of LDL, in the hyperlipemic serum and do not necessarily preclude HDL uptake. Human arterial 'intimacytes' can take up ^{131}I-labeled LDL from the growth medium but when the lipoproteins were doubly labeled with ^{3}H-cholesterol and ^{131}I-protein, uptake of the former was much greater [234, 235]. Studies of the relative rates of uptake of ^{125}I-protein-labeled lipoproteins by homologous cultures of aortic smooth muscle cells in the stationary growth phase showed that total cellular incorporation of VLDL and HDL ranged from 0.1 to 0.5% and was proportional to the lipoprotein concentration in the medium [19]. Pulse-chase experiments with labeled and unlabeled HDL indicated that the cells were also capable of catabolizing HDL. In similar studies, the uptake of ^{125}I-protein-labeled VLDL was compared with that of cholesterol-rich remnant particles produced by the *in vitro* action of postheparin plasma on ^{125}I-labeled VLDL, by rat aortic smooth muscle cells in stationary growth phase [18]. After 48 h of incubation, uptake of labeled remnant particles (average diameter 234 Å) from the medium was more than twice that for VLDL (average diameter 425 Å). Although the smaller remnant particles would be expected to enter cells more readily than VLDL, the fact that the incorporation of 'remnants' exceeded that of HDL (diameter 75–110 Å) suggests that size is not the only criterion which determines cellular uptake. Differences in surface charge [95], the high content of B protein, the presence of the arginine-rich peptide, or an excess of free cholesterol and phospholipid in the remnant may all be related to its apparently avid uptake by the cells. Recent studies [306a] on the uptake of LDL suggest that lipoproteins containing the B apoprotein may be selectively taken up by cells.

There are indications of the presence of cellular binding sites for both VLDL, VLDL 'remnants', and HDL on the aortic smooth muscle cells [18, 19]. Such binding sites may be intimately involved in the regulation of cellular lipid metabolism and the lack of feedback inhibition of HMG CoA reductase by LDL in cultured skin fibroblasts from patients with homozygous type II hypercholesterolemia may be due to deficient or abnormal cellular binding sites for LDL [39].

Studies on the cellular uptake of 'core' lipoprotein lipids, especially cholesterol esters, by a variety of cells, including those derived from rabbit aortic smooth muscle [49] and in culture [13, 246] are indicative of the uptake of whole lipoproteins, but the interpretation of these findings is far less definitive than that of studies in which the lipoproteins were labeled in the protein moiety. Decreases in the esterified cholesterol content of the medium [13, 49] together with smaller increases in the extracellular free sterol content may be viewed from two extremes: (1) cellular uptake of lipoproteins with subsequent egression of some of the free cholesterol produced by intracellular hydrolysis or, alternatively, (2) hydrolysis of the cholesterol esters as the lipoproteins are bound to the cellular plasma membrane by esterases contained therein [232]. Since the decreases noted in the cholesterol ester content of the medium are severalfold greater than the corresponding values observed for cellular uptake of ^{125}I-protein-labeled lipoproteins, both mechanisms are probably operative, but the hydrolysis of cholesterol esters on the plasma membrane is chiefly responsible for decreases in the concentration of these lipids in the culture medium.

The precise mechanism of cellular uptake remains to be elucidated, but incorporation of lipoproteins by pinocytosis seems highly likely. Morphological studies on cultured smooth muscle cells from guinea pig aorta [240, 241] support this view; the cells contained numerous cytoplasmic plasmalemmal vesicles 400–1,000 Å in diameter which resemble those seen in medial smooth muscle cells from arteries fixed *in vivo* [195, 231, 302]. Plasmalemmal vesicles seem to be derived from inpocketings of the plasma membrane, termed caveoli, which 'bud off' and subsequently traverse into the cell. Since the exterior layer of the plasma membrane gives rise to the interior of the vesicle, lipoprotein molecules bound to the interstitial or exterior surface of the cell may be incorporated into the vesicle without previous release from their binding sites. Alternatively, the molecules also are incorporated as constituents of the plasma contained in caveoli before vesicular entrapment. As we shall discuss later, pinocytotic uptake of intact lipoproteins by endothelial cells is probably involved in the entry of these molecules into the normal arterial wall *in vivo*.

F. Penetration of Lipoproteins into the Arterial Wall

Passage of lipoproteins from the vascular lumen into the arterial wall has long been considered a major source of lesion lipids [192]. This filtration theory of atherogenesis is supported by reports which have documented immunologically the immunological presence of LDL in both normal and atherosclerotic arteries. Although these reports have focused primarily on human tissue [106, 108, 133, 137, 290, 291, 297, 330, 331, 345] antigens similar to plasma LDL have also been demonstrated in the arteries of hyperlipemic rhesus monkeys [108, 275, 342] and rabbits [274, 329]. Fluorescein-labeled antibodies further demonstrated the genetically determined human Lpa lipoprotein in lesions from patients with this variant lipoprotein in their plasma but not in lesions of other Lpa-negative individuals [328]. Similar techniques have also been used to document the presence of apo lipoprotein C antigens [328] in arteries from both normal and hyperlipemic human subjects. In the latter, however, it cannot be ascertained whether the peptides represent the entry of VLDL or HDL into the artery, since both contain the C peptides. In view of the scarcity of information pertaining to the uptake mechanisms of lipoproteins by arterial tissue in nonhuman primates, a brief discussion of recent observations, mainly in the rat, will be presented. It is anticipated that much of the work on the rat may be extended to nonhuman primates.

1. Transport of Macromolecules across the Endothelium

The rate of passage through uninjured endothelium determines the entry of lipoproteins and other macromolecules into the subendothelial tissue. The permeability of capillary endothelium and its capacity to transport substances from the lumen into the vessel wall or extracellular space has been widely studied [40, 135, 279, 280]; similar work on arterial endothelium has, however, been less prolific. Electron microscopic studies [84, 120] have examined the mechanisms of transport of horseradish peroxidase (diameter 50Å) across the aortic endothelium of the rat. The transendothelial passage occurred through both the intercellular junctions and via plasmalemmal vesicles. In contrast, ferritin (110 Å diameter) was transported exclusively by the latter vesicles [120]; this finding agrees with those for muscle capillaries [40]. However, the peroxidase detected in the intercellular junctions may in fact be derived from plasmalemmal vesicles emptying into this space rather than from passage along an intercellular channel [84, 120, 271]. The demonstration of peroxidase-labeled caveoli opening into the intercellular spaces [271] provides experimental support for this theory as do recent rapid-pulse studies on the transport of whale myoglobin (diameter 33Å) through the capillaries of rat diaphragm [280].

The rates of transendothelial transport of plasmalemmal vesicles are rapid. For instance in samples of rat and mouse aorta examined 1–2 min after the injection of horseradish peroxidase into the vascular compartment [120, 271, 305], peroxidase activity was demonstrated within vesicles in the endothelial cells, in caveoli emptying into the subendothelial space, and in the subendothelial space itself. Short-term studies on the uptake of myoglobin [280] have indicated that the total transit time for plasmalemmal vesicles in the capillary endothelium of rat diaphragm is 45–60 sec: about 15 sec for the vesicular uptake of myoglobin from the lumen, 10–25 sec for transcellular movement of the vesicle, and about 15 sec for vesicular fusion and discharge of the abluminal surface.

It may be anticipated from the preceding discussion that if molecules of 80–110 Å diameter such as ferritin [84] or lactoperoxidase [305] cross the endothelium exclusively via plasmalemmal vesicles, then this should also be the pathway followed by lipoproteins of similar or larger dimensions. That this is indeed the case is indicated by recent radioautographic studies on the transport of ^{125}I-labeled human lipoproteins in rat aorta [307]. In these experiments, rat heart and aorta were perfused *in vitro* with ^{125}I-protein-labeled VLDL, LDL, HDL, delipidated HDL apoprotein, or rat HDL; and the uptake of label was evaluated by both biochemical and radioautographic criteria. After a 30-min perfusion with VLDL, LDL, or HDL, the radioautographic reaction was seen over the endothelium, the subendothelial space, and the inner media. In the latter, the label was mainly extracellular rather than in medial smooth muscle cells. An area of unlabeled media was interposed between the labeled inner surface and label derived from the vasa vasorum of the adventitia. Although numerous silver grains were seen over plasmalemmal vesicles of endothelial cells, little labeling of the intercellular junctions was observed. Transendothelial transport of all three lipoproteins was therefore considered to occur via the plasmalemmal vesicles rather than via passage through intercellular channels. On a quantitative basis the total uptake and distance of penetration of lipoproteins into the arterial wall were inversely related to their size and proceeded in the order HDL > LDL > VLDL. Delipidated ^{125}I-labeled apoHDL, which contains at least five peptides ranging in molecular weight from 7,000 to 25,000, was taken up far more extensively than any of the intact lipoproteins and was uniformly distributed over the entire media. Although transendothelial transport of protein-labeled lipoproteins *in vivo* has been demonstrated in the rabbit [189], dog [74], and man [273], the studies centered on total incorporation of lipoproteins rather than on their mechanism of uptake. Impractically huge amounts of isotope would be required for whole body experiments designed to follow *in vivo* the uptake by arterial tissue of labeled lipoproteins from the total plasma pool in order to examine this uptake at the cellular level by electron microscopic radioautography.

In addition to this pathway of transendothelial transport of lipoproteins via plasmalemmal vesicles, various pharmacological agents [57, 160, 238] or a large single dose of dietary cholesterol [276] reportedly promote the contraction of endothelial cells with subsequent widening of the intercellular spaces between the cells. The vasoactive peptide angiotensin II was found to be particularly effective in eliciting the contraction of rat aortic endothelial cells [238]. When either ^{3}H-cholesterol-labeled VLDL or LDL was injected simultaneously with angiotensin II, both were incorporated into the medial layers of the aorta. In contrast, when the lipoproteins were injected 8 min after the infusion of angiotensin II, at which time the interendothelial gaps were closed, transendothelial transport into the media was not observed. Similar effects have also been observed by other workers [274, 276]. The increased transport of these macromolecular tracers through contracted endothelial cells is presumed to occur through intercellular channels; however, the possibility that such transport is mediated by increases in the number and rates of transit of plasmalemmal vesicles across the endothelium cannot be totally discounted. Such an effect on the transendothelial transport of ferritin has been described in hypertensive and normal rats [120].

2. A Defective Endothelial Barrier

In the normal artery the transport of macromolecules, including lipoproteins, into the subendothelial layers is clearly limited by their rate of passage across the endothelium. Similarly, the egression of material from the subendothelial tissue to the vascular lumen may be determined by the rate of transfer from the abluminal to the luminal surfaces of the endothelial cells; however, studies designed to follow such transfer have been lacking. The response of the arterial wall to injury or removal of the endothelium has been studied in various species including the rabbit [24, 26] and pigtailed macaque *(M. nemestrina)* [243]. As previously discussed, studies on cultured aortic smooth muscle cells would lead us to indicate that the de-endothelialization of the artery *in vivo* is followed by intimal thickening and deposition of extra- and intracellular lipid droplets. After this initial de-endothelialization, lipoproteins rapidly enter the arterial wall. The uptake of ^{3}H-cholesterol-labeled lipoproteins into such injured areas of rabbit aorta was some 15 times greater than into nearby areas in which the endothelium was intact; lipid was initially present extracellularly but during regrowth of the endothelium abundant intracellular lipid droplets became evident [27]. Earlier studies have also demonstrated these injured areas to be characterized by an increased permeability to plasma albumin [23]. As previously discussed, the intracellular lipidosis of aortic smooth muscle cells may result from direct pinocytotic uptake of lipoproteins from the extracellular matrix or from an alteration in the normal exchange-diffusion equilibrium which would favor the intracellular accretion of cholesterol. Six months after de-endothelialization with a balloon catheter, the thickened intimal lesions induced in the iliac arteries of normocholesterolemic pigtailed macaques *(M. nemestrina)* had regressed to two to three cell layers (re-endothelialization takes place after seven to 14 days [243]. In contrast, similar lesions in animals with diet-induced hypercholesterolemia failed to regress and resembled the lipid-rich fibrous plaques seen in man [245]. Although de-endothelialization clearly represents an extremely unphysiological condition, similar but less pronounced effects on arterial permeability and atherosclerosis have been induced by various techniques including immunological injury [111, 168] and diet-induced hypercholesterolemia [301]. Both these factors appear to damage the endothelial barrier and result in an increased permeation of plasma constituents into the arterial wall. A twofold increase was observed in the aorta-to-serum ratio after injection of ^{131}I-labeled LDL into rabbits fed cholesterol for 24 h [189]. The increase was even more marked (fourfold) in those animals with visible atherosclerosis. In preliminary studies in our laboratories [221a], control and hyperlipemic rabbits and squirrel monkeys were injected with either LDL or HDL biosynthetically labeled with both ^{3}H- leucine and ^{3}H-cholesterol; the results indicated that the number of lipoprotein molecules that enter the aorta of atherosclerotic animals is considerably greater than that in controls. It has not been established whether this increase in permeability, which is reflected by increases in the rate of entry of lipoproteins, occurs through widened intercellular gaps between injured cells or as the result of increased vesicular transport.

3. The Form of Lipoproteins in the Arterial Wall

The presence of lipoproteins in the arterial wall is well established. The predilection of LDL to specifically accumulate in the arterial wall could imply either that the HDL do not enter the wall or that LDL are selectively retained therein. STEIN *et al.* have clearly shown that the former hypothesis is not valid [307]. Their data indicate that the rate of entry of HDL into the arterial wall exceeds that of LDL. The specific accumulation of LDL would, therefore, appear to be attributable either to the selective trapping of this lipoprotein in the extracellular matrix of the arterial wall or to the inability of the aortic smooth muscle cells to catabolize it.

The localization of LDL in the arterial wall has been studied by immunofluorescent microscopic techniques but the results vary somewhat. KNIERIEM *et al.* [137], for example, found virtually all LDL in human arterial lesions to be intracellular, while a predominantly extracellular localization was found by WALTON and WILLIAMSON [330]. In fatty streaks, immunofluorescence was seen predominantly in the extracellular matrix [106, 345], which supports the findings of WALTON and WILLIAMSON [330], who failed to detect any fluorescence over fat-filled cells. In thickened fibrous areas of human aorta, fluorescence was localized extracellularly and seemed to be associated with lipid droplets [106, 330]. The predominantly extracellular localization of LDL can also be inferred from chemical and immunoelectrophoretic studies on minced samples of human arterial tissue [291, 294] previously separated by microdissection. These studies (for review see SMITH and SLATER [291, 295, 296]) have indicated that the concentrations

Table VIII. LDL and fibrinogen/fibrin-related antigen (FRA) concentrations in human aortic lesions[1]

	Concentration in lesion[2] (% of normal intima)		Ratio LDL/albumin in areas of intima[3]	
	LDL	FRA	upper layer	lower layer
Normal intima	100	100	7.4	44.3
Fatty streak	22	62	8.6	
Gelatinous thickenings	179	171	10.8	8.4
Gelatinous periphery of plaques	308	425	10.4	18.5
Small plaques	92	175	9.0	16.9

[1] From SMITH and co-workers [294, 296, 297].

[2] Determined by quantitative immunoelectrophoresis into antibody-containing agarose gels.

[3] The ratio is in terms of volumes of plasma from individual patients. The upper (luminal) section contained two-thirds of the intima, the lower section, one-third.

of both LDL and fibrinogen and fibrin-related antigens increase appreciably in gelatinous thickenings and small fibrous plaques, but that in fatty streak lesions the concentrations are less than for corresponding samples of normal intima (table VIII). It is also apparent that in the lower layers of fibrous plaques, the retention of LDL in relation to albumin exceeds that in other lesions or in normal intima (table VIII). The fatty acid profile of the cholesterol esters from gelatinous thickenings of fibrous plaques also resembles those from plasma LDL and have a linoleate/oleate ratio of 1.5 to 2. In contrast, the cholesterol ester fatty acids from fatty streak lesions are predominantly oleic acid (linoleate/oleate ratio 0.2 to 0.3) and resemble those formed by cholesterol esterification *in situ* [295, 296]. What therefore is the basis for the observed retention of LDL and to a lesser extent fibrinogen and fibrin-related antigens in the extracellular matrix of aortic tissue?

The extracellular matrix of the arterial wall contains four classes of macromolecules: collagen, elastin, proteoglycans (protein-acidic glycosaminoglycan complexes), and glycoproteins [233]. Two of these, glycosaminoglycans and elastin, have both been implicated in the arterial sequestration of extracellular LDL and lipids. The ability of a wide variety of anionic sulphated glycosaminoglycans (mucopolysaccharides) such as heparin, to form complexes with serum, LDL and VLDL is well-known [45]. Despite the fact that LDL is an acidic molecule with an isoelectric point of pH 5.7, there appears to be an uneven charge distribution so that local areas at the surface behave as cations. Chemical modification of free amino groups, e.g. arginine and lysine, which contribute to the basic areas of LDL, with a resultant alteration of the charge profile, has been shown to completely inhibit the binding of this lipoprotein to either amylopectin sulfate [148] or dextran sulfate [186]. These findings, together with the known susceptibility of lipoprotein acid mucopolysaccharide complexes to dissociate at high ionic strength [299] or high pH [148], indicate that the binding forces are primarily electrostatic. The presence of divalent cations, e.g. calcium or magnesium, further enhances complex formation. The ability of glycosaminoglycans extracted from human aortas to bind to homologous LDL *in vitro* has been demonstrated [21]. Other workers have also succeeded in isolating glycosaminoglycan-LDL complexes from human aortic lesions [20, 299, 323]. Apparently, these latter complexes are similar to those found *in vitro* and are held together mainly by ionic binding forces. Formation of complexes between fibrinogen and glycosaminoglycans, analogous to those observed for serum LDL, has also been demonstrated [2]. Thus, the selective retention (table VIII) of both fibrinogen and LDL in arterial tissue is probably at least partly attributable to specific electrostatic interactions between these macromolecules and negatively charged glycosaminoglycans present in the extracellular matrix. In addition to this selective retention, a focal accumulation of extracellular lipids along the elastic lamellae is frequently observed in atherosclerotic lesions. Elastic fibers contain two morphological components and are composed of a series of microfibrillar glycoproteins covalently bound to an amorphous matrix of elastin molecules [127, 242]. The very high proportions of apolar amino acids they contain has been related to their resistance to aqueous base-catalyzed hydrolysis. Elastin preparations isolated from normal arterial tissue have been shown to contain about 1% lipid [139] composed mostly of cholesterol esters, with lesser amounts of free cholesterol, phospholipids, and triglycerides. In tissue from atherosclerotic plaques, however, the lipid content of elastin was found to increase in proportion to the severity of the lesions and in severe fibrous fatty plaques constituted 37% of the total weight of elastin. The cholesterol that was bound to elastin accounted for 20–34% of the total cholesterol of the plaque [139]. The increased lipid-binding affinity of elastin from plaques appears to be related to specific changes in the amino acid composition. Plaque elastin was shown to contain increased proportions of polar amino acids – e.g. aspartic acid, serine, leucine – concomitant with a decrease in the content of the cross-linking amino acids desmosine, isodesmosine, and lysinonor-

leucine [139]. More recent *in vitro* studies [141] have examined the ability
of lipoproteins to bind and transfer lipids to delipidated samples of elastin
derived from both normal and atherosclerotic areas of human aorta. In the
presence of elastin from either source, a net transfer of lipid, composed
primarily of cholesterol esters, from VLDL and LDL, but not from HDL,
to the delipidated elastin was observed. The extent of this transfer, however,
was some five to six times greater with elastin derived from atherosclerotic
plaques than with normal tissue. Despite the marked affinity of elastin for
apolar lipids, little binding of the protein moiety of LDL was observed. The
chemical basis for increased binding of apolar lipids to plaque elastin appears
to be an alteration in the amino acid composition, but the causative mechanism responsible for this change is not known. However, the marked
age-related increases in the concentrations of esterified cholesterol in connective tissue, e.g. dura mater and psoas tendon [60], suggest that the mechanisms of lipid accumulation is similar in these nonvascular tissues and
in the arterial elastin.

From the preceding discussion, it is clear that the extracellular matrix
of the arterial intima-media constitutes a milieu favorable to the entrapment
of lipoproteins, specifically those which contain the B apoprotein, and to
fibrinogen. The glycosaminoglycans and elastin may well act in a concerted
synergistic way. Thereby, initial entrapment of LDL by electrostatic binding
to acidic glycosaminoglycans would precede denaturation and subsequent
hydrophobic binding of cholesterol esters, derived from LDL, to elastin.
The fate of the other lipids, and of LDL apoprotein derived from denatured
LDL, may involve cellular uptake, binding to other matrix constituents,
extracellular precipitation of apoprotein or crystallization of cholesterol or,
in the case of phospholipids and free cholesterol, incorporation into cellular
membranes. The cholesterol clefts frequently seen in atherosclerotic lesions
may well be derived from such denatured lipoproteins, a theory supported
by the relatively slower rate of removal of small implants of this lipid than
of phospholipids, palmitic acid, or triglycerides from the aortic intima-media
of dogs [38].

IV. Conclusions about Mechanisms of Change in Arterial Composition

One of us [204, 206] suggested an hypothesis to explain the development
of atherosclerosis in squirrel monkeys. It was based in part on the observations that during the induction of atherosclerosis, the initial increase in free
cholesterol and sphingomyelin was followed by an increase in lecithin and

cholesteryl esters in the intima plus inner media. Intermediate between those two phases was a marked increase in the lycolecithin content. Aging alone resulted in a striking increase in the sphingomyelin content of the intima plus inner media. A number of observations on the metabolism of these compounds, their precursors and products, and on the activities and quantities of subcellular constituents isolated by differential and density gradient centrifugation of homogenates of arteries were also integrated into the hypothesis. There seemed to be two interrelated stages: (1) a change in the cellular constituents, particularly in the smooth muscle plasma membrane, and (2) the accumulation of plasma LDL or altered lipoproteins.

Within the last few years, considerable new information has appeared on the mechanisms involved in the second stage of experimental atherosclerosis. For example, lesions in the aorta of rabbits and pigtailed macaques *(M. nemestrina)* were induced by mechanical disruption of the endothelium [26, 243], and atherosclerosis was produced by immune injury [102, 167]. These studies have shown that once the endothelium is damaged, hyperlipemia is not necessary for the development of lesions. Some lesions induced by mechanical means or by hyperlipemia in experimental animals eventually regress with low plasma lipoprotein concentrations. In man, the concentration of extracellular LDL increases with age and in mature lesions, but not in fatty streaks. There is a very low level of LDL in the arterial wall of normal experimental animals, but this could result from a rapid clearance of lipoprotein from the arterial wall rather than from a lack of penetration. Some new evidence indicates that under normal circumstances lipoproteins cross the endothelium in plasmalemmal vesicles, although in certain instances it has been suggested that they may also enter via the intercellular gaps. It is unknown as yet whether the number or the transport activity of these vesicles increases during atherogenesis. If lipoproteins in the lumina of vesicles move across the endothelial cell as a normal process, an increase in lipoprotein concentration in the plasma would presumably lead to an increased rate of transport even without a change in the number of vesicular carriers. The smooth muscle cell also has a network of vesicles originating from the plasmalemma but their function is unclear. The finding that age *per se* and hyperlipemia in monkeys result in more plasma membrane per cell [211] may indicate an increase in the number of plasmalemmal vesicles per smooth muscle cell. There is proliferation of arterial smooth muscle cells with hyperlipemia *in vivo* and a similar effect of lipoproteins on isolated smooth muscle cells. Furthermore, isolated smooth muscle cells take up remnants of VLDL metabolism [18].

The accumulation of intact lipoprotein or lipoprotein fragments after lesions are initiated has been well documented, but the identification of the initial changes leading to the alteration of the metabolic barrier against lipid accumulation [347] is still very unclear. We [206] have assumed that this 'metabolic' barrier is a permeability barrier and that the increase in the plasma membrane fraction from smooth muscle cells of the intima plus inner media [211] and in its sphingomyelin and free cholesterol content during the first stage of lesion formation [206, 213] reveals something about the change of the artery wall from a tissue that did not accumulate cholesteryl esters to one that did. The first increase in intracellular cholesteryl oleate probably may reflect, at least partly, the esterification of the increased free cholesterol of the plasma membrane which was first carried into the cell as caveoli.

The early compositional changes in the intima plus inner media were observed in tissues in which the number of medial smooth muscle cells was much greater than that of endothelial cells. We assume, therefore, that such changes are prominent in the former cell type. Thus, the 'metabolic' barrier to accumulation of lipoproteins and their products must apply to the smooth muscle cell and its ability to deal with lipoproteins.

The early increase in sphingomeylin and free cholesterol in the smooth muscle cell plasma membrane is, we think, a response to an increase in the concentration and change in the composition of LDL in the extracellular space which are one component of a normal exchange diffusion equilibrium. In human LCAT deficiency, the lipoproteins are unusually rich in free cholesterol and marked atherosclerosis is observed [95a]. Susceptibility to cholesterol accumulation in this condition, which is characterized by low plasma cholesteryl esters and B lipoproteins, may be an extreme example of a perturbed diffusion equilibrium.

The large increase in lysolecithin in the plasma membrane of smooth muscle cells is the result of the greater total synthesis (including that produced by LCAT in plasma) of lysolecithin in a freely miscible body pool. The increase of these lipid moieties may stimulate the formation of plasma membrane including plasmalemmal vesicles. The free cholesterol in the matrices of these caveoli is then esterified; plasma cholesteryl esters are also transported as VLDL fragments or other forms in the lumina of these vesicles. These conditions result in the proliferation not only of plasma membrane per cell but of the number of cells. The permeability of the endothelium may not be altered (i.e. may not admit an increased volume of plasma into the intercellular space) until the lesion becomes large enough to physically distort the anatomy of the intima.

Because of the heterogeneity of even the most carefully dissected layers of normal and atherosclerotic arteries from experimental animals and of the limited amount of available tissues, chemical studies must be used cautiously and in conjunction with morphological approaches to explain the pathogenesis of atherosclerosis. The availability of methods for studying the composition and metabolism of small quantities of arterial lipids has, however, helped to make some progress possible. Numerous techniques for characterizing small quantities of plasma and tissue lipoproteins and of tissue membranes [326], and for working with isolated cells and organs should further widen the scope of biochemical approaches to the arterial wall.

Note added in proofs. Between the time that we completed this review in early 1974 and corrected proofs in May, 1975, a number of important new papers on arterial metabolism appeared. Some of these advances were anticipated in a number of abstracts of papers presented at the 3rd International Congress on Atherosclerosis in Berlin in October, 1973, and at the 27th meeting of the Council on Arteriosclerosis in Atlantic City in November, 1973, and which are included in this review. Smith [289a] recently presented an important review of the factors influencing the concentrations of plasma lipoproteins in the arterial wall.

One of the most significant developments in atherosclerosis research in recent years [96a] was the recognition that low density lipoproteins have specific binding sites on human fibroblasts. The binding of those lipoproteins depresses cholesterol synthesis, and enhances cholesterol esterification and catabolism of the low density lipoprotein apoprotein by the fibroblast. Fibroblasts from patients with homozygous Type II hyperlipoproteinemia lack these binding sites and control mechanisms.

Another important new approach to the pathogenesis of atherosclerosis was the report of Small and Shipley [286a] on the physical-chemical basis of lipid deposition. These investigators used concentrations of a four-component system with water, cholesterol, cholesteryl esters, and lecithin to predict the existence of one or more possible physical phases, lamellar liquid crystal, crystal, or oily liquid. Crystals of free cholesterol would theoretically be the most resistant form to mobilize from the arterial wall.

We [Portman, manuscript in preparation] have evidence for the role of lysolecithin as a ligand between plasma low density lipoproteins and cellular membranes of the arterial wall of squirrel and rhesus monkeys. The enhanced binding of intact lipoproteins was associated with inhibition of free cholesterol exchange between lipoproteins and cellular membranes.

References

1 Adams, C. W. M. and Bayliss, O. B.: Histochemical observations on the localization and origin of sphingomyelin, cerebroside and cholesterol in the normal and atherosclerotic artery. J. Path. Bact. *85:* 113–119 (1963).

2 Anderson, A. J.: The formation of chondromucoprotein-fibrinogen and chondromucoprotein-β-lipoprotein complexes. Biochem. J. *88:* 460–469 (1963).

3 ANDRUS, S. B. and PORTMAN, O. W.: Comparative studies of spontaneous and experimental atherosclerosis in primates; in FIENNES Some recent developments in comparative medicine, pp. 161–177 (Academic Press, London 1966).

4 ANDRUS, S. B.; PORTMAN, O. W., and RIOPELLE, A. J.: Comparative studies of spontaneous and experimental atherosclerosis in primates. II. Lesions in chimpanzees including myocardial infarction and cerebral aneurysms. Prog. biochem. pharmacol., vol. 4, pp. 393–419 (Karger, Basel 1968).

5 ARMSTRONG, M. L. and MEGAN, M. D.: Lipid depletion in atheromatous coronary arteries in rhesus monkeys after regression diets. Circulation Res. 30: 675–680 (1972).

6 ARMSTRONG, M. L.; WARNER, E. D., and CONNER, W. E.: Regression of coronary atheromatosis in rhesus monkeys. Circulation Res. 27: 59–67 (1970).

7 ASHWORTH, L. A. E. and GREEN, C.: Plasma membranes: phospholipid and sterol content. Science 151: 210–211 (1966).

8 AVIGAN, J.: A method for incorporating cholesterol and other lipids into serum lipoproteins in vitro. J. biol. Chem. 234: 787–790 (1959).

9 AZARNOFF, D. L.: Species differences in cholesterol biosynthesis by arterial tissue. Proc. Soc. exp. Biol. Med. 98: 680–683 (1958).

10 BAGDAGE, J. D. and WAYS, P. O.: Erythrocyte membrane lipid composition in exogenous hypertriglyceridemia. J. Lab. clin. Med. 75: 53–60 (1970).

11 BAILEY, J. M.: Lipid metabolism in cultured cells. IV. Serum alpha globulins and cellular cholesterol exchange. Expl Cell Res. 37: 175–182 (1965).

12 BAILEY, J. M.: Cellular lipid nutrition and lipid transport; in ROTHBLAT and KRITCHEVSKY Lipid metabolism in tissue culture cells, pp. 85–113 (Wistar Institute Press, Philadelphia 1967).

13 BAILEY, J. M.: Regulation of cell cholesterol content; in Atherosclerosis: initiating factors. Ciba Symp. No. 12, pp. 63–92 (Elsevier, Amsterdam 1973).

14 BAILEY, J. M.; HOWARD, B. V., and TILLMAN, S. F.: Lipid metabolism in cultured cells. XI. Utilization of serum triglycerides. J. biol. Chem. 248: 1240–1247 (1973).

15 BAILEY, P. J. and KELLER, D.: The deposition of lipids from serum into cells cultured in vitro. Atherosclerosis 13: 333–343 (1971).

16 BANGHAM, A. D.: Lipid bilayers and biomembranes. Annu. Rev. Biochem. 41: 753–776 (1972).

17 BELL, F. P. and SCHWARTZ, C. J.: Exchangeability of cholesterol between swine serum lipoproteins and erythrocytes in vitro. Biochim. biophys. Acta 231: 553–557 (1971).

18 BIERMAN, E. L.; EISENBERG, S.; STEIN, O., and STEIN, Y.: Very low density lipoprotein 'remnant' particles: uptake by aortic smooth muscle cells in culture. Biochim. biophys. Acta 329: 163–169 (1973).

19 BIERMAN, E. L.; STEIN, O., and STEIN, Y.: Lipoprotein uptake and metabolism by rat aortic smooth muscle cells in tissue culture. Circulation Res. 35: 136–150 (1974).

20 BIHARI-VARGA, M.; SIMON, J., and GERO, S.: Identification of glycosaminoglycan-β-lipoprotein complexes in the atherosclerotic aortic intima by thermoanalytical methods. Acta biochim. biophys. hung. 3: 375–383 (1968).

21 BIHARI-VARGA, M. and VÉGH, M.: Quantitative studies on the complexes formed between aortic mucopolysaccharides and serum lipoproteins. Biochim. biophys. Acta 144: 202–210 (1967).

22 BJÖRKERUD, S.: Synthesis of lipids from glucose and incorporation of palmitate in atherosclerotic and normal rabbit aorta; in JONES Proc. 2nd Int. Symp. Atherosclerosis, pp. 126–129 (Springer, Berlin 1970).

23 BJÖRKERUD, S. and BONDJERS, G.: Arterial repair and atherosclerosis after mechanical injury. I. Permeability and light microscopic characteristics of endothelium in non-atherosclerotic and atherosclerotic lesions. Atherosclerosis 13: 355–363 (1971).

24 BJÖRKERUD, S. and BONDJERS, G.: Uptake, accumulation and removal of cholesterol in the normal and atherosclerotic arterial wall. Nutr. Metabol. 15: 27–36 (1973).

25 BLATON, V. H. and PEETERS, H.: Integrated approach to plasma lipid and lipoprotein analysis; in NELSON Blood lipids and lipoproteins, quantitation, composition and metabolism, pp. 275–314 (Wiley-Interscience, New York 1972).

26 BONDJERS, G. and BJÖRKERUD, S.: Arterial repair and atherosclerosis after mechanical injury. IV. Uptake and composition of cholesteryl ester in morphologically defined regions of atherosclerotic lesions. Atherosclerosis 15: 273–284 (1972).

27 BONDJERS, G. and BJÖRKERUD, S.: Arterial repair and atherosclerosis after mechanical injury. III. Cholesterol accumulation and removal in morphologically defined regions of aortic atherosclerotic lesions in the rabbit. Atherosclerosis 17: 85–94 (1973).

28 BORENSZTAJN, J.; GETZ, G. S., and WISSLER, R. W.: The in vitro incorporation of [³H] thymidine into DNA and of ³²P into phospholipids and RNA in the aorta of rhesus monkeys during early atherogenesis. Atherosclerosis 17: 269–280 (1973).

29 BOSCH, H. VAN DEN; GOLDE, L. M. G. VAN, and DEENEN, L. L. M. VAN: Dynamics of phosphoglycerides. Ergebn. Physiol. 66: 13–145 (1972).

30 BÖTTCHER, C. J. F.: Phospholipids of atherosclerotic lesions in the human aorta; in JONES Evolution of the atherosclerotic plaque, pp. 109–116 (Univ. of Chicago Press, Chicago 1963).

31 BÖTTCHER, C. J. F.: Chemical constituents of human atherosclerotic lesions. Proc. R. Soc. Med. 57: 792–795 (1964).

32 BÖTTCHER, C. J. and GENT, C. M. VAN: Changes in the composition of phospholipids and of phospholipid fatty acids associated with atherosclerosis in the human aortic wall. J. Atheroscler. Res. 1: 36–46 (1961).

33 BÖTTCHER, C. J. F.; WOODFORD, F. P.; HAAR ROMENY-WACHTER, C. C. TER; BOELSMA VAN HOUSE, E., and GENT, C. M. VAN: Fatty-acid distribution in lipids of the aortic wall. Lancet i: 1378–1383 (1960).

34 BOWYER, D. E.; HOWARD, A. N.; GRESHAM, G. A.; BATES, D., and PALMER, B. V.: Aortic perfusion in experimental animals. A system for the study of lipid synthesis and accumulation. Prog. biochem. Pharmacol., vol. 4, pp. 235–243 (Karger, Basel 1968).

35 BRECHER, P.; KESSLER, M.; CLIFFORD, C., and CHOBANIAN, A. V.: Cholesterol ester hydrolase in aortic tissue. Biochim. biophys. Acta 316: 386–394 (1973).

36 BRECHER, P. I.; TERCYAK, A., and CHOBANIAN, A. V.: Increased cholesterol esterification by different enzyme systems in atherosclerotic aortic tissue. Circulation 48: suppl. IV, p. 78 (1973).

37 BRETSCHER, M. S.: Membrane structure: some general principles. Science 181: 622–629 (1973).

38 BROWN, B. G.; PIERCE, J. E., and FRY, D. L.: The fate of surgically implanted crystalline lipids in the arterial wall of dogs. Circulation 48: suppl. IV, p. 78 (1973).

39 Brown, M. S. and Goldstein, J. L.: Defective regulation of 3-hydroxy-3 methyl glutaryl coenzyme A reductase activity in familial hypercholesterolemia. Circulation *48:* suppl. IV, p. 69 (1973).

40 Bruns, R. R. and Palade, G. E.: Studies on blood capillaries. II. Transport of ferritin molecules across the wall of muscle capillaries. J. Cell Biol. *37:* 277–299 (1968).

41 Buck, R. C. and Rossiter, R. J.: Lipids of normal and atherosclerotic aortas; a chemical study. Archs Path. *51:* 224–237 (1951).

42 Bullock, B. C.: Effects of age and diet on coronary artery atherosclerosis in *Cebus albifrons.* Fed. Proc. Fed. Am. Socs exp. Biol. *26:* 371 (1967).

43 Bullock, B. C.; Clarkson, T. B.; Lehner, N. D. M.; Lofland, H. B., jr., and St. Clair, R. W.: Atherosclerosis in *Cebus albifrons* monkeys. III. Clinical and pathological studies. Expl molec. Path. *10:* 39–62 (1969).

44 Burns, C. H. and Rothblatt, G. H.: Cholesterol excretion by tissue culture cells: effect of serum lipids. Biochim. biophys. Acta *176:* 616–625 (1969).

45 Burstein, M. and Scholnick, H. R.: Lipoprotein-polyanion-metal interactions. Adv. Lipid Res. *13:* 67–108 (1973).

46 Camejo, G.; Bosch, V.; Arreaza, C., and Mendez, H. C.: Early changes in plasma lipoprotein structure and biosynthesis in cholesterol-fed rabbits. J. Lipid Res. *14:* 61–68 (1973).

47 Campbell, D. J.; Day, A. J.; Skinner, S. L., and Tume, R. K.: The effect of hypertension on the accumulation of lipids and the uptake of [3H] cholesterol by aorta of normal-fed and cholesterol-fed rabbits. Atherosclerosis *18:* 301–319 (1973).

48 Castelli, W. P.; Nickerson, R. J.; Newell, J. M., and Rutstein, D. D.: Serum NEFA following fat, carbohydrate and protein ingestion and during fasting as related to intracellular lipid deposition. J. Atheroscler. Res. *6:* 328–341 (1966).

49 Chen, R. M.: Effects of hyperlipidemic rabbit serum and its lipoproteins on proliferation and lipid metabolism of rabbit aortic medial cells *in vitro* · Ph. D. thesis, Chicago (1973).

50 Chen, R.; Dzoga, K.; Borensztajn, J., and Wissler, R. W.: Effect of hyperlipemic lipoproteins on the lipid accumulation of rabbit aortic medial cells *in vitro.* Circulation *46:* suppl. II, p. 253 (1972).

51 Chen, R. M.; Getz, G. S.; Fisher-Dzoga, K., and Wissler, R. W.: Effects of hyperlipemic serum on lipid biosynthesis and efflux of rabbit aortic medial cells *in vitro.* Circulation *48:* suppl. IV, p. 152 (1973).

52 Chernick, S.; Srere, P. A., and Chaikoff, I. L.: The metabolism of arterial tissue. II. Lipide syntheses: the formation *in vitro* of fatty acids and phospholipides by rat artery with C^{14} and P^{32} as indicators. J. biol. Chem. *179:* 113–118 (1949).

53 Chobanian, A. V.: Sterol synthesis in the human arterial intima. J. clin. Invest. *47:* 595–603 (1968).

54 Chobanian, A. V. and Hollander, W.: Phospholipid synthesis in the human arterial intima. J. clin. Invest. *45:* 932–938 (1966).

55 Chobanian, A. V. and Lille, R. D.: Effects of atherosclerosis on lipid and protein synthesis in human aorta; in Jones Proc. 2nd Int. Symp. Atherosclerosis, pp. 282–285 (Springer, Berlin 1970).

56 Cliff, W. J.: The aortic tunica media in growing rats studied with the electron microscope. Lab. Invest. *17:* 599–615 (1967).

57 CONSTANTINIDES, P. and ROBINSON, M.: Ultrastructural injury of arterial endothelium I. Effects of pH, anoxia, and temperature. II. Effects of vasoactive amines. Archs Path. *88:* 99–105, 106–112 (1969).

58 COOPER, R. A.: Anemia with spur cells: a red cell defect acquired in serum and modified in the circulation. J. clin. Invest. *48:* 1820–1831 (1969).

59 COOPER, R. A.: Lipids of human red cell membranes: normal composition and variability in disease. Sem. Hemat. *7:* 296–322 (1970).

60 CROUSE, J. R.; GRUNDY, S. M., and AHRENS, E. H., jr.: Cholesterol distribution in the bulk tissues of man: variation with age. J. clin. Invest. *51:* 1292–1296 (1972).

61 DALY, M. M.; DEMING, Q. B.; RAEFF, V. M., and BRUN, L. M.: Cholesterol concentration and cholesterol synthesis in aortas of rats with renal hypertension. J. clin. Invest. *42:* 1606–1612 (1963).

62 DAOUD, A. S.; Fritz, K. E.; JARMOLYCH, J., and AUGUSTIN, J. M.: Use of aortic medial explants in the study of atherosclerosis. Expl molec. Path. *18:* 177–189 (1973).

63 DAY, A. J. and TUME, R. K.: Incorporation of [^{14}C] oleic acid into lipid by foam cells and by other fractions separated from rabbit atherosclerotic lesions. Atherosclerosis *11:* 291–299 (1970).

64 DAY, A. J. and WAHLQVIST, M. L.: Uptake and metabolism of ^{14}C-labeled oleic acid by atherosclerotic lesions in rabbit aorta. Circulation Res. *23:* 779–788 (1968).

65 DAY, A. J. and WAHLQVIST, M. L.: Cholesterol ester and phospholipid composition of normal aortas and of atherosclerotic lesions in children. Expl molec. Path. *13:* 199–216 (1970).

66 DAY, A. J. and WILKINSON, G. K.: Incorporation of ^{14}C-labeled acetate into lipid by isolated foam cells and by atherosclerotic arterial intima. Circulation Res. *21:* 593–600 (1967).

67 DAYTON, S.: Turnover of cholesterol in the artery walls of normal chickens. Circulation Res. *7:* 468–475 (1959).

68 DAYTON, S.: Decline in rate of cholesterol synthesis during maturation of chicken aorta. Proc. Soc. exp. Biol. Med. *108:* 257–261 (1961).

69 DAYTON, S. and HASHIMOTO, S.: Movement of labeled cholesterol between plasma lipoprotein and normal arterial wall across the intimal surface. Circulation Res. *19:* 1041–1049 (1966).

70 DEMPSEY, M. D.; RITTER, M. C., and LUX, S. E.: Functions of a specific plasma apoprotein in cholesterol biosynthesis. Fed. Proc. Fed. Am. Socs exp. Biol. *31:* 430 (1972).

71 DEVAUX, P. and MCCONNELL, H. M.: Lateral diffusion in spinlabeled phosphatidylcholine multilayers. J. Am. chem. Soc. *94:* 4475–4481 (1972).

72 D'HOLLANDER, F. and CHEVALLIER, F.: in PEETERS Protides of biological fluids, pp. 237–241 (Pergamon Press, Oxford 1972).

73 DOD, B. J. and GRAY, G. M.: The lipid composition of rat-liver plasma membranes. Biochim. biophys. Acta *150:* 397–404 (1968).

74 DUNCAN, L. E.; BUCK, K., and LYNCH, A.: Lipoprotein movement through canine aortic wall. Science *142:* 972–973 (1963).

75 EISENBERG, S.; RACHMILEWITZ, D.; Stein, O., and STEIN, Y.: Metabolic non-homogeneity of lecithin and cholesterol in aortae. Biochim. biophys. Acta *231:* 198–207 (1971).

76　Eisenberg, S.; Stein, Y., and Stein, O.: The role of lysolecithin in phospholipid metabolism of human umbilical and dog carotid arteries. Biochim. biophys. Acta *137:* 211–231 (1967).

77　Eisenberg, S.; Stein, Y., and Stein, O.: Phospholipases in arterial tissue. II. Phosphatide acyl-hydrolase and lysophosphatide acyl-hydrolase activity in human and rat arteries. Biochim. biophys. Acta *164:* 205–214 (1968).

78　Eisenberg, S.; Stein, Y., and Stein, O.: Phospholipases in arterial tissue. III. Phosphatide acyl-hydrolase, lysophosphatide acyl-hydrolase and sphingomyelin choline phosphohydrolase in rat and rabbit aorta in different age groups. Biochim. biophys. Acta *176:* 557–569 (1969).

79　Eisenberg, S.; Stein, Y., and Stein, O.: Phospholipases in arterial tissue. IV. The role of phosphatide acyl hydrolase, lysophosphatide acyl hydrolase, and sphingomyelin choline phosphohydrolase in the regulation of phospholipid composition in the normal human aorta with age. J. clin. Invest. *48:* 2320–2329 (1969).

80　Eisley, N. F. and Pritham, G. E. H.: Arterial synthesis of cholesterol *in vitro* from labeled acetate. Science *122:* 121 (1955).

81　Field, H., jr.; Swell, L.; Schools, P. E., jr., and Treadwell, C. R.: Dynamic aspects of cholesterol metabolism in different areas of the aorta and other tissues in man and their relationship to atherosclerosis. Circulation *22;* 547–558 (1960).

82　Fisher-Dzoga, K.; Chen, R., and Wissler, R. W.: Effects of serum lipoproteins on the morphology, growth and metabolism of arterial smooth muscle cells. Adv. expl. Med. *43:* 299–310 (1974).

83　Fisher-Dzoga, K.; Jones, R. M.; Vesselinovitch, D., and Wissler, R. W.: Ultrastructural and immunohistochemical studies of primary cultures of aortic medial cells. Expl molec. Path. *18:* 162–176 (1973).

84　Florey, L. and Sheppard, B. L.: The permeability of arterial endothelium to horseradish peroxidase. Proc. R. Soc. B *174:* 435–443 (1970).

85　Foote, J. L. and Coles, E.: Cerebrosides of human aorta: isolation, identification of the hexose, and fatty acid distribution. J. Lipid Res. *9:* 482–486 (1968).

86　Fowler, S. and Duve, C. De: Digestive activity of lysosomes. III. The digestion of lipids by extracts of rat liver lysosomes. J. biol. Chem. *244:* 471–481 (1969).

87　Garlick, D. G. and Courtice, F. C.: The composition of the lipoproteins in the plasma of rabbits with hypercholesterolaemia or Triton WR-1339-induced hyperlipaemia. Q. J. expl. Physiol. *47:* 211–220 (1962).

88　Geer, J. C.; Catsulis, C.; McGill, H. C., jr., and Strong, J. P.: Fine structure of the baboon aortic fatty streak. Am. J. Path. *52:* 265–286 (1968).

89　Geer, J. C.; McGill, H. C., jr.; Strong, J. P., and Holman, R. L.: Electron microscopy of human atherosclerotic lesions. Fed. Proc. Fed. Am. Socs exp. Biol. *19:* 15 (1960).

90　Geer, J. C. and Malcolm, G. T.: Cholesterol ester fatty acid composition of human aorta fatty streaks and normal intima. Expl molec. Path. *4:* 500–507 (1965).

91　Geer, J. C.; Panganamala, R. V., and Cornwell, D. G.: Position of double bonds in the fatty acids of cholesterol esters from human aorta. Atherosclerosis *12:* 63–74 (1970).

92　Getz, G. S.; Wissler, R. W.; Hughes, R. H.; Graber, C., and Tantra, S.: The lipid composition and biosynthesis of lecithin in atherosclerotic aorta of rhesus monkeys fed three food fats. Circulation *36:* II–13 (1967).

93 GETZ, G. S.; WISSLER, R. W.; HUGHES, R. H.; GRABER, C., and TANTRA, S.: Metabolism of lipids in the atherosclerotic rhesus monkey aorta. Proc. Inst. Med., Chicago *27:* 106–107 (1968).

94 GEYER, R. P.: Uptake and retention of fatty acids by tissue culture cells; in ROTHBLAT and KRITCHEVSKY Lipid metabolism in tissue culture, pp. 33–47 (Wistar Inst. Press, Philadelphia 1967).

95 GHOSH, S.; BASU, M. K., and SCHWEPPE, J. S.: The isoelectric point and surface charge density of lipo-proteins in normal and hyperlipemic individuals. Circulation *44:* Suppl. II, p. 6 (1971).

95a GJONE, E.: Familial LCAT deficiency. Acta med. scand. *194:* 353–356 (1973).

96 GLOMSET, J. A. and NORUM, K. R.: The metabolic role of lecithin: cholesterol acyltransferase: perspectives from pathology. Adv. Lipid Res. *13:* 1–65 (1973).

96a GOLDSTEIN, J. L. and BROWN, M. S.: Hyperlipidemia in coronary heart disease: a biochemical genetic approach. J. Lab. clin. Med. *85:* 15–25 (1975).

97 GOODMAN, D. S.: The metabolism of chylomicron cholesterol ester in the rat. J. clin. Invest. *41:* 1886–1896 (1962).

98 GOULD, R. G.; WISSLER, R. W., and JONES, R. J.: The dynamics of lipid deposition in arteries; in JONES Evolution of the atherosclerotic plaque, pp. 205–214 (Univ. of Chicago Press, Chicago 1963).

99 GREEN, J. R.; EDWARDS, P. A., and GREEN, C.: Optical-rotatory-dispersion studies of compounds related to cholesterol in liposomes and the membranes of erythrocyte 'ghosts'. Biochem. J. *135:* 63–71 (1973).

100 GRESHAM, G. A. and HOWARD, A. N.: Vascular lesions in primates. Ann. N. Y. Acad. Sci. *127:* 694–701 (1965).

101 GURD, F. R. N.: Some naturally occurring lipoprotein systems; in HANAHAN Lipide Chemistry, pp. 260–325 (J. Wiley & Sons, Chichester 1960).

102 HARDIN, N. J.; MINICK, C. R., and MURPHY, G. E.: Experimental induction of atheroarteriosclerosis by the synergy of allergic injury to arteries and lipid-rich diet. III. The role of earlier acquired fibromuscular intimal thickening in the pathogenesis of later developing atherosclerosis. Am. J. Path. *73:* 301–326 (1973).

103 HARVEY, M. S.; WIRTZ, K. W. A.; KAMP, H. H.; ZEGERS, B. J. M., and DEENEN, L. L. M. VAN: A study in phospholipid exchange proteins present in the soluble fractions of beef liver and brain. Biochim. biophys. Acta *323:* 234–239 (1973).

104 HASHIMOTO, S.; DAYTON, S., and ALFIN-SLATER, R. B.: Esterification of cholesterol by homogenates of atherosclerotic and normal aortas. Life Sci. *12:* 1–12 (1973).

105 HAUSHEER, L. und BERNHARD, K.: Cerebroside aus menschlichen Aorten. Hoppe-Seyler's Z. physiol. Chem. *331:* 41–44 (1963).

106 HAUST, M. D.: Electron microscopic and imuno-histochemical studies of fatty streaks in human aorta. Prog. biochem. Pharmacol., vol. 4, pp. 429–437 (Karger, Basel 1968).

107 HELLER, M. and SHAPIRO, B.: Enzymic hydrolysis of sphingomyelin by rat liver. Biochem. J. *98:* 763–769 (1966).

108 HOLLANDER, W.: Recent advances in experimental and molecular pathology. Influx synthesis and transport of arterial lipoproteins in atherosclerosis. Expl molec. Path. *7:* 248–258 (1967).

109 Howard, A. N.; Blaton, V.; Gresham, G. A.; Vandamme, D., and Peeters, H.: The lipoproteins in hyperlipidaemic primates as a model for human atherosclerosis; in Peeters Protides of the biological fluids, pp. 341–344 (Pergamon Press, Oxford 1972).

110 Howard, A. N.; Blaton, V.; Vandamme, D.; Landschoot, N. van, and Peeters, H.: Lipid changes in the plasma lipoproteins of baboons given an atherogenic diet. 3. A comparison between lipid changes in the plasma of the baboon and chimpanzee given atherogenic diets and those in human plasma lipoproteins of Type II hyperlipoproteinemia. Atherosclerosis 16: 257–272 (1972).

111 Howard, A. N.; Patelski, J.; Bowyer, D. E., and Gresham, G. A.: Atherosclerosis induced in hypercholesteroalemic baboons by immunological injury; and the effects of intravenous polyunsaturated phosphatidyl choline. Atherosclerosis 14: 17–29 (1971).

112 Howard, C. F., jr.: De novo synthesis and elongation of fatty acids by subcellular fractions of monkey aorta. J. Lipid Res. 9: 254–261 (1968).

113 Howard, C. F., jr.: Lipogenesis from glucose-2-^{14}C and acetate-1-^{14}C in aorta. J. Lipid Res. 12: 725–730 (1971).

114 Howard, C. F., jr.: Aortic lipogenesis during aerobic and hypoxic incubation. Atherosclerosis 15: 359–369 (1972).

115 Howard, C. F., jr.: Spontaneous diabetes in Macaca nigra. Diabetes 21: 1077–1090 (1972).

116 Howard, C. F., jr.: Atherosclerosis in spontaneously diabetic monkeys. Circulation 48: Suppl. IV, p. 41 (1973).

117 Howard, C. F., jr.: Lipogenesis from [2-^{14}C] glucose in squirrel monkeys. Proc. Soc. exp. Biol. Med. 142: 660–665 (1973).

118 Howard, C. F., jr. and Bonnett, L.: Effects of hypoxic and aerobic incubations on lipogenesis from [U-^{14}C] glucose in sections of normal and atherosclerotic aorta from the rabbit. Atherosclerosis 18: 469–477 (1973).

119 Howard, C. F., jr. and Portman, O. W.: Hydrolysis of cholesteryl linoleate by a high-speed preparation of rat and monkey aorta. Biochim. biophys. Acta 125: 623–626 (1966).

120 Huttner, I.; More, R. H., and Rona, G.: Fine structural evidence of specific mechanism for increased endothelial permeability in experimental hypertension. Am. J. Path. 61: 395–412 (1970).

121 Illingworth, D. R. and Portman, O. W.: Exchange of phospholipids between low and high density lipoproteins of squirrel monkeys. J. Lipid Res. 13: 220–227 (1972).

122 Illingworth, D. R. and Portman, O. W.: Independence of phospholipid and protein exchange between plasma lipoproteins in vivo and in vitro. Biochim. biophys. Acta 280: 281–289 (1972).

123 Illingworth, D. R. and Portman, O. W.: The uptake and metabolism of plasma lysophosphatidylcholine in vivo by the brain of squirrel monkeys. Biochem. J. 130: 557–567 (1972).

124 Illingworth, D. R.; Portman, O. W.; Robertson, A. L., jr., and Magyar, W. A.: The exchange of phospholipids between plasma lipoproteins and rapidly dividing human cells grown in tissue culture. Biochim. biophys. Acta 306: 422–436 (1973).

125 INSULL, W., jr. and BARTSCH, G. E.: Cholesterol, triglyceride, and phospholipid content of intima, media, and atherosclerotic fatty streak in human thoracic aorta. J. clin. Invest. *45:* 513–523 (1966).

126 JACKSON, R. L. and GOTTO, A. M.: Phospholipids in biology and medicine. New Engl. J. Med. *290:* 24–29, 87–93 (1974).

127 JACOTOT, B.: BEAUMONT, J. L.; MONNIER, G.; SZIGETI, M.; ROBERT, B., and ROBERT, L.: Role of elastic tissue in cholesterol deposition in the arterial wall. Nutr. Metabol. *15:* 46–58 (1973).

128 JARMOLYCH, J.; DAOUD, A. S.; LANDAU, J.; FRITZ, K. E., and McELVENE, E.: Aortic medial explants. Cell proliferation and production of mucopolysaccharides, collagen, and elastic tissue. Expl. molec. Path. *9:* 171–188 (1968).

129 JENSEN, J.: A further study of the kinetics of cholesterol uptake at the endothelial cell surface of the rabbit aorta *in vitro,* biochim. biophys. Acta *173:* 71–77 (1969).

130 KAGAWA, Y.; JOHNSON, L. W., and RACKER, E.: Activation of phosphorylating vesicles by net transfer of phosphatidyl choline by phospholipid transfer protein. Biochem. biophys. Res. Commun. *50:* 245–251 (1973).

131 KAMAT, V. B. and WALLACH, D. F. H.: Separation and partial purification of plasma-membrane fragments from Ehrlich ascites carcinoma microsomes. Science *148:* 1343–1345 (1965).

132 KANFER, J. N.; YOUNG, O. M.; SHAPIRO, D., and BRADY, R. O.: The metabolism of sphingomyelin. I. Purification and properties of a sphingomyelin-cleaving enzyme from rat liver tissues. J. biol. Chem. *241:* 1081–1084 (1966).

133 KAO, V. C. Y. and WISSLER, R. W.: A study of the immunohistochemical localization of serum lipoproteins and other plasma proteins in human atherosclerotic lesions. Expl molec. Path. *4:* 465–479 (1965).

134 KAO, V. C. Y.; WISSLER, R. W., and DZOGA, K.: The influence of hyperlipemic serum on the growth of medial smooth muscle cells of rhesus monkey aorta *in vitro.* Circulation *38:* Suppl. VI, p. 12 (1968).

135 KARNOVSKY, M. J.: The ultrastructural basis of capillary permeability studied with peroxidase as a tracer. J. Cell Biol. *35:* 213–236 (1967).

136 KARRER, H. E.: An electron microscope study of the aorta in young and in aging mice. J. Ultrastruct. Res. *5:* 1–27 (1961).

137 KNIERIEM, H. J.; KAO, V. C. Y., and WISSLER, R. W.: Actomyosin and myosin and the deposition of lipids and serum lipoproteins. Archs Path. *84:* 118–129 (1968).

138 KORNBERG, R. D. and McCONNELL, H. M.: Inside-outside transitions of phospholipids in vesicle membranes. Biochemistry *10:* 1111–1120 (1971).

139 KRAMSCH, D. M.; FRANZBLAU, C., and HOLLANDER, W.: The protein and lipid composition of arterial elastin and its relationship to lipid accumulation in the atherosclerotic plaque. J. clin. Invest. *50:* 1666–1677 (1971).

140 KRAMSCH, D. and HOLLANDER, W.: Occlusive atherosclerotic disease of the coronary arteries in monkey *(Macaca irus)* induced by diet. Expl molec. Path. *9:* 1–22 (1968).

141 KRAMSCH, D. M. and HOLLANDER, W.: The interaction of serum and arterial lipoproteins with elastin of the arterial intima and its role in the lipid accumulation in atherosclerotic plaques. J. clin. Invest. *52:* 236–247 (1973).

142 KRESSE, H.; FILIPOVIC, I., und BUDDECKE, E.: Untersuchungen zur Chemie der Arterienwand XIV. Gesteigerte ^{14}C Inkorporation in die Triacylglycerine (Triglyceride)

des Arteriengewebes bei Sauerstoffmangel. Hoppe-Seyler's Z. physiol. Chem. *350:* 1611–1618 (1969).

143 KÜNNERT, B. and KRUG, H.: The composition of cholesterol esters in fatty streaks and atherosclerotic plaques of the human aorta. Histochromatographic investigation. Atherosclerosis *13:* 93–101 (1971).

144 LANDS, W. E. M.: Metabolism of glycerolipids. II. The enzymatic acylation of lysolecithin. J. biol. Chem. *235:* 2233–2237 (1960).

145 LANG, P. D. and INSULL, W., jr.: Lipid droplets in atherosclerotic fatty streaks of human aorta. J. clin. Invest. *49:* 1479–1488 (1970).

146 LECUYER, H. and DERVICHIAN, D. G.: Structure of aqueous mixtures of lecithin and cholesterol. J. molec. Biol. *45:* 39–57 (1969).

147 LESLIE, R. B.: Some physical and physico-chemical approaches to the structure of serum high density lipoproteins (HDL); in Plasma lipoproteins, Biochem. Soc. Symp. No. 33, pp. 47–86 (Academic Press, New York 1971).

148 LEVY, R. S. and DAY, C. E.: Low density lipoprotein structure and its relation to atherogenesis; in JONES Proc. 2nd Int. Symp. Atherosclerosis, pp. 186–189 (Springer, Berlin 1970).

149 LOFLAND, H. B., jr. and CLARKSON, T. B.: Certain metabolic patterns of atheromatous pigeon aortas. Archs Path. *80:* 291–296 (1965).

150 LOFLAND, H. B.; CLARKSON, T. B., and ARTOM, C.: Lipid synthesis in aorta preparations from atherosclerosis susceptible or resistant pigeons. Archs Biochem. *88:* 105–109 (1960).

151 LOFLAND, H. B., jr ; CLARKSON, T. B.; ST. CLAIR, R. W., and LEHNER, N. D. M.: Studies on the regulation of plasma cholesterol levels in squirrel monkeys of two genotypes. J. Lipid Res. *13:* 39–47 (1972).

152 LOFLAND, H. B., jr.; MOURY, D. M.; HOFFMAN, C. W., and CLARKSON, T. B.: Lipid metabolism in pigeon aorta during atherogenesis. J. Lipid Res. *6:* 112–118 (1965).

153 LOFLAND, H. B., jr.; ST. CLAIR, R. W.; CLARKSON, T. B.; BULLOCK, B. C., and LEHNER, N. D. M.: Atherosclerosis in cebus monkeys. II. Arterial metabolism. Expl molec. Path. *9:* 57–70 (1968).

154 MACA, R. D. and ROSE, K. D.: The role of beta-lipoprotein, cholesterol, and various sera in tissue culture intracellular lipidosis. Proc. Soc. exp. Biol. Med. *136:* 457–460 (1971).

155 McCANDLESS, E. L. and ZILVERSMIT, D. B.: The effect of cholesterol on the turnover of lecithin, cephalin and sphingomyelin in the rabbit. Archs Biochem. *62:* 402–410 (1956).

156 McGILL, H. C., jr.; STRONG, J. P.; HOLMAN, R. L., and WERTHESSEN, N. T.: Arterial lesions in the Kenya baboon. Circulation Res. *8:* 670–679 (1960).

157 MacKENZIE, C. G.; MacKENZIE, J. B., and REISS, O. K.: Regulation of cell lipid metabolism and accumulation. V. Quantitative and structural aspects of triglyceride accumulation caused by lipogenic substances; in ROTHBLAT and KRITCHEVSKY Lipid metabolism in tissue culture cells, pp. 63–83 (Wistar Institute Press, Philadelphia 1967).

158 McMURRAY, W. C. and DAWSON, R. M. C.: Phospholipid exchange reactions within the liver cell. Biochem. J. *112:* 91–108 (1969).

159 MacNintch, J. E.; St. Clair, R. W.; Lehner, N. D. M.; Clarkson, T. B., and Lofland, H. B.: Cholesterol metabolism and atherosclerosis in Cebus monkeys in relation to age. Lab. Invest. *16:* 444–452 (1967).

160 Majno, G.; Shea, S. M., and Leventhal, M.: Endothelial contraction induced by histamine-type mediators. An electron microscopic study. J. Cell Biol. *42:* 647–672 (1969).

161 Mann, G. V. and Andrus, S. B.: Xanthomatosis and atherosclerosis produced by diet in an adult rhesus monkey. J. Lab. clin. Med. *48:* 533–550 (1956).

162 Mann, G. V.; Andrus S. B.; McNally, A., and Stare, F. J.: Experimental atherosclerosis in Cebus monkeys. J. exp. Med. *98:* 195–217 (1953).

163 Mateu, L.; Tardieu, A.; Luzzati, V.; Aggerbeck, L., and Scanu, A. M.: On the structure of human serum low density lipoproteins. J. molec. Biol. *70:* 105–116 (1972).

164 Meldolesi, J.; Jamieson, J. D., and Palade, G. E.: Composition of cellular membranes in the pancreas of the guinea pig. II. Lipids. J. Cell Biol. *49:* 130–149 (1971).

165 Mellors, A. and Tappel, A. L.: Hydrolysis of phospholipids by a lysosomal enzyme. J. Lipid Res. *8:* 479–485 (1967).

166 Middleton, C. C.; Clarkson, T. B.; Lofland, H. B., and Pritchard, R. W.: Atherosclerosis in the squirrel monkey. Naturally occurring lesions in the aorta and coronary arteries. Archs Path. *78:* 16–23 (1964).

167 Minick, C. R. and Murphy, G. E.: Experimental induction of athero-arteriosclerosis by the synergy of allergic injury to arteries and lipid-rich diet. II Effect of repeatedly injected foreign protein in rabbits fed a lipid-rich, cholesterol-poor diet Am J. Path. *73:* 265–300 (1973).

168 Minick, C. R.; Murphy, G. E., and Campbell, W. G., jr.: Experimental induction of athero-arteriosclerosis by the synergy of allergic injury to arteries and lipid-rich diet. I. Effect of repeated injections of horse serum in rabbits fed a dietary cholesterol supplement. J. exp. Med. *124:* 635–652 (1966).

169 Morin, R. J.: Phospholipid composition and synthesis in male rabbit aortas. Metabolism *17:* 1051–1058 (1968).

170 Morris, M. D. and Lee, J. A.: Serum lipids and low density lipoproteins: spontaneous hypercholesterolemia vs. cholesterol feeding in rhesus monkeys; in Schettler and Weizel Atherosclerosis III, pp. 374–376 (Springer, Berlin 1974).

171 Moscowitz, M. S.: Fatty acid-induced steatosis in monolayer cell cultures; in Rothblat and Kritchevsky Lipid metabolism in tissue culture cells, pp. 49–62 (Wistar Institute Press, Philadelphia 1967).

172 Murphy, J. R.: Erythrocyte metabolism. III. Relationship of energy metabolism and serum factors to the osmotic fragility following incubation. J. Lab. clin. Med. *60:* 86–109 (1962).

173 Murphy, J. R.: Erythrocyte metabolism. IV. Equilibration of cholesterol-4-C^{14} between erythrocytes and variously treated sera. J. Lab. clin. Med. *60:* 571–578 (1962).

174 Myasnikov, A. L. and Block, Y. E.: Influence of some factors on lipoidosis and cell proliferation in aorta tissue cultures of adult rabbits. J. Atheroscler. Res. *5:* 33–42 (1965).

175 MYASNIKOV, A. L.; BLOCK, Y. E., and PAVLOV, V. M.: Influence of lipemic serum of patients with atherosclerosis on tissue cultures of adult human aortas. J. Atheroscler. Res. *6:* 224–231 (1966).

176 NAKATANI, M.; SASAKI, T.; MIYAZAKI, T., and NAKAMURA, M.: Synthesis of phospholipids in arterial walls. I. Incorporation of ^{32}P into phospholipids of aortas and coronary arteries of various animals. J. Atheroscler. Res. *7:* 747–757 (1967).

177 NEERHOUT, R. C.: Erythrocyte stromal lipids in hyperlipemic states. J. Lab. clin. Med. *71:* 448–454 (1968).

178 NEWMAN, H. A. I.; DAY, A. J., and ZILVERSMIT, D. B.: *In vitro* phospholipid synthesis in normal and atheromatous rabbit aortas. Circulation Res. *19:* 132–138 (1966).

179 NEWMAN, H. A. I.; MCCANDLESS, E. L., and ZILVERSMIT, D. B.: The synthesis of C^{14}-lipids in rabbit atheromatous lesions. J. biol. Chem. *236:* 1264–1268 (1961).

180 NEWMAN, H. A. I.; SETH, S. K., and HUNT, W. T.: Sphingomyelin influx and biosynthesis in rabbit atheroma. Fed. Proc. Fed. Am. Socs exp. Biol. *32:* 293 (1973).

181 NEWMAN, H. A. I. and ZILVERSMIT, D. B.: Quantitative aspects of cholesterol flux in rabbit atheromatous lesions. J. biol. Chem. *237:* 2078–2084 (1962).

182 NEWMAN, H. A. I. and ZILVERSMIT, D. B.: Accumulation of lipid and nonlipid constituents in rabbit atheroma. J. Atheroscler. Res. *4:* 261–271 (1964).

183 NEWMAN, H. A. I. and ZILVERSMIT, D. B.: Uptake and release of cholesterol by rabbit atheromatous lesions. Circulation Res. *18:* 293–303 (1966).

184 NEWMAN, W. P., III; EGGEN D. A., and STRONG, J. P.: Comparison of lesions and serum lipids in rhesus and spider monkeys on an egg-butter diet. Fed. Proc. Fed. Am. Socs exp. Biol. *26:* 371 (1967).

185 NICHOLS, A. V. and SMITH, L.: The effect of very low-density lipoproteins on lipid transfer in incubated serum. J. Lipid Res. *6:* 206–210 (1965).

186 NISHIDA, T. and COGAN, U.: Nature of the interaction of dextran sulfate with low density lipoproteins of plasma. J. biol. Chem. *245:* 4689–4697 (1970).

187 NORUM, K. R. and GJONE, E.: The influence of plasma from patients with familial lecithin:cholesterol acyl transferase deficiency on the lipid pattern of erythrocytes. Scand. J. clin. Lab. Invest. *22:* 94–98 (1968).

188 OCKNER, R. K.; MANNING, J. A., POPPENHAUSEN, R. B., and HO, W. K. L.: A binding protein for fatty acids in cytosol of intestinal mucosa, liver, myocardium, and other tissues. Science *177:* 56–58 (1972).

189 OKISHIO, T.: Studies on the transfer of ^{131}I-labelled serum lipoproteins into the aorta of rabbits with experimental atherosclerosis. Med. J. Osaka Univ. *11:* 367–381 (1961).

190 OSTWALD, R. and SHANNON, A.: Composition of tissue lipids and anaemia of guinea pigs in response to dietary cholesterol. Biochem. J. *91:* 146–154 (1964).

191 OSTWALD, R.; YAMANAKA, W., and LIGHT, M.: The phospholipids of liver, plasma, and red cells in normal and cholesterol-fed anemic guinea pigs. Proc. Soc. exp. Biol. Med. *134:* 814–820 (1970).

192 PAGE, I. H.: Atherosclerosis. An introduction. Circulation *10:* 1–27 (1954).

193 PANGANAMALA, R. V.; GEER, J. C., and CORNWELL, D. G.: Long-chain bases in the sphingolipids of atherosclerotic human aorta. J. Lipid Res. *10:* 445–455 (1969).

194 PARKER, F.: An electron microscopic study of experimental atherosclerosis. Am. J. Path. *36:* 19–53 (1960).

195 PARKER, F. and ODLAND, G. F.: A correlative histochemical, biochemical and electron microscopic study of experimental atherosclerosis in the rabbit aorta with special reference to the myo-intimal cell. Am. J. Path. *48:* 197–239 (1966).

196 PARKER, F.; ORMSBY, J. W.; PETERSON, N. F.; ODLAND, C. F., and WILLIAMS, R. H.: *In vitro* studies of phospholipid synthesis in experimental atherosclerosis: possible role of myo-intimal cells, Circulation Res. *19:* 700–710 (1966).

197 PARKER, F.; SCHIMMELBUSCH, W., and WILLIAMS, R. H.: The enzymatic nature of phospholipid synthesis in normal rabbit and human aorta. Results of in vitro studies. Diabetes *13:* 182–188 (1964).

198 PATELSKI, J.; BOWYER, D. E.; HOWARD, A. N., and GRESHAM, G. A.: Changes in phospholipase A, lipase and cholesterol esterase activity in the aorta in experimental atherosclerosis in the rabbit and rat. J. Atheroscler. Res. *8:* 221–228 (1968).

199 PEETERS, H.; BLATON, V.; DECLERCQ, B.; HOWARD, A. N., and GRESHAM, G. A.: Lipid changes in the plasma lipoproteins of baboons given an atherogenic diet. Atherosclerosis *12:* 283–290 (1970).

200 PETERS, T. J.; MULLER, M., and DUVE, C. DE: Lysosomes of the arterial wall. I. Isolation and subcellular fractionation of cells from normal rabbit aorta. J. exp. Med. *136:* 1117–1139 (1972).

201 PETERS, T. J.; TANAKO, T., and DUVE, C. DE: Subcellular fractionation studies on the cells isolated from normal and atherosclerotic aorta; in Atherosclerosis: initiating factors. Ciba Symp. No. 12, pp. 197–214 (Elsevier, Amsterdam 1973).

202 PETERSON, J. A. and RUBIN, H.: The exchange of phospholipids between cultured chick embryo fibroblasts and their growth medium. Expl Cell Res. *58:* 365–378 (1969).

203 PETERSON, M.; DAY, A. J.; TUME, R. K., and EISENBERG, E.: Ultrastructure, fatty acid content, and metabolic activity of foam cells and other fractions separated from rabbit atherosclerotic lesions. Expl molec. Path. *15:* 157–169 (1971).

204 PORTMAN, O. W.: Incorporation of fatty acids into phospholipids by cell-free and subcellular fractions of squirrel monkey and rat aorta. Importance of endogenous lysophosphatidylcholine. J. Atheroscler. Res. *7:* 617–628 (1967).

205 PORTMAN, O. W.: Atherosclerosis in nonhuman primates. Sequences and possible mechanisms of change in phospholipid composition and metabolism. Ann. N.Y. Acad. Sci. *162:* 120–136 (1969).

206 PORTMAN, O. W.: Arterial composition and metabolism. Esterified fatty acids and cholesterol. Adv. Lipid Res. *8:* 41–114 (1970).

207 PORTMAN, O. W. and ALEXANDER, M.: Lipid composition of aortic intima plus inner media and other tissue fractions from fetal and adult rhesus monkeys. Archs Biochem. *117:* 357–365 (1966).

208 PORTMAN, O. W. and ALEXANDER, M.: Lysolecithin concentrations and metabolism in aorta. Circulation *38:* Suppl. VI, p. 19 (1968).

209 PORTMAN, O. W. and ALEXANDER, M.: Lysophosphatidylcholine concentrations and metabolism in aortic intima plus inner media: effect of nutritionally induced atherosclerosis. J. Lipid Res. *10:* 158–165 (1969).

210 PORTMAN, O. W. and ALEXANDER, M.: Metabolism of sphingolipids by normal and atherosclerotic aorta of squirrel monkeys. J. Lipid Res. *11:* 23–30 (1970).

211 PORTMAN, O. W. and ALEXANDER, M.: Changes in arterial subfractions with aging and atherosclerosis. Biochim. biophys. Acta *260;* 460–474 (1972).

212 PORTMAN, O. W.; ALEXANDER, M., and MARUFFO, C. A.: Composition of subcellular constituents of aortic intima plus inner media isolated by differential and density gradient centrifugation. Archs Biochem. *122:* 344–353 (1967).

213 PORTMAN, O. W.; ALEXANDER, M., and MARUFFO, C. A.: Nutritional control of arterial lipid composition in squirrel monkeys. Major ester classes and types of phospholipids. J. Nutr. *91:* 35–46 (1967).

214 PORTMAN, O. W.; ALEXANDER, M., and OSUGA, T.: Heterogeneity of lipid composition of microsome subfractions from aorta and liver. Biochim. biophys. Acta *187:* 435–438 (1969).

215 PORTMAN, O. W. and ANDRUS, S. B.: Comparative evaluation of three species of New World monkeys for studies of dietary factors, tissue lipids and atherogenesis. J. Nutr. *87:* 429–438 (1965).

216 PORTMAN, O. W.; BEHRMAN, R. E., and SOLTYS, P.: Transfer of free fatty acids across the primate placenta. Am. J. Physiol. *216:* 143–147 (1968).

217 PORTMAN, O. W.; SOLTYS, P.; ALEXANDER, M., and OSUGA, T.: Metabolism of lysolecithin *in vivo.* Effects of hyperlipemia and atherosclerosis in squirrel monkeys. J. Lipid Res. *11:* 596–604 (1970).

218 PORTMAN, O. W. and ILLINGWORTH, D. R.: Lysolecithin binding to human and squirrel monkey plasma and tissue components. Biochim. biophys. Acta *326:* 34–42 (1973).

219 PORTMAN, O. W. and ILLINGWORTH, D. R.: Factors determining the concentrations of lysolecithin in plasma and tissues. Scand. J. clin. Lab. Invest. *33:* Suppl. 137, 49–55 (1974).

220 PORTMAN, O. W. and ILLINGWORTH, D. R.: Metabolism of lysolecithin *in vivo* and *in vitro* with particular emphasis on the arterial wall. Biochim. biophys. Acta *348:* 136–144 (1974).

221 PORTMAN, O. W.; ILLINGWORTH, D. R., and ALEXANDER, M.: Lysolecithin and sphingosinephosphorylcholine in the metabolism of brain phospholipids of the rhesus monkey *(Macaca mulatta).* Effects of development. J. Neurochem. *20:* 1659–1667 (1973).

221a PORTMAN, O. W.; ILLINGWORTH, D. R., and ALEXANDER, M.: The effects of hyperlipidemia on lipoprotein metabolism in squirrel monkeys and rabbits. Biochim. biophys. Acta (in press).

222 PORTMAN, O. W. and SUGANO, M.: Factors influencing the level and fatty acid specificity of the cholesterol esterification activity in human plasma. Archs Biochem. *105:* 532–540 (1964).

223 PRIES, C.; AUMONT, A., and BÖTTCHER, C. J. F.: Analysis of phospholipids. Biochim. biophys. Acta *125:* 277–287 (1966).

224 PROUDLOCK, J. W. and DAY, A. J.: Cholesterol esterifying enzymes of atherosclerotic rabbit intima. Biochim. biophys. Acta *260:* 716–723 (1972).

225 PROUDLOCK, J. W.; DAY, A. J., and TUME, R. K.: Cholesterol-esterifying enzymes of foam cells isolated from atherosclerotic rabbit intima. Atherosclerosis *18:* 451–457 (1973).

226 PUCAK, G. J.; LEHNER, N. D. M.; CLARKSON, T. B.; BULLOCK, B. C., and LOFLAND, H. B.: Spider monkeys (Ateles sp.) as animal models for atherosclerosis research. Expl molec. Path. *18:* 32–49 (1973).

227 QUARFORDT, S. H.; BOSTON, F., and HILDERMAN, H: Transfer of triglyceride between isolated human lipoproteins. Biochim. biophys. Acta *231*: 290–294 (1971).

228 QUARFORDT, S. H. and HILDERMAN, H. L.: Quantitation of the *in vitro* free cholesterol exchange of human red cells and lipoproteins. J. Lipid Res. *11*: 528–535 (1970).

229 RACHMILEWITZ, D.; EISENBERG, S.; STEIN, Y., and STEIN, O.: Phospholipases in arterial tissue. I. Sphingomyelinase activity in human, dog, guinea pig, rat and rabbit arteries. Biochim. biophys. Acta *144*: 624–632 (1967).

230 REED, C. F.: Phospholipid exchange between plasma and erythrocytes in man and the dog. J. clin. Invest. *47*: 749–760 (1968).

231 RHODIN, J. A. G.: Fine structure of vascular wall in mammals, with special reference to the smooth muscle components. Physiol. Rev. *42*: suppl. 5, pp. 48–81 (1962).

232 RIDDLE, M. and GLOMSET, J. A.: Cholesterol ester hydrolysing enzymes of rat livers. Fed. Proc. Fed. Am. Socs exp. Biol. *32*: 363 (1973).

233 ROBERT, L.: The macromolecular matrix of the arterial wall: collagen, elastin, mucopolysaccharies; in JONES Proc. 2nd Int. Symp. Atherosclerosis, pp. 59–68 (Springer, Berlin 1970).

234 ROBERTSON, A. L., jr.: Intracellular incorporation of plasma lipoproteins by arterial intima in relation to early stages of intravascular thrombosis; in SAWYER Biophysical mechanisms in vascular homeostasis and intravascular thrombosis, pp. 267–274 (Appleton Century Crofts, New York 1965).

235 ROBERTSON, A. L., jr.: Transport of plasma lipoproteins and ultrastructure of human arterial intimacytes in culture; in ROTHBLAT and KRITCHEVSKY Lipid metabolism in tissue culture cells, pp. 115–125 (Wistar Institute Press, Philadelphia 1967).

236 ROBERTSON, A. L., jr.: in WOLF The artery and the process of atherosclerosis: pathogenesis. Advances in experimental medicine and biology, vol. 16 A, pp. 104–107 (Plenum Publishing, New York 1971).

237 ROBERTSON, A. L. jr. and INSULL, W., jr.: Dissection of normal and atherosclerotic human artery with proteolytic enzymes *in vitro*. Nature, Lond. *214*: 821–823 (1967).

238 ROBERTSON, A. L., jr. and KHAIRALLAH, P. A.: Arterial endothelial permeability and vascular disease. The 'trap door' effect. Expl molec. Path. *18*: 241–260 (1971).

239 ROHEIM, P. S.; HAFT, D. E.; GIDEZ, L. I.; WHITE, A., and EDER, H. A.: Plasma lipoprotein metabolism in perfused rat livers. II. Transfer of free and ester cholesterol into the plasma. J. clin. Invest. *42*: 1277–1285 (1963).

240 ROSS, R.: The smooth muscle cell. II. Growth of smooth muscle in culture and formation of elastic fibers. J. Cell Biol. *50*: 172–186 (1971).

241 ROSS, R.: The arterial smooth muscle cell; in WISSLER and GEER The pathogenesis of atherosclerosis, pp. 147–163 (Williams & Wilkins, Baltimore 1972).

242 ROSS, R.: The elastic fiber. A review. J. Histochem. Cytochem. *21*: 199–208 (1973).

243 ROSS, R. and GLOMSET, J. A.: Atherosclerosis and the arterial smooth muscle cell. Science *180*: 1332–1339 (1973).

244 ROSS, R.; GLOMSET, J. A.; KARIYA, B., and HARKER, L.: A platelet-dependent serum factor that stimulates the proliferation of arterial smooth muscle cells *in vitro*. Proc nat. Acad. Sci. USA *71*: 1207–1210 (1974).

245 ROSS, R.; WIGHT, T., and GLOMSET, J. A.: Lesion formation in hypercholesteremic primates following intravascular endothelialization; in SCHETTER and WEIZEL Atherosclerosis III, pp. 860 (Springer, Berlin 1974).

246 ROTHBLAT, G. H.: Lipid metabolism in tissue culture cells. Adv. Lipid Res. *7:* 135–163 (1969).

247 ROTHBLAT, G. H.; BUCHKO, M. K., and KRITCHEVSKY, D.: Cholesterol uptake by L5178Y tissue culture cells: studies with delipidized serum. Biochim. biophys. Acta *164:* 327–338 (1968).

248 ROTHBLAT, G. H.; HARTZELL, R.; MIALHE, M., and KRITCHEVSKY, D.: Cholesterol metabolism in tissue culture cells; in ROTHBLAT and KRITCHEVSKY Lipid metabolism in tissue culture cells, pp. 129–149 (Wistar Institute Press, Philadelphia 1967).

249 ROTHBLAT, G. H. and KRITCHEVSKY, D.: The metabolism of free and esterified cholesterol in tissue culture cells. A review. Expl molec. Path. *8:* 314–329 (1968).

250 ROUSER, G. and SOLOMON, R. D.: Changes in phospholipid composition of human aorta with age. Lipids *4:* 232–234 (1969).

251 RUBENSTEIN, B. and RUBINSTEIN, D.: Interrelationship between rat serum very low density and high density lipoproteins. J. Lipid Res. *13:* 317–324 (1972).

252 RUDEL, L. L. and LOFLAND, H. B., jr.: Circulating lipoproteins in nonhuman primates. Prim. Med., vol. 9, pp. 224–266 (Karger, Basel 1976).

253 RUTSTEIN, D. D.; CASTELLI, W. P., and NICKERSON, R. J.: Effect of carbohydrate ingestion in humans on intracellular lipid deposition in tissue culture. Am. J. clin. Nutr. *20:* 98–107 (1967).

254 RUTSTEIN, D. D.; CASTELLI, W. P.; SULLIVAN, J. C.; NEWELL, J. M., and NICKERSON, R. J.: Effects of fat and carbohydrate ingestion in human beings on serum lipids and intracellular lipid deposition in tissue cultures. New Engl. J. Med. *271:* 1–7 (1964).

255 RUTSTEIN, D. D.; INGENITO, E. F.; CRAIG, J. M., and MARTINELLI, M.: Effects of linolenic and stearic acids on cholesterol-induced lipoid deposition in human aortic cells in tissue-culture. Lancet *i:* 545–552 (1958).

256 ST. CLAIR, R. W. and LOFLAND, H. B., jr.: Uptake and esterification of exogenous cholesterol by organ cultures of normal and atherosclerotic pigeon aorta. Proc. Soc. exp. Biol. Med. *138:* 632–637 (1971).

257 ST. CLAIR, R. W.; LOFLAND, H. B., jr., and CLARKSON, T. B.: Composition and synthesis of fatty acids in atherosclerotic aortas of the pigeon. J. Lipid Res. *9:* 739–747 (1968).

258 ST. CLAIR, R. W.; LOFLAND, H. B., and CLARKSON, T. B.: Influence of atherosclerosis on the composition, synthesis and esterification of lipids in aortas of squirrel monkeys *(Saimiri sciureus).* J. Atheroscler. Res. *10:* 193–201 (1969).

259 ST. CLAIR, R. W.; LOFLAND, H. B., and CLARKSON, T. B.: Influence of duration of cholesterol feeding on esterification of fatty acids by cell-free preparation of pigeon aorta. Studies on the mechanism of cholesterol esterification. Circulation Res. *27:* 213–225 (1970).

260 ST. CLAIR, R. W.; LOFLAND, H. B., jr.; PRICHARD, R. W., and CLARKSON, T. B.: Synthesis of squalene and sterols by isolated segments of human and pigeon arteries. Expl molec. Path. *8:* 201–215 (1968).

261 SARDET, C.; HANSMA, H., and OSTWALD, R.: Characterization of guinea pig plasma lipoproteins: the appearance of new lipoproteins in response to dietary cholesterol. J. Lipid Res. *13:* 624–639 (1972).

262 Sardet, C.; Hansma, H., and Ostwald, R.: Effects of plasma lipoproteins from control and cholesterol-fed guinea pigs on red cell morphology and cholesterol content: an *in vitro* study. J. Lipid Res. *13:* 705–715 (1972).

263 Scandella, C. J.; Devaux, P., and McConnell, H. M.: Rapid lateral diffusion of phospholipids in rabbit sarcoplasmic reticulum. Proc. nat. Acad. Sci. USA *69:* 2056–2060 (1972).

264 Scanu, A. M.: Structural studies on serum lipoproteins. Biochim. biophys. Acta *265:* 471–508 (1972).

265 Scanu, A. M.: The structure of human serum low and high density lipoproteins; in Atherosclerosis: initiating factors. Ciba Symp. No. 12, pp. 223–246 (Elsevier, Amsterdam (1973).

266 Scanu, A. M.; Edelstein, C.; Vitello, L.; Jones, R., and Wissler, R.: The serum high density lipoproteins of macacus rhesus. I. Isolation and composition. J. biol. Chem. *248:* 7648–7652 (1973).

267 Scanu, A. M. and Wisdom, C.: Serum lipoproteins: structure and function. Annu. Rev. Biochem. *41:* 703–730 (1972).

268 Schneider, H.; Morrod, R. S.; Colvin, J. R., and Tattrie, N. H.: The lipid core model of lipoproteins. Chem. Phys. Lipids *10:* 328–353 (1973).

269 Schönheimer, R.: Zur Chemie der gesunden und der atherosklerotischen Aorta. I. Über die quantitativen Verhältnisse des Cholesterins und der Cholesterinester. Hoppe-Seyler's Z. Physiol. Chem. *160:* 61–76 (1926).

270 Schönheimer, R.: Zur Chemie der gesunden und der atherosklerotischen Aorta, Hoppe-Seyler's Z. Physiol. Chem. *177:* 143–157 (1928).

271 Schwartz, S. M. and Benditt, E. P.: Studies on aortic intima. I. Structure and permeability of rat thoracic aortic intima. Am. J. Path. *66:* 241–264 (1972).

272 Schwenk, E. and Werthessen, N. T.: Studies on the biosynthesis of cholesterol. III. Purification of C^{14}-cholesterol from perfusions of livers and other organs. Archs Biochem. *40:* 334–341 (1952).

273 Scott, P. J. and Hurley, P. J.: The distribution of radio-iodinated serum albumin and low density lipoprotein in tissues and the arterial wall. Atherosclerosis *11:* 77–103 (1970).

274 Shimamoto, T.; Kobayashi, M., and Numano, F.: Infiltration of γ-globulin, fibrinogen and β-lipoprotein into blood vessel wall by atherogenic stress visualized by immunofluorescence. Proc. jap. Acad. *48:* 336–341 (1972).

275 Shimamoto, T. and Numano, F.: β-Lipoprotein entry into the arterial wall and its preventionin; in Schettler and Weizel Atherosclerosis III, pp. 89–92 (Springer, Berlin 1974).

276 Shimamoto, T. and Sunaga, T.: Contraction of endothelial cells as a key mechanism in atherogenesis. Proc. jap. Acad. *48:* 633–639 (1972).

277 Shohet, S. B.: Release of phospholipid fatty acid from human erythrocytes. J. clin. Invest. *49:* 1668–1678 (1970).

278 Shore, M. L.; Zilversmit, D. B., and Ackerman, R. F.: Plasma phospholipide deposition and aortic phospholipide synthesis in experimental atherosclerosis. Am. J. Physiol. *181:* 527–531 (1955).

279 SIMIONESCU, N.; SIMIONESCU, M., and PALADE, G. E.: Permeability of intestinal capillaries. Pathway followed by dextrans and glycogens. J. Cell Biol. *53:* 365–392 (1972).

280 SIMIONESCU, N.; SIMIONESCU, M., and PALADE, G. E.: Permeability of muscle capillaries to exogenous myoglobin. J. Cell Biol. *57:* 424–452 (1973).

281 SINGER, S. J.: The molecular organization of biological membranes; in ROTHFIELD Structure and function of biological membranes, pp. 145–222 (Academic Press, New York 1971).

282 SIPERSTEIN, M. D.; CHAIKOFF, I. L., and CHERNICK, S. S.: Significance of endogenous cholesterol in arteriosclerosis: synthesis in arterial tissue. Science *113:* 747–749 (1951).

283 SKIPSKI, V. P.: Lipid composition of lipoproteins in normal and diseased states; in NELSON Blood lipids and lipoproteins, quantitation, composition and metabolism; pp. 471–583 (Wiley-Interscience, New York 1972).

284 SKIPSKI, V. P.; BARCLAY, M.; ARCHIBALD, F. M.; TERABUS-KEKISH, P.; REID, E. S., and GOOD, J. J.: Lipid composition of rat liver cell membranes. Life Sci. *4:* 1673–1680 (1965).

285 SLAKEY, L. L.; NESS, G. C.; QURESHI, N., and PORTER, J. W.: Occurrence of the enzymes effecting the conversion of acetyl coA to squalene in homogenates of hog aorta. J. Lipid Res. *14:* 485–494 (1973).

286 SMALL, D. M.: The physical state of lipids of biological importance: cholesterol esters, cholesterol and triglycerides. Adv. Exp. Med. Biol. *7:* 55–83 (1970).

286a SMALL, D. M. and SHIPLEY, G. G.: Physical-chemical basis of lipid deposition in atherosclerosis. Science *185:* 222–229 (1974).

287 SMITH, E. B.: Intimal and medial lipids in human aortas. LANCET *i:* 799–803 (1960).

288 SMITH, E. B.: The influence of age and atherosclerosis on the chemistry of aortic intima. 1. The lipids. J. Atheroscler. Res. *5:* 224–240 (1965).

289 SMITH, E. B.: The origin and significance of the changes in the lipids of vascular tissue with age. J. Atheroscler. Res. *8:* 197–199 (1968).

289a SMITH, E. B.: The relationship between plasma and tissue lipids in human atherosclerosis. Adv. Lipid Res. *12:* 1–49 (1974).

290 SMITH, E. B. and SLATER. R. S.: The chemical and immunological assay of low density lipoproteins extracted from human aortic intima. Atherosclerosis *11:* 417–438 (1970).

291 SMITH, E. B. and SLATER, R. S.: The lipoproteins of the lesions; in JONES Proc. 2nd Int. Symp. Atherosclerosis pp. 42–49 (Springer, Berlin 1970).

292 SMITH, E. B. and SLATER, R. S.: The microdissection of large atherosclerotic plaques to give morphologically and topographically defined fractions for analysis. Atherosclerosis *15:* 37–56 (1972).

293 SMITH, E. B. and SLATER, R.S.: Lipids and lipoproteins in ageing aortic intima. Proc. R. Soc. Med. *65:* 675–677 (1972).

294 SMITH, E. B. and SLATER, R. S.: Relationship between low density lipoproteins in aortic intima and serum lipid levels. Lancet *i:* 463–469 (1972).

295 SMITH, E. B. and SLATER, R. S.: Relationship between plasma lipids and arterial tissue lipids. Nutr. Metabol. *15:* 17–26 (1973).

296 SMITH, E. B. and SLATER, R. S.: Lipids and low-density lipoproteins in intima in relation to its morphological characteristics; in Atherosclerosis: initiating factors. Ciba Symp. No. 12, pp. 39–52 (Elsevier, Amsterdam 1973).

297 SMITH, E. B.; SLATER, R. S., and HUNTER, J. A.: Quantitative studies on fibrinogen and low density lipoproteins in human aortic intima. Atherosclerosis *18:* 479–487 (1973).

298 SMITH, S. C.; STROUT, R. G.; DUNLOP, W. R., and SMITH, E. C.: Fatty acid composition of cultured aortic cells from white Carneau and show racer pigeons. J. Atheroscler. Res. *5:* 379–387 (1965).

299 SRINIVASAN, S. R.; DOLAN, P.; RADHAKRISHNAMURTHY, B., and BERENSON, G. E.: Isolation of lipoprotein-acid mucopolysaccharide complexes from fatty streaks of human aortas. Atherosclerosis *16:* 95–104 (1972).

300 STEELE, J. M. and KAYDEN, H. L.: The nature of the phospholipids in human serum and atheromatous vessels. Trans. Ass. Am. Physns *68:* 249–254 (1955).

301 STEFANOVICH, V. and GORE, I.: Cholesterol diet and permeability of rabbit aorta. Expl molec. Path. *14:* 20–29 (1971).

302 STEIN, O.; EISENBERG, S. and STEIN, Y.: Morphologic and biochemical changes in smooth muscle cells of aortas in growth-restricted rats. Lab. Invest. *25:* 149–157 (1971).

302a STEIN, O.; EISENBERG, S., and STEIN, Y.: Aging of aortic smooth muscle cells in rats and rabbits. A morphological and biochemical study. Lab. Invest. *21:* 386–397 (1969).

303 STEIN, O.; RACHMILEWITZ, D.; EISENBERG, S., and STEIN, Y.: Aortic phospholipids. A biochemical, radioautographic and morphological study of phospholipid localization in normal adult aorta. Israel J. med. Sci. *6:* 53–66 (1970).

303a STEIN, O. and STEIN, Y.: Lecithin synthesis, intracellular transport, and secretion in rat liver. IV. A radioautographic and biochemical study of choline-deficient rats injected with choline-³H. J. Cell Biol. *40:* 461–483 (1969).

304 STEIN, O. and STEIN, Y.: Lipid synthesis and transport in the normal and atherosclerotic aorta. An autoradiographic study of rat and rabbit aortae incubated and perfused with choline-³H and oleic acid-³H. Lab. Invest. *23:* 556–566 (1970).

305 STEIN, O. and STEIN, Y.: An electron microscopic study of the transport of peroxidases in the endothelium of mouse aorta. Z. Zellforsch. *133:* 211–222 (1972).

306 STEIN O. and STEIN, Y.: The removal of cholesterol from Landschütz ascites cells by high-density apolipoprotein. Biochim. biophys. Acta *326:* 232–244 (1973).

306a STEIN, O. and STEIN, Y.: Comparative uptake of rat and human serum low-density and high-density lipoproteins by rat aortic smooth muscle cells in culture. Circulations Res. *36:* 436–443 (1975).

307 STEIN, O.; STEIN, Y. and EISENBERG, S.: A radioautographic study of the transport of ¹²⁵I-labeled serum lipoproteins in rat aorta. Z. Zellforsch. mikrosk. Anat. *138:* 223–237 (1973).

308 STEIN, Y. and STEIN, O.: Incorporation of fatty acids into lipids of aortic slices of rabbits, dogs, rats and baboons. J. Atheroscler. Res. *2:* 400–412 (1962).

309 STEIN, Y. and STEIN, O.: Metabolism of lysolecithin by the perfused rat heart. Biochim. biophys. Acta *106:* 527–539 (1965).

310 STEIN, Y. and STEIN, O.: Transport of lipids in the arterial wall: a biochemical and radioautographic study. Expos. Ann. biochim. Med. *31:* 97–110 (1972).

311 STEIN, Y.; EISENBERG, S., and STEIN, O.: Metabolism of lysolecithin by human umbilical and dog carotid arteries. Prog. biochem. Pharmacol. vol. 4, pp. 253–259 (Karger, Basel New York 1968).

312 STEIN, Y.; STEIN, O., and SHAPIRO, B.: Enzymic pathways of glyceride and phospholipid synthesis in aortic homogenates. Biochim. biophys. Acta 70: 33–42 (1963).

313 STEIN, Y.; WIDNELL, C., and STEIN, O.: Acylation of lysophosphatides by plasma membrane fractions of rat liver. J. Cell Biol. 39: 185–192 (1968).

314 STOFFEL, W. and GRETEN, H.: Studies on lipolytic activities of rat liver lysosomes. Hoppe-Seyler's Z. physiol. Chem. 348: 1145–1150 (1967).

315 STOUT, R. W., BIERMAN, E. L., and ROSS, R.: Effects of insulin on the proliferation of cultured primate arterial smooth muscle cells. Circulation Res. 35: 319–327 (1975).

316 STOUT, R. W.; BUCHANAN, K. D., and VALLANCE-OWEN, J.: Arterial lipid metabolism in relation to blood glucose and plasma insulin in rats with streptozotocin-induced diabetes. Diabetologia 8: 398–401 (1972).

317 STRONG, J. P. and TAPPEN, N. C.: Naturally occurring arterial lesions in African monkeys. Comparison of sudanophilic aortic lesions. Archs Path. 79: 199–205 (1965).

318 SUGANO, M. and PORTMAN, O. W.: Fatty acid specificities and rates of cholesterol esterification in vivo and in vitro. Archs Biochem. 107: 341–351 (1964).

319 SUGANO, M. and PORTMAN, O. W.: Essential fatty acid deficiency and cholesterol esterification activity of plasma and liver in vitro and in vivo. Archs Biochem. 109: 302–315 (1965).

320 TAYLOR, C. B.; COX, G. E.; HALL-TAYLOR, B. J., and NELSON, L. G.: Atherosclerosis in areas of vascular injury in monkeys with mild hypercholesterolemia. Circulation 10: 613 (1954).

321 TAYLOR, C. B.; COX, G. E.; MANALO-ESTERELLA, P.; SOUTHWORTH, J.; PATTON, D. E., and CATHCART, C.: Atherosclerosis in rhesus monkeys. II. Arterial lesions associated with hypercholesterolemia induced by dietary fat and cholesterol. Archs Path. 74: 16–34 (1962).

322 TEAL, S. W. and GAMBLE, W.: Effect of p-chlorophenoxyisobutyric acid, ethyl p-chlorophenoxyisobutyrate, and atromid on the synthesis of nonsaponifiable material and cholesterol in the bovine aorta. Biochem. Pharmacol. 14: 896–898 (1965).

323 TRACY, R. E.; DZOGA, K. R., and WISSLER, R. W.: Sequestration of serum low-density lipoproteins in the arterial intima by complex formation. Proc. Soc. exp. Biol. Med. 118: 1095–1098 (1965).

324 URRUTIA, G.; BEAVAN, D. W., and CAHILL, G. F., jr.: Metabolism of glucose-U-C^{14} in rat aorta in vitro. Metabolism 11: 530–534 (1962).

325 VOST, A.: Uptake and metabolism of circulating chylomicron triglyceride by rabbit aorta. J. Lipid Res. 13: 695–703 (1972).

326 WALLACH, D. F. H. and LIA, P. S.: A critical evaluation of plasma membrane fractionation. Biochim. biophys. Acta 300: 211–254 (1973).

327 WALSH, M. R.; TEAL, S. W., and GAMBLE, W.: Biosynthesis of cholesterol by the bovine aorta and the mechanism of action of α-para-chlorophenoxyisobutyric acid. Archs Biochem. 130: 7–18 (1969).

328 WALTON, K. W.: Identification of lipoproteins involved in human atherosclerosis. in SCHETTLER and WEIZEL Atherosclerosis III, pp. 93–95 (Springer, Berlin 1974).

329 WALTON, K. W.: Studies on experimental atherosclerosis by immunofluorescence. in SCHETTLER and WEIZEL Atherosclerosis III, pp. 863–864 (Springer, Berlin 1974).

330 WALTON, K. W. and WILLIAMSON, N.: Histological and immunofluorescent studies on the evolution of the human atheromatous plaque. J. Atheroscler. Res. *8:* 599–624 (1968).

331 WATTS, H. F.: Role of lipoproteins in the formation of atherosclerotic lesions; in: JONES Evolution of the atherosclerotic plaque, pp. 117–132 (Univ. of Chicago Press, Chicago 1964).

332 WAYS, P.; REED, C. F., and HANAHAN, D. J.: Red cell and plasma lipids in acanthocytosis. J. clin. Invest. *42:* 1248–1260 (1963).

333 WEINHOUSE, S. and HIRSCH, E. F.: Chemistry of atherosclerosis; lipid and calcium content of intima and media of aorta with and without atherosclerosis. Archs Path. *29:* 31–41 (1940).

334 WERTHESSEN, N. T.; MILCH, L. J.; REDMOND, R. F.; SMITH, L. L., and SMITH, E. C.: Biosynthesis and concentration of cholesterol by the intact surviving bovine aorta *in vitro*. Am. J. Physiol. *178:* 23–29 (1954).

335 WHEREAT, A. F.: Lipid biosynthesis in aortic intima from normal and cholesterol-fed rabbits. J. Atheroscler. Res. *4:* 272–282 (1964).

336 WHEREAT, A. F.: Fatty acid synthesis in cell-free system from rabbit aorta. J. Lipid Res. *7:* 671–677 (1966).

337 WHEREAT, A. F.: Fatty acid biosynthesis in aorta and heart. Adv. Lipid Res. *9:* 119–159 (1971).

338 WHEREAT, A. F. and ORISHIMO, M. W.: The effect of cholesterol administration on lipid biosynthesis by rabbit aortic mitochondria. Expl molec. Path. *9:* 230–241 (1968).

339 WIRTZ, K. W. A.; KAMP, H. H., and DEENEN, L. L. M. VAN: Isolation of a protein from beef liver which specifically stimulates the exchange of phosphatidylcholine. Biochim. biophys. Acta *274:* 606–617 (1972).

340 WIRTZ, K. W. A. and ZILVERSMIT, D. B.: Participation of soluble liver proteins in the exchange of membrane phospholipids. Biochim. biophys. Acta *193:* 105–116 (1969).

341 WISSLER, R. W.; FRAZIER, L. E.; HUGHES, R. H., and RASMUSSEN, R. A.: Atherogenesis in the cebus monkeys. I. A comparison of three food fats under controlled dietary conditions. Archs Path *74:* 312–322 (1962).

342 WISSLER, R. W. and VESSELINOVITCH, D.: Comparative pathogenic patterns in atherosclerosis. Adv. Lipid Res. *6:* 181–206 (1968).

343 WISSLER, R. W.; VESSELINOVITCH, D.; GETZ, G. S., and HUGHES, R. H.: Aortic lesions and blood lipids in Rhesus monkeys fed three food fats. Fed. Proc. Fed. Am Socs exp. Biol. *26:* 371 (1967).

344 WISSLER, R. W.; VESSELINOVITCH, D., and HUGHES, R. H.: A comparison of the aortic atherosclerotic response in the rhesus and stump tailed monkey using rations rich in vegetable fat with or without added cholesterol. Am. J. Path. *66:* 92a (1972)

345 WOOLF, N. and PILKINGTON, T. R. E.: The immunohistochemical demonstration of lipoproteins in vessel walls. J. Path. Bact. *90:* 459–463 (1965).

346 ZILVERSMIT, D. B.: A proposal linking atherogenesis to the interaction of endothelial lipoprotein lipase with triglyceride-rich lipoproteins. Circulation Res. *33:* 633–638 (1973).

347 Zilversmit, D. B. and Newman, H. A. I.: Does a metabolic barrier to circulating cholesterol protect the arterial wall? Circulation *33:* 7 (1966).

348 Zilversmit, D. B.; Shore, M. L., and Ackerman, R. F.: The origin of aortic phospholipid in rabbit atheromatosis. Circulation *9:* 581–585 (1954).

349 Zilversmit, D. B.; Sweeley, C. C., and Newman, H. A. I.: Fatty acid composition of serum and aortic intimal lipids in rabbits fed low- and high-cholesterol diets. Circulation Res. *9:* 235–241 (1961).

Dr. Oscar W. Portman, Oregon Regional Primate Research Center, 505 N.W. 185th Avenue, *Beaverton, OR 97005* (USA)

Prim. Med., vol. 9, pp. 224–266 (Karger, Basel 1976)

Circulating Lipoproteins in Nonhuman Primates[1]

Lawrence L. Rudel and Hugh B. Lofland †

Arteriosclerosis Research Center,
Departments of Comparative Medicine and Pathology,
Bowman Gray School of Medicine, Wake Forest University, Winston-Salem, N. C.

Contents

I. Introduction

The purpose of this chapter is to review the available information on distribution, structure, and metabolism of plasma[2] lipoproteins of nonhuman primates. For purposes of comparison, a brief description of knowledge about lipoproteins in other species will be included since most of our present understanding about these macromolecules has been derived from work carried out in other species. Comprehensive study of lipoproteins in

[1] Supported in part by USPH, NIH grant HL 14164.

[2] In this chapter, no differentiation will be made between plasma lipoproteins and serum lipoproteins.

nonhuman primates is just beginning. The interest in plasma lipoproteins in primates stems from the fact that significant elevations in the concentration of many plasma lipoproteins have been described in primates under circumstances in which the prevalence of atherosclerosis is high. It is not certain that there is a cause and effect relationship between elevated lipoprotein levels and atherosclerosis, but much evidence would suggest that this may be the case for some forms of the disease. The complete answer will be known only after much more investigation of controlling factors of lipoprotein metabolism and of lipoprotein interactions with tissues.

Due to their close phylogenetic relationship, nonhuman primates are important models for research on human atherosclerosis. Development of understanding of lipoprotein metabolism in these animals may lead to a more complete understanding of the relationship between atherosclerosis and circulating lipoproteins in human beings. Thus, this presentation is offered in the hope that it will stimulate gathering of knowledge in this area of research.

II. Techniques of Analysis

Plasma lipoproteins are circulating macromolecules which consist of varying amounts of lipid, protein, and carbohydrate. The major component in most lipoproteins is lipid, including phospholipids, glycerides, cholesteryl esters, cholesterol, and unesterified fatty acids. Several proteins, or apolipoproteins, are part of the lipoprotein macromolecules and one of their functions is to maintain the lipid soluble in water; some apolipoproteins are glycoproteins [56]. The ratio of lipid to protein determines the density of the lipoprotein. The plasma lipoproteins are routinely named according to their densities; the very low density lipoproteins (VLDL) are the least dense class (d < 1.006); the low density lipoproteins (LDL) have a density between 1.006 and 1.063, and the high density lipoproteins (HDL) have a density between 1.063 and 1.21. An additional class of d < 1.006 lipoproteins, the chylomicra, appear in plasma during fat absorption after a meal. In most species, these are the major classes commonly found in plasma, although some heterogeneity within each class has been described.

Classes of lipoproteins are routinely separated by sequential ultracentrifugation of plasma at appropriate densities by a technique similar to that of HAVEL et al. [26]. Isolated classes can then be analyzed in the analytical ultracentrifuge to determine the distribution of particles according to their respective flotation rates [39]. Within each density class, there is a distribution of closely related particles which differ slightly in lipid to protein ratio, hence in density. One can determine the distribution according to flotation rate, i.e. S_f (defined as Svedburg flotation units or negative Svedburg sedimentation units, 10^{-13} cm/sec·dyne·g, at d 1.063) or F (at densities higher than 1.063 g/ml). The characteristic S_f values for lipoproteins are: chylomicra, $S_f > 400$; VLDL, S_f 20–400; LDL, S_f 0–20; and HDL F 0–10. The concentration of lipoproteins in serum can also be calculated from the analytical ultracentrifuge data (see Lindgren and Jensen [39] for details).

A new method for preparatively separating and characterizing plasma lipoproteins has recently been described by Rudel et al. [54]. In this technique, all lipoproteins are quantitatively separated according to size using agarose column chromatography. The elution profile is monitored at 280 nm and is recorded, thereby providing a description of the size distribution of lipoproteins in the plasma sample. This method provides a relatively quick way to determine lipoprotein distribution while, at the same time, each fraction is purified for further analysis. The major density classes as described by sequential ultracentrifugation are easily identified as individual peaks which are eluted according to their size. Significant variation in lipoprotein distribution among individuals is often found using this technique. One of the more common findings in primates is the presence of a fraction intermediate in size between VLDL and LDL, which in this communication will be termed intermediate-size low density lipoprotein (IDL) (fig. 1a).

Although these types of analyses are excellent, they are not available on a routine basis to many laboratories. A quick and simple method of lipoprotein separation uses anionic polysaccharides (heparin, dextran sulfate) and divalent metal ions (Ca^{++}, Mn^{++}) to precipitate VLDL and LDL, since both contain the apolipoprotein of LDL [11, 13, 34]. This technique is extremely useful when large numbers of samples are to be evaluated. The products of this type of separation are the HDL (α-lipoproteins) which remain in the supernatant solution, and the LDL + VLDL (β- + pre-β-lipoproteins) which aggregate as part of a complex with the divalent metal and anionic polysaccharide. Although this procedure can be used to determine the amounts of HDL and LDL + VLDL in the original plasma sample, the type of information generated is limited due to the fact that LDL and VLDL are not separated, and no estimate of size distribution is obtained. This technique has been developed using human serum primarily. Its applicability in other species should always be verified for the species in question since results from dextran sulfate-Ca^{++} precipitation in sera from some nonhuman

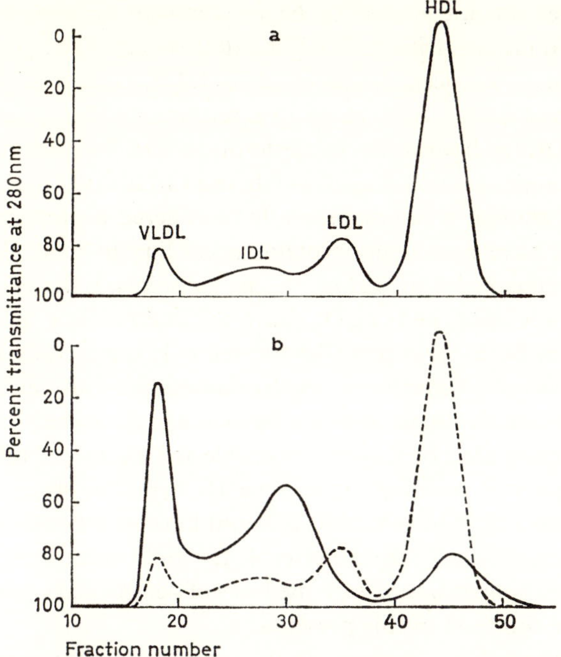

Fig. 1. Lipoprotein size distribution in *M. mulatta* as determined by agarose column chromatography. Blood was collected after a 16- to 18-hour fasting period. All lipoproteins were simultaneously isolated from plasma by ultracentrifugation and were then separated according to size by agarose gel chromatography. The extent of separation was determined by continuously monitoring and recording the percent transmittance (%T) of the column eluate at 280 nm. Turbidity present in the large VLDL component decreased the %T out of proportion to the concentration of lipoprotein in this fraction. Shown in *a* is the separation of plasma lipoproteins from a representative male rhesus monkey (plasma cholesterol concentration 160 mg/dl) which was fed control diet A (table IV). In *b*, the solid line represents the separation of lipoproteins of a male rhesus monkey (plasma cholesterol concentration 805 mg/dl) which was fed test diet B (table IV). The dotted line in *b* is the lipoprotein separation of the control animal of *a* and is presented to illustrate clearly the difference in relative size of the peaks of the control and cholesterol-fed samples; each fraction contained the same volume (3.3 ml) and for any peak, elution volume (Ve) is a function of lipoprotein size, when it is assumed that all molecules are spherical. For more details of the method refer to RUDEL *et al.* [54].

primate species do not appear to be in complete agreement with results from analytical ultracentrifugation. FREDRICKSON and LEVY [19] recommend a combination of ultracentrifugation at d 1.006, combined with heparin-manganese precipitation, as a means to determine the distribution of VLDL, LDL, and HDL (as lipoprotein cholesterol) in sera from man.

Electrophoretic analysis is also widely used to identify classes of plasma lipoproteins although it is not generally considered a preparative method. Some authors have named lipoproteins according to their electrophoretic mobility when compared to that of plasma globulins. LDL have β-mobility, HDL have α-mobility, and VLDL have an intermediate 'pre-β-mobility', thus the names β-, α-, and pre-β-lipoproteins. It is generally true that this relationship between density and electrophoretic mobility exists. However, there are notable exceptions, as might be expected since the properties which determine density are not those responsible for electrophoretic migration. VLDL may show β mobility, as in type III hyperlipoproteinemia [19], in Tangiers disease [18], and in rabbits [12] and in rhesus monkeys [RUDEL and CLARKSON, unpublished] fed cholesterol. It would seem that presently the least redundant nomenclature is on the basis of density class, and this nomenclature will be followed in this presentation.

Many attempts have been made to quantitate lipoprotein distribution after densitometric scanning of electrophoretic separations [28, 40]. Such procedures can be useful, provided the user remains aware of the pitfalls, including loss or alteration of material during staining, lack of correlation between lipoprotein density and electrophoretic mobility, and variation in staining properties with changing composition. However, it is more meaningful if electrophoretic analysis can be combined with a direct quantitation procedure before interpretations are finalized. Accordingly, in the present chapter, we will place more emphasis on studies in which this has been done.

III. Distribution and Composition of Lipoproteins

The distribution of lipoproteins among animal species, as well as within any one species, is quite variable so that generalization is not without risk. In addition, many evironmental factors affect lipoprotein distribution. For example, the amount of fat and cholesterol in the diet [48] are both factors that increase LDL concentration. During fat absorption after a meal, the lipoprotein distribution changes significantly [25] with VLDL and chylomicra increasing in concentration during absorption, and with HDL possibly de-

Table I. Normal human serum lipoprotein cholesterol distribution[1]

Age	Sex	N	Mean serum cholesterol mg/dl	Cholesterol distribution, %		
				VLDL	LDL	HDL
0–19	M	72	166	7.2	62.0	30.7
	F	71	169	7.1	60.9	32.0
20–29	M	47	180	8.9	63.9	27.2
	F	41	180	8.9	60.0	31.1
30–39	M	38	195	9.7	66.7	23.6
	F	29	191	7.8	62.3	29.8
40–49	M	25	212	10.4	67.0	22.6
	F	34	211	8.1	61.1	30.8
50–59	M	14	223	11.2	70.0	18.8
	F	10	242	10.7	69.0	20.2

[1] Data from FREDERICKSON *et al.* [19].

creasing in concentration. Lipoprotein distribution between adult human males and females is also different. More HDL is present in females than in males, and the percentage of total lipoprotein present as LDL is lower in females [3, 14]. Lipoprotein distribution is also age-related. In older individuals, more VLDL and LDL are present, and both absolute amount and percentage of total lipoprotein present as VLDL and LDL are increased [38]. These facts demonstrate the importance of determining the exact experimental conditions under which results are obtained before any attempt is made to generalize or to extrapolate from human studies in which diet is rarely well controlled, to studies in experimental animals in which diet is usually fixed. With these reservations in mind, this discussion of lipoprotein distribution in primates is presented. Included is a brief discussion of human plasma lipoproteins to serve as a point of reference.

A. Human Beings

In human beings, there have been many studies to determine the normal distribution of lipoproteins. Generally, these studies have been carried out on fasting individuals. Table I presents an average distribution of lipoprotein cholesterol in normal individuals, as determined by FREDRICKSON and LEVY [19]. LDL is present in highest concentration and contains about 60–70% of the plasma total cholesterol. As the concentration of cholesterol in plasma increases with age, the percentages present in VLDL and LDL rise, and the percentage in HDL decreases; thus, the LDL/HDL ratio increases. The real significance of a shift in LDL/HDL ratio is unknown, although some workers have suggested that the decreased prevalence and incidence of atherosclerosis in premenopausal human females may in some way be related to this ratio [7]. It is important to note that when the percentage in any one lipoprotein class increases with age, the absolute concentration of that lipoprotein also increases, since plasma total cholesterol increases. It is difficult to ascertain whether the absolute concentration of LDL is a more important consideration than the shift in the LDL/HDL ratio; both may be important in atherosclerosis. For example, in human type II hyperlipoproteinemia [21], the concentration of LDL is increased (some 5x normal in homozygous individuals) and the percentage distribution of cholesterol is shifted so that 90% is in LDL. Type II patients appear to be among the human populations most highly predisposed to premature death from atherosclerosis.

The distribution of lipoproteins in human serum has been determined in several laboratories using analytical ultracentrifugation (for review see BARCLAY [2]). Representative data from NICHOLS [47] is shown in table II. In normal premenopausal adult females, the HDL are present in higher concentration and the VLDL in lower concentration than in normal males. This observation is similar to what was seen when studying the cholesterol distribution (table I). However, analytical ultracentrifugation is a more sensitive technique for the study of differences in serum lipoprotein distribution, since it permits one to determine a shift in mean density within any one lipoprotein class. We will return to the importance of this fact in the discussion of alterations in nonhuman primates.

General agreement as to the chemical composition of lipoproteins has been reached among most laboratories [62]. The data in table III show the composition of lipoproteins obtained by RUDEL *et al.* [54] in three normocholesterolemic individuals. The values represent the means for duplicate determinations on lipoproteins which were separated by two methods for each individual.

In the protein moiety of each lipoprotein, several specific apolipoproteins have been identified using polyacrylamide gel electrophoresis. Individual apolipoproteins have now been separated using sephadex and DEAE-ion exchange chromatography, and some of their chemical properties have been

Table II. Serum lipoprotein distribution in normal human beings[1]

Sex	Age	Serum concentration mg/dl						
		VLDL		LDL		HDL		
		S_f 100–400	S_f 20–100	S_f 10–20	S_f 0–12	HDL_1	HDL_2	HDL_3
Male	30–39	51	91	40	322	23	37	217
		142		362		277		
Female	30–49	14	55	41	330	22	83	237
		69		372		342		

[1] Data from NICHOLS [7]. Serum was collected from individuals who had fasted overnight. The concentration of lipoprotein subclasses was determined by analytical ultracentrifugation.

Table III. Percentage composition of human lipoproteins

| | Lipoprotein | | |
	VLDL	LDL	HDL
Chemical[1]			
Protein	12.5	20.8	43.8
Lipid	87.5	79.2	56.2
Lipid			
Triglyceride	55.4	6.6	8.7
Chol. ester	14.5	51.8	37.5
Free chol.	6.6	11.5	5.5
Phospholipid	23.4	29.2	48.2
Protein[2]			
apoLP-gln I	±	±	65–75
apoLP-gln II	±	±	20–25
apoLDL	40–45	>90	±
apoLP-ser	8–10	<1	2–4
apoLP-glu	8–10	<1	2–4
apoLP-ala	30	<1	5–10

[1] Data from RUDEL *et al.* [54]. The values are means of duplicate determinations on lipoproteins isolated from four individual plasma samples obtained from three people. The values shown are mean values obtained for lipoproteins isolated by agarose column chromatography and by preparative ultracentrifugation.

[2] Data from FREDERICKSON *et al.* [20]. It should be recognized that only those apolipoproteins which have been consistently found and were chemically characterized are listed here; this list is incomplete at present.

identified, including the complete amino acid sequence of apoLP-ser [61], apoLP-gln II [5], and apoLP-ala [6]. Nomenclature of the apolipoproteins has been complicated by the fact that the individual laboratories studying apolipoprotein composition have not agreed on a single system. The names for apolipoproteins which we present in table III follow the recommendations of FREDRICKSON *et al.* [20], and percentages shown are taken from their work.

Detailed knowledge of lipoprotein structure has contributed to a partial understanding of lipoprotein metabolism. The primary tissue of synthesis appears to be the liver [44] although the intestine also contributes significantly [66]. Cells secrete primarily VLDL and HDL into the circulation [43]. The plasma LDL apparently originate mainly as products of metabolism of VLDL (and chylomicra) from the action of lipoprotein lipase within the capillaries [43]. The plasma enzyme, lecithin: cholesterol acyltransferase has also been implicated in the intravascular metabolism of VLDL [58], although its role seems highly speculative at present. The catabolism of plasma LDL and HDL is not well understood, although some recent evidence suggests that the liver is the site of HDL degradation [52]. It seems logical that the liver may also degrade LDL, since this is the only tissue in the body which could catabolize the large amounts of cholesterol present in these lipoproteins, but supporting evidence seems lacking at present. Unfortunately, it is accurate to say that our present knowledge does not define the control mechanisms by which the body maintains normal lipoprotein concentrations.

B. New World Monkeys

Lipoprotein distribution is incompletely known for most New World primate species. The aspect of lipoprotein distribution which has been studied most thoroughly is the partitioning of plasma cholesterol between HDL and LDL + VLDL after anionic polysaccharide-metal precipitation of the latter. Some available data are shown in table IV. Samples taken after capture in the wild in cebus, marmoset, and squirrel monkeys all had about 25% of the plasma cholesterol present in the HDL fraction, and 75% in the LDL + VLDL [42]. Although lower in absolute concentration, this distribution of lipoprotein cholesterol appears to be similar to that found in man. It is interesting to note that in all wild caught animals, the triglyceride concentration was within the normal human range although the distribution of this lipid among lipoproteins has not been reported. Wild caught spider

Table IV. New World monkey plasma lipoprotein cholesterol distribution[1]

Species	Sex	N	Diet[2]	Mean serum cholesterol, mg/dl	Mean serum triglyceride, mg/dl	HDL cholesterol %	mg/dl	Reference number
Ateles (Spider)								
geoffroy	mixed	29	native	130	90	9.2	12	42
geoffroy	mixed	9	m. chow	143	–	25.0	36	63
sp.	mixed	7	C	200	–	13.9	28	51
sp.	mixed	9	D	290	–	10.3	30	51
Cebus								
albifrons	mixed	57	native	90	62	23.3	21	42
apella	mixed	21	native	98	97	23.4	23	42
Lagothrix (woolly)								
lagothrica	mixed	20	native	133	108	8.2	11	42
lagothrica	males	4	A	111	–	14.0	16	8
lagothrica	males	3	B	106	–	13.0	14	8
Sanguinus (marmoset)								
nigrocollis	mixed	46	native	69	95	26.1	18	42
Saimiri (squirrel)								
sciureus	males	110	native	103	100	31.1	32	42
sciureus	females	110	native	106	101	27.4	29	42
sciureus	males	7	A	179	–	24.0	43	8
sciureus	males	9	B	469	–	17.0	80	8
sciureus	mixed	14	m. chow	182	–	42.0	76	63

[1] Plasma cholesterol concentration is shown as the percent distribution between HDL and LDL + VLDL. The latter were precipitated using anionic polysaccharide-divalent metal complexing techniques. In most cases, the LDL + VLDL were removed from plasma, after which the concentration of the remaining HDL cholesterol was determined.

[2] The diets shown include: native diet, for which the exact composition is unknown since the animals were studied after capture in the wild; diet A is a semipurified diet which contains 45 % of calories as lard and 0.05 mg cholesterol per Cal; diet B is a semipurified diet with the same ingredients as A, but with 1 mg cholesterol per Cal. Diet C is a natural products diet of monkey chow plus 39 % of calories as equal quantities of butter and coconut oil, with a low cholesterol content; diet D is a natural products diet similar to C but with 1 mg cholesterol per Cal. The brand or composition of the monkey chow was not given in the reference.

and woolly monkeys had less cholesterol in the HDL fraction, showing > 90% present in LDL + VLDL. The HDL cholesterol percentage in spider and squirrel monkeys fed monkey chow was apparently higher than when the same species were studied in the wild, and the plasma cholesterol concentrations reported for these monkeys were also higher [63]. When fed experimental diets with and without added cholesterol, the percentage distribution of cholesterol for spider monkeys [51] and for squirrel monkeys [8] seemed to be similar, indicating that cholesterol feeding did not cause redistribution of cholesterol among lipoproteins even though the concentration of cholesterol in plasma increased. That is to say, the concentration of HDL and LDL + VLDL cholesterol was higher when these animals were fed cholesterol. This was not found to be the case in 30 squirrel monkeys whose cholesterol values were determined after capture, during shipment, and after relocation at the laboratory [65]. In these cases, the plasma cholesterol varied from 116 mg/dl in the wild to 194 mg/dl during shipment, and the percent of cholesterol in LDL + VLDL, determined after dextran sulfate precipitation, increased from 65% to 80%. The increase in LDL + VLDL cholesterol accounted for all of the rise in cholesterol concentration; the increase in this group of animals was attributed to the stress associated with capture and transportation.

In a study on the distribution of lipoproteins in five spider monkeys, SRINIVASAN et al. [64] noted an increase in both HDL and LDL + VLDL cholesterol after the animals were placed on a diet containing cholesterol for 20 days. The agarose electrophoresis patterns were analyzed using a densitometer, an increase of HDL concentration and of the HDL to LDL + VLDL ratio was observed.

In other studies on New World monkeys, the lipoprotein distribution has been determined by analytical ultracentrifugation. Table V presents a summary of information gathered by LOFLAND et al. [unpublished]. Cebus and squirrel monkeys fed monkey chow appeared to have a similar pattern of lipoprotein distribution, although the HDL/LDL ratio is higher for squirrel monkeys than for either *Cebus albifrons* or *Cebus apella*. The HDL_2 (d 1.063–1.125) to HDL_3 (d 1.125–1.21) ratio is also higher in squirrel monkeys than in cebus. Interestingly enough, the woolly monkeys (which had been fed cholesterol) had a much lower concentration of HDL and a slightly higher concentration of LDL than for either squirrel monkeys or cebus monkeys fed monkey chow, although the plasma lipoprotein and cholesterol concentrations are comparable. By comparison, the woolly monkeys also had a low HDL_2 to HDL_3 ratio.

Table V. Serum lipoprotein distribution of some New World Monkeys[1]

Species	Diet	Chol.	Serum concentration mg/dl				
			VLDL	LDL		HDL$_2$	HDL$_3$
			S_f 20–400	S_f 12–20	S_f 0–12		
Cebus							
albifrons	m. chow[2]	169	13	20	227	231	194
				260		426	
apella	m. chow[2]	139	12	10	187	148	184
				209		221	
Lagothrix							
lagothrica	1.0 mg/Cal	125	0	4	301	26	73
	(B)			305		99	
Saimiri							
sciureus	m. chow[2]	~140[3]	8	7	147	302	113
				162		415	
	0.05 mg/Cal	160	10	9	170	306	134
	(A)			189		440	
	0.3 mg/Cal	197	14	18	234	324	113
				266		437	
	1.0 mg/Cal	290	40	93	229	388	177
	(B)			362		565	
	2.0 mg/Cal	~500[3]	138	168	318	128[4]	600
				623		795	

[1] Unpublished data of LOFLAND, H. B., jr. *et al.* Analytical ultracentrifugation kindly performed by Dr. FRANK LINDGREN. Analyses were performed on lipoproteins isolated from pooled serum samples from young animals which had been fasted overnight. We are indebted to Mrs BETTY MASKET of the NHLI for assisting in these analyses.

[2] The diet is designated as m. chow. for monkey chow, or with a number indicating how much cholesterol was added to a semipurified diet which had 40% of calories as fat. Diets A and B are the same as described in table IV.

[3] The cholesterol concentration has been estimated where indicated, since a value on the pooled sample was not obtained.

[4] This sample had a HDL$_1$ concentration of 67 mg/dl.

The control mechanisms by which normal lipoprotein concentrations are maintained apparently differ among species. Although we do not understand the biochemical processes which regulate lipoprotein levels, it seems clear that they are influenced by exogenous cholesterol metabolism. The results shown in table V for squirrel monkeys fed increasingly larger amounts of cholesterol illustrate this point. As the amount of cholesterol in the diet was increased, the concentration of each subclass of LDL (including VLDL) increased. Proportionally, the concentration of S_f 12–20 and 20–400 lipoproteins increased the most. The HDL_2 and HDL_3 also increased in concentration up to a point. Beyond this point, HDL_2 decreased in concentration and HDL_3 increased so that the ratio of HDL_2 to HDL_3 changed from 3 to 0.2. The reason for such a marked shift in lipoprotein distribution is not known, but this shift would appear to be a fruitful area for further study of the regulation of lipoprotein concentration and its interaction with exogenous cholesterol metabolism.

One unexplained problem is that the data in tables IV and V do not appear to be in agreement in cases where the same species has been fed a comparable diet. The distribution of cholesterol among lipoprotein classes in squirrel monkeys fed monkey chow, for example, shows that only 42% of lipoprotein cholesterol was present in HDL (table IV). On the other hand, the lipoprotein distribution as determined by analytical ultracentrifugation (table V) indicates that $2^1/_2$ times more HDL is present than LDL + VLDL. A similar discrepancy is also apparent for cebus monkeys. Although the complete chemical compositions for squirrel monkey lipoproteins (or for any of the other New World monkey lipoproteins) are not known, a correction for the fact that HDL probably contain relatively less cholesterol as a percent of total mass than LDL + VLDL would not bring the data into agreement (compare, for example, the human data in tables I and II). As more animals are studied, these apparent discrepancies will hopefully be resolved. Presently it would appear that some of the HDL of some New World monkeys precipitate by the dextran sulfate-Ca^{++} precipitation method.

PORTMAN and ANDRUS [50] found that the LDL concentration in cebus and woolly monkeys increased when the dietary fat was changed from corn oil to coconut oil. The distribution within LDL subclasses also changed in both species; the percent of LDL present in the S_f 12–400 class in animals on a coconut oil diet was 45% in cebus monkeys and 34% in woolly monkeys. Comparable values were much lower in animals on a corn oil diet. When cholesterol was added to either diet, the percentage and absolute amount in the S_f 12–400 class was low. The authors make no statement concerning a possible relationship between the alteration in lipoprotein distribution and concentration, and the observed incidence of aortic atherosclerosis, but they do indicate that a significant correlation was found between increased plasma cholesterol concentration and aortic atherosclerosis in these animals. Presumably, the lipoprotein distribution shift was related to plasma cholesterol increase. No data were given for normal concentrations of HDL nor for the effect of diet upon this lipoprotein class. The effect of cholesterol feeding on lipoprotein distribution in this study does not appear to be quite the same as described above for squirrel monkeys.

CLARKSON and LOFLAND [unpublished] examined the effect of age and sex on lipoprotein distribution in a group of eight cholesterol-fed cebus monkeys[3]. No sex difference was seen for lipoprotein distribution between two young males and two young females (less than 18 months old) with serum cholesterol concentrations near 325 mg/dl. In adult animals (about five years old) fed the same diet, males had an average serum cholesterol concentration of 445 mg/dl which was higher than that of young animals and higher than the average of 305 mg/dl in adult females. The percentage distribution of cholesterol among the lipoproteins separated by ultracentrifugation was apparently related to age and sex. An average of 55% of serum cholesterol was present in LDL in young male and female animals, whereas 64% was present in LDL of adult males and females. In adult males 15% of plasma cholesterol was present in VLDL, and 20% was in HDL, whereas in adult females 6% was present in VLDL and 30% was present in HDL. Although the numbers of animals in these studies were small, the results suggest that further work may elucidate the mechanisms by which differences in lipoprotein distribution occur. The percentage distribution of lipoprotein cholesterol in these animals was not dissimilar to that in human beings (table I). It is interesting to note that in a study of the effect of age and sex on coronary artery atherosclerosis in cholesterol-fed cebus monkeys, BULLOCK *et al.* [10] found an increased incidence of the disease which was age related but apparently not dependent on sex.

Information on composition and metabolism of lipoproteins in New World monkeys is limited. ILLINGWORTH and PORTMAN [30] have reported that phospholipids, including lecithin and sphingomyelin, readily exchange between LDL and HDL of squirrel monkeys, as was also found to be the case in rabbits and in human beings [31]. Interpretation of these data is made

[3] Appreciation is expressed to Drs. ROBERT FURMAN and PETER ALAUPOVIC of the Oklahoma Medical Research Foundation for carrying out these analyses.

difficult since the value for the amount of lecithin and/or sphingomyelin in either HDL or LDL was not published.

The question of whether some of the apolipoproteins of VLDL are transferred to HDL during VLDL metabolism has been studied in squirrel monkeys [24, 29, 57]. In the study of SCHONFELD et al. [57], human VLDL were injected into squirrel monkeys, and the appearance in plasma of the apolipoproteins of human VLDL was followed by immunoassay using antisera specific for individual apolipoproteins of human VLDL. Transfer of specific human apolipoproteins onto squirrel monkey HDL was documented. Together with the earlier report of this group [24], in which human VLDL were found to yield LDL after catabolism in the squirrel monkey, these data indicate that in the squirrel monkey, metabolism of VLDL may proceed in the manner postulated for human beings. In this scheme, it is proposed that during or immediately after the hydrolysis and removal of VLDL triglycerides by the enzyme lipoprotein lipase, plasma LDL are formed which contain the apoLDL of VLDL, and the apoLP-glu and apoLP-ala of VLDL are transferred to the HDL fraction in plasma. It is important to note, however, that although no chemical compositions were determined, the polyacrylamide disc gel electrophoresis patterns of apolipoproteins of squirrel monkey HDL suggest that there are basic apolipoprotein differences between human and squirrel monkey HDL.

Additional information on the apolipoprotein patterns of squirrel monkey lipoproteins was presented by ILLINGWORTH and PORTMAN [29]. These authors separated the apolipoproteins of squirrel monkey HDL and VLDL on polyacrylamide gel electrophoresis and compared the patterns to those of human HDL and VLDL. The patterns for both VLDL and HDL appeared quite different between squirrel monkeys and humans. In addition, the authors injected radioactive leucine into squirrel monkeys to study the appearance of apoLDL in plasma LDL by transfer from VLDL. Their results suggested that LDL is formed during VLDL catabolism. Thus, their conclusions, based on studies of squirrel monkey lipoproteins *per se* were similar to those of SCHONFELD et al. [57] and GULBRANDSEN et al. [24] using human lipoproteins in squirrel monkeys.

Some aspects of the metabolism of lipoproteins in squirrel monkeys were studied by GREENE and co-workers [23, 41]. Juvenile squirrel monkeys were fed diets which contained cholesterol (0.75 mg/Cal) and in which the dietary fat was either butter or safflower oil, both fed as 40% of calories. After 12 months, lipoprotein cholesterol distribution was determined by heparin-manganese precipitation. The fatty acid composition of lipoprotein phos-

Table VI. Distribution of lipoprotein cholesterol in plasma by heparin-manganese precipitation (means $\pm$ SE)

Phenotype and diet	Cholesterol concentration, mg/dl			
	total	LDL + VLDL	HDL	(%)[1]
Hyperresponder[2] butter	412 ± 29	280 ± 38	133 ± 12	(33)
Hyporesponder[3] butter	272 ± 21	174 ± 18	99 ± 6	(37)
Hyperresponder safflower	327 ± 18	213 ± 13	114 ± 8	(35)
Hyporesponder safflower	245 ± 9	144 ± 9	101 ± 7	(41)

[1] Percent of total cholesterol in HDL.
[2] Progeny of hyperresponder parents.
[3] Progeny of hyporesponder parents.

Table VII. Average fatty acid compositions of lipoprotein phospholipids

Lipoprotein fraction, g/ml	Diet group[1]	Fatty acids[2] (% of total fatty acids)[3]				
		16:0	18:0	18:1	18:2	20:4
d <1.006	safflower	30	24	7	39	0[4]
	butter	47	27	16	6	4
d 1.019–1.063	safflower	33	25	7	29	6
	butter	38	25	23	8	6
d 1.063–1.21	safflower	27	19	12	32	10
	butter	35	24	26	10	6

[1] Includes both hypo- and hyperresponders.
[2] Fatty acids designated by carbon chain length:number of unsaturated bonds.
[3] Percent of these five fatty acids.
[4] Not determined.

pholipids and cholesterol esters was determined by gas-liquid chromatography. During the last 84 days of the experiment, the turnover rate of plasma cholesterol was determined following a single intravenous pulse label of cholesterol-4-^{14}C.

Table VI shows the distribution of the plasma lipoprotein cholesterol. The monkeys were of different phenotypes, depending on the responsiveness of the parents to cholesterol feeding, and the results indicate that there are both dietary and phenotypic differences among the groups. In these studies, when plasma cholesterol concentrations were elevated, the elevation was primarily in the LDL + VLDL fraction.

Table VIII. Average fatty acid composition of lipoprotein cholesteryl esters

Lipoprotein fraction, g/ml	Diet group[1]	Fatty acids[2] (% of total fatty acids)[3]			
		16:0	18:0	18:1	18:2
d <1.006	safflower	10	2	15	72
	butter	29	10	56	5
d 1.019–1.063	safflower	14	4	20	62
	butter	23	6	52	18
d 1.063–1.21	safflower	10	2	23	65
	butter	24	5	54	17

[1] Includes both hypo- and hyperresponders.
[2] Expressed as chain length:number of double bonds.
[3] Percent of these four fatty acids.

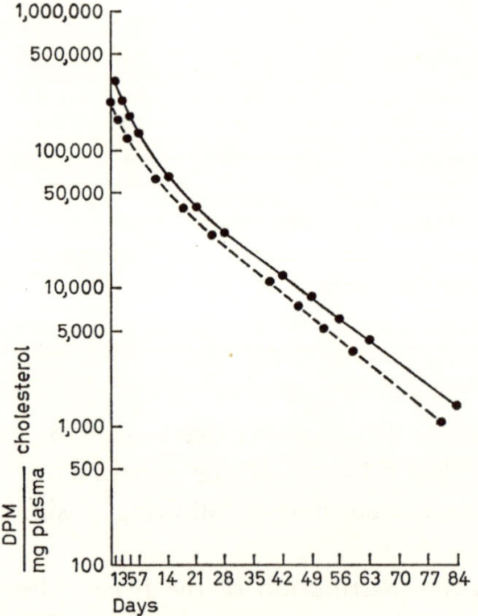

Fig. 2. A computer-fitted plot of plasma cholesterol specific activity versus time after an injection of radioactive cholesterol. The radioactive cholesterol was administered intravenously to ten monkeys on a semipurified diet containing safflower oil (solid line) and ten monkeys on a diet containing butter (broken line). Both diets contained 0.75 mg of cholesterol per cal, and the fat in the diet was 40% of the calories.

Table VII shows the fatty acid composition of plasma lipoprotein phospholipid, and table VIII shows the fatty acid composition of lipoprotein cholesteryl esters. Major compositional differences were seen, especially in the 18:2 fraction of the cholesteryl esters, as would be expected when fats are fed which differ so strikingly in the degree of unsaturation. The fatty acid pattern of the plasma lipoproteins was influenced by that of the diet, suggesting that the lipoproteins may differ in size and in steric configuration.

In this context, it is of interest that isotopically determined turnover rates of plasma cholesterol were very similar. The average disappearance curves obtained on all the animals on either diet are shown graphically in figure 2. Turnover times in days for the four groups were, respectively, hyperresponders (butter), 67 days; hyporesponders (butter), 61 days; hyperresponders (safflower oil), 60 days; hyporesponders (safflower oil), 59 days. Thus, in spite of the chemical (and possibly steric) differences in the plasma lipoproteins, the overall rate of catabolism of plasma cholesterol occurred at approximately the same rate.

C. Old World Monkeys

1. Lipoprotein Distribution in Rhesus Monkeys

The distribution of plasma lipoproteins has been more completely studied in rhesus monkeys *(Macaca mulatta)* than in any other species of Old World monkeys, having been examined under several dietary conditions, some of which are known to produce atherosclerosis. A variety of methods has been used so that techniques can also be compared.

Several reports have appeared which describe lipoprotein distribution in rhesus monkeys fed monkey chow. SCANU *et al.* [55] have reported the total lipoprotein distribution in juvenile males with plasma cholesterol concentration less than 150 mg/dl is 6.5% VLDL, 22% LDL, 45% HDL$_2$, and 23% HDL$_3$. In these studies, lipoproteins were isolated by sequential ultracentrifugation. VLDL were isolated as the d < 1.006 fraction; LDL, d 1.019–1.063; HDL$_2$, d 1.063–1.125; and HDL$_3$, d 1.125–1.21. LEE and MORRIS [36] have reported the cholesterol distribution among lipoproteins for two normal adult male rhesus monkeys with an average serum cholesterol of 165 mg/dl to be 1% in VLDL (d < 1.006), 1% in LDL$_1$ (d 1.006–1.019), 37% in LDL$_2$ (d 1.019–1.063), and 60% in HDL (d 1.063–1.21). Analytical ultracentrifuge data on serum lipoprotein distribution in a pooled sample obtained from young male rhesus monkeys is shown as part of table IX. Almost all of the LDL were S$_f$ 0–12 material and represented 34% of total plasma lipoproteins.

Table IX. Serum lipoprotein distribution of some Old World monkeys[1]

Genus/species	Diet[2]	Chol.	Serum concentration, mg/dl				
			VLDL S_f 20–400	LDL S_f 12–20	S_f 0–12	HDL₂	HDL₃
Macaca							
mulatta	m. chow	154	2	3	266	192	254
(rhesus)				231			446
	2.0 mg/cal	882	520	767	562	23	102
				1,849			125
arctoides	0.05 mg/cal	184	35	85	114	161	250
(stumptailed)	(A)			233			411
	1 mg/cal	515	316	429	275	22	134
	(B)			1,021			156
Malaysian							
fascicularis	m. chow	121	6	5	134	139	177
(cynomologus)				145			316
Philippine	m. chow	135	7	6	187	178	178
fascicularis (cynomologus)				200			356
nemestrina	m. chow	129	13	9	134	238	122
(pigtailed)				156			360
Cercopithecus							
aethiops	1 mg/cal	367	100	208	408	79	225
(African green)	(B)			716			304

[1] Unpublished data of Lofland, H. B., jr. *et al.* Analytical ultracentrifugation kindly performed by Dr. Frank Lindgren. Analyses were performed as described in the legend to table V.

[2] Diet designation is described in the legend to table IV.

In contrast to the data of Scanu *et al.* [55], the HDL₂ were 28 % of the total lipoprotein and were present in lower concentration than HDL₃ which were 38 % in this study. The results from these two laboratories agree in showing that the predominant lipoprotein class present in normal male rhesus monkeys fed monkey chow is HDL. In contrast, VLDL were found to be present in only very small amounts. This distribution is clearly different from that of normal human beings (about 30 and 10 % of lipoprotein cholesterol in HDL and VLDL, respectively) but the human diet is also quite different.

Table X. The effect of cholesterol feeding on the relative size and cholesterol distribution among plasma lipoproteins of *M. mulatta*[1]

Animals	N	Mean cholesterol, mg/dl	Lipoprotein	Cholesterol		Ve/V_T[2]
				%	mg/dl	
Control diet	2	165	VLDL	1.7	3	0.309
			IDL	15.4	25	0.506
			LDL	30.8	51	0.595
			HDL	50.1	83	0.760
Test diet	2	836	VLDL	7.6	64	0.321
			IDL	12.7	106	0.438
			LDL	74.4	622	0.555
			HDL	3.3	28	0.788[3]

[1] The plasma lipoproteins were separated as described in the legend for figure 1. Individual peaks were combined and the cholesterol in each was determined and is expressed here as a percent of the total of all cholesterol eluted from the column.

[2] The relative size of the lipoprotein fraction was determined by measuring the elution volume (Ve) at the peak (fig. 1) and dividing it by the total column volume (V_T).

[3] A double peak was consistently seen for HDL, and the Ve/V_T is given for the highest portion of the peak (fig. 1 b).

RUDEL and CLARKSON [unpublished] have examined the distribution of plasma cholesterol among lipoproteins separated on agarose column chromatography in two adult male rhesus monkeys fed a diet with 45% of the calories as fat, and with 0.05 mg cholesterol/Cal. This diet more closely resembles the 'normal' human diet, although the amount of cholesterol is still lower in the diet of the monkeys. The average plasma cholesterol in these monkeys was 165 mg/dl. The lipoproteins were separated using a Bio-Gel A-15m (4%) agarose chromatography column according to the method of RUDEL et al. [54]. A size distribution of lipoproteins for one of the animals is shown in figure 1a. The three major fractions seen in human plasma, i.e. VLDL, LDL, and HDL, were present, as well as a fourth population having a size intermediate between VLDL and LDL. Since the size of the latter is intermediate, it has been termed 'IDL'. After combining the material under each peak, the cholesterol distribution was determined and is shown in table X. The mean percentage distribution for plasma cholesterol in these animals was 1.7% in VLDL, 15.4% in IDL, 30.8% in LDL, and 50.1% in HDL. The fat and the small amount of cholesterol in the diet did not appear to change the lipoprotein cholesterol distribution when compared to that reported by LEE and MORRIS [36], although the percent in HDL may be slightly lower. Only small numbers of animals are represented in these experiments, but the values appear to be in general agreement between laboratories, and are likely to be representative.

MORRIS and FITCH [46] have reported naturally-occurring hyperbetalipoproteinemia in two male rhesus monkeys. These animals maintained serum cholesterol concentrations of 449 and 495 mg/dl, respectively, which is more than three times higher than in the control animals (mean of 144 mg/dl). The diet contained less than 1 mg of digitonin-precipitable sterol per 100 g. The plasma lipoproteins of both monkeys were separated at d 1.063 by ultracentrifugation and by heparin-manganese precipitation. Using both techniques, more than 90% of the plasma cholesterol was present in the fraction which contains LDL, whereas only 45% was present in this fraction in the control animals of this study. More recently, LEE and MORRIS [36] have fractionated the plasma lipoproteins of these two hyperbetalipoproteinemic monkeys by preparative ultracentrifugation. The animals had cholesterol concentrations of 725 and 705 mg/dl, respectively. The cholesterol distribution among lipoproteins was, respectively, 15 and 26% in VLDL (d < 1.006), 24 and 20% in LDL_1 (d 1.006–1.019), 57 and 50% in LDL_2 (d 1.019–1.063), 3 and 3% in HDL (d 1.063–1.21). The rather high values in VLDL and LDL were interpreted as a lipoprotein disorder similar to the human type IIb as described by FREDRICKSON and LEVY [19]. It is interesting to note that this percentage distribution is not unlike that seen in cholesterol-fed monkeys by LOFLAND *et al.* [unpublished observations, table IX] and RUDEL and CLARKSON [unpublished observations, table X].

Lipoprotein distribution between HDL and VLDL + LDL has been determined in rhesus monkeys using anionic polysaccharide-divalent metal precipitation methods. MORRIS and GREER [45] have examined the change in lipoprotein distribution at varying intervals after animals were shifted from a monkey chow diet to a diet which contained cholesterol. These workers used heparin-Mn^{++} to precipitate LDL + VLDL. The HDL cholesterol concentration increased from an average of 72 mg/dl at zero time to an average of 119 mg/dl between 7 and 28 days of the study. It then decreased, even though the plasma total cholesterol concentration continued to increase. HDL cholesterol eventually dropped significantly below its concentration at zero time. Interestingly, there were two monkeys in the group of 13 that responded differently. The concentration of HDL cholesterol increased upon cholesterol feeding in these animals, and reached a value of almost 150 mg/dl. The level of HDL cholesterol then remained elevated and did not

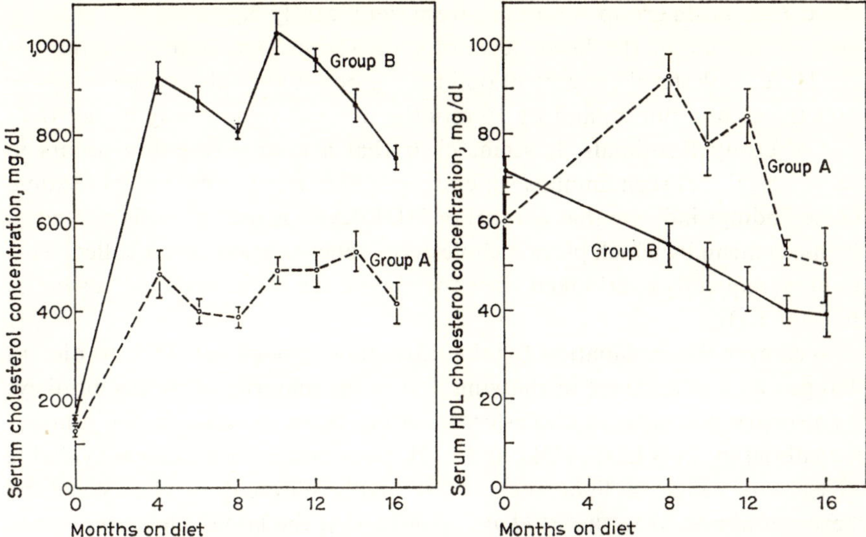

Fig. 3. The response of serum cholesterol and HDL cholesterol to test diet B (table IV) in young male hyporesponding (group A) and in young male hyperresponding (group B) *M. mulatta.* The 13 animals in group A and 19 animals in group B were ingesting monkey chow until day zero, at which time the diet was changed. Fasting blood samples were collected and HDL cholesterol was determined after heparin-manganese precipitation of the VLDL + LDL. Bars represent standard effort.

decrease for the 71 days of the experiment. At the same time, total plasma cholesterol concentrations in these two animals increased only half as much as in the other animals. The mechanism which controls these differences is unknown, but it seems to be more than an inability to form more LDL in response to cholesterol feeding, since the HDL also reacts in a unique way.

LOFLAND and CLARKSON [unpublished] have also examined the prolonged changes in distribution of lipoprotein cholesterol which occur after cholesterol feeding (fig. 3). All animals in this study were fed the same semi-purified diet, which consisted of 1.0 mg cholesterol per Cal, with 45% of the calories as lard. The 13 animals in group A responded to the diet with an increased plasma cholesterol concentration which averaged about 450 mg/dl during the 16 months of the study, whereas the 19 animals in group B responded with significantly higher values averaging about 890 mg/dl, nearly twice that of group A. Heparin and manganese were used to precipitate the VLDL + LDL in the serum, and the level of HDL cholesterol remaining in serum was monitored. Throughout most of the study the HDL cholesterol concentrations in group A animals were significantly higher than in group B animals in spite of the lower level of total serum cholesterol in group A. The HDL cholesterol concentration in the group A animals dropped steadily after the eighth month, and by 16 months was not significantly higher than that of group B animals. It seems clear that a unique response occurs in plasma HDL between animals which react differently to cholesterol feeding. These findings indicate that control of HDL levels is part of, or is related to the mechanism by which plasma cholesterol concentration is controlled. This point is frequently overlooked since the magnitude of the response is smaller than for LDL.

Whatever the explanation for the intriguing response of HDL to dietary changes, all studies agree in showing that in the majority of rhesus monkeys the primary response to cholesterol feeding is an increase in the amount of material in the VLDL, IDL, and LDL fractions. Information is available on the size range of lipoprotein most highly elevated in cholesterol-fed rhesus monkeys. ARMSTRONG et al. [1] found that the largest elevation was in the S_f 12–20 fraction, with large increases also in the S_f 0–12 and 20–100 size ranges. No rise occurred in the S_f 100–400 fraction. The data in table IX show the analytical ultracentrifuge results of LOFLAND et al. [unpublished] to be in general agreement with ARMSTRONG's findings. In addition to the large increase in concentration of LDL, this data also shows an absolute decrease in the concentration of HDL, with a proportionally larger decrease in HDL_2 than in HDL_3.

RUDEL and CLARKSON [unpublished] have also demonstrated a major alteration in size distribution of lipoproteins in cholesterol-fed adult male rhesus monkeys (fig. 1b). A large increase in the relative amount of material in the VLDL, IDL and LDL region was found; each of these lipoprotein fractions was found to have β mobility on agarose electrophoresis. Proportionally, a large decrease in HDL occurred, and the skewed curve for HDL indicates heterogeneity of size. Table X shows the extent of change between test diet fed animals and controls. The percent of plasma cholesterol in LDL doubled after cholesterol feeding, whereas the percentage of cholesterol in HDL decreased by a factor of 10; thus, the change in LDL/HDL ratio was massive. In terms of absolute cholesterol concentration, the order of relative increase due to cholesterol feeding was VLDL > LDL > IDL. HDL decreased in absolute concentration. The last column of table X indicates the relative size of the lipoprotein fractions. Larger lipoproteins elute from the column first and have smaller Ve/VT values. Using Bio-Gel A-15m (4% agarose), the VLDL appear essentially at the void volume of the column and differences

Table XI. Percentage composition of serum LDL and HDL of *M. mulatta*

Lipoprotein	Chemical		Lipid			
	protein	lipid	trigly-ceride	cholesteryl ester	free cholesterol	phospho-lipid
LDL[1]	22.9	77.1	12.2	42.5	10.1	30.0
LDL[2]	22.9	77.1	13.5	46.3	9.0	31.2
HDL_2	45.0	55.0	6.7	33.3	9.4	50.5
HDL_3[3]	53.0	47.0	6.1	28.3	8.0	57.6

[1] From LEE and MORRIS [36]. Values are means for LDL isolated from five normal adult males fed monkey chow.

[2] From LEE and MORRIS [36]. Values are means for several determinations performed on two type IIb hyperlipoproteinemic adult males fed monkey chow.

[3] From SCANU *et al.* [55]. Values are from juvenile males fed monkey chow.

in size of this fraction would not be detected. IDL and LDL of cholesterol-fed animals were clearly larger than IDL and LDL, respectively, of control animals. This difference in size is quite marked and suggests that the composition of these lipoproteins may also be abnormal in cholesterol-fed animals. The next section describes such a difference in composition (table XII).

LASSER and ALLEBAUGH [35] have recently reported a study of the size distribution of lipoproteins in cholesterol fed rhesus monkeys and concluded that the primary change caused by cholesterol feeding was the appearance of a LDL of lower than normal density. These results appear to be consistent with the size change shown in figure 1b. Thus, the results from several laboratories agree in showing an increase in concentration of all subfractions of LDL and a decrease in HDL during long-term cholesterol feeding of rhesus monkeys. It is important to emphasize that the increase in LDL is not simply an increase in normal LDL, but is instead an increase in abnormally large LDL and IDL, and in VLDL. It is possible that normal LDL may not increase in concentration at all. In contrast, the spontaneous hyperbetalipoproteinemic rhesus monkeys of MORRIS and co-workers [36, 46] have an absolute increase in the concentration of apparently normal LDL.

The effect of different types of dietary fat on lipoprotein distribution in three groups of cholesterol fed rhesus monkeys was reported by FRASER *et al.* [17]. Diets consisted of powdered monkey chow plus 25%, by weight, of corn oil, coconut oil, or milk fat and 2% by weight of crystalline cholesterol. The only significant effect that appeared to be attributable to type of fat was that the HDL cholesterol was lower in concentration in animals fed corn oil for 30 days than in animals fed the other types of fats. In animals fed each fat there was an average of 75% of plasma cholesterol present in LDL (S_f 0–20). It is somewhat disturbing that an average of only 82% of plasma cholesterol was recovered in the isolated lipoprotein fractions. The authors do not comment on where the loss might occur. A selective loss of almost 20% somewhere in the procedure could influence the interpretation of the results.

These same authors also studied the lipoprotein distribution in the thoracic duct lymph of their groups of animals. Lymph duct cannulations were performed 6 h after a meal. About 8 ml of lymph was collected for 2 h after cannulation. Lymph was then placed in the preparative ultracentrifuge and separated into chylomicra (S_f > 400), VLDL, LDL, and HDL, and the distribution of cholesterol among fractions was determined. The chylomicra contained more cholesterol than any other fraction in monkeys fed all types of fats. The average distribution of lipoprotein cholesterol for all diets was 50% in chylomicra, 24% in VLDL, 21.5% in LDL, and 4.5% in HDL with the mean lymph cholesterol concentration being 250 mg/dl. The effect of the type of fat was not readily apparent in most cases, although, on the average, the corn oil plus cholesterol group contained more chylomicra and less LDL and HDL cholesterol, as a percent of the total, than animals on the other diets.

2. Chemical Composition and Metabolism of
Lipoproteins in Rhesus Monkeys

The chemical composition of LDL and HDL in male rhesus monkeys fed monkey chow has been reported [36, 55], and is shown in table XI. In LDL, the percentage of cholesteryl ester is lower and that of triglyceride is higher than for LDL of human beings (table III). SCANU et al. [55] separated HDL into an HDL_2 (d 1.063–1.125) and HDL_3 (d 1.125–1.21) fraction. The percentage composition of these HDL subfractions is much like those in human beings. The composition of VLDL in monkey chow fed rhesus monkeys has not been reported.

SCANU et al. [55] have recently characterized many of the physical properties of HDL_2 and HDL_3 of monkey chow-fed juvenile male rhesus monkeys fed monkey chow and have found that in most instances the rhesus HDL are indistinguishable from human HDL. When these workers analyzed the structure of the apolipoproteins of HDL, significant differences between rhesus monkeys and human beings were identified [15]. Two major poly-

Table XII. Effect of cholesterol feeding on plasma lipoprotein composition in *M. mulatta*[1]

Diet	Lipo-protein	Chemical		Lipid			
		protein	lipid	trigly-cerides	choles-teryl ester	free choles-terol	phospho-lipid
Control (A)	VLDL	13.1	86.9	46.4	25.3	10.0	18.2
	IDL	18.6	81.4	9.0	51.4	11.0	28.4
	LDL	18.3	81.7	5.1	51.7	12.7	30.3
	HDL	40.5	59.5	4.0	35.4	7.6	53.0
Cholesterol-added (B)	VLDL	11.3	88.7	9.2	66.5	7.4	16.8
	IDL	13.0	87.0	3.0	57.0	12.4	27.5
	LDL	12.2	87.8	1.4	57.3	12.5	28.7
	HDL	48.4	51.6	2.4	34.7	5.8	57.1

[1] Unpublished data of RUDEL and CLARKSON. The same animals and diets are described in the legend to figure 1. The values are means for duplicate determinations on lipoproteins isolated from two animals.

peptides were identified in rhesus HDL, which had similar immunochemical spectral properties (circular dichroism), and molecular weights to apoLP-gln I and apoLP-gln II of human HDL. However, the two major apolipo-proteins from monkey HDL contained amino acid differences which distinguished them from their human counterpart. The rhesus monkey counterpart to human apoLP-gln I was very similar in amino acid composition except that it contained more glutamic acid and less methionine and arginine. The rhesus monkey counterpart to human apoLP-gln II contained no cysteine, but instead was found to contain a serine in position 6, which is the position occupied by cysteine in human apoLP-gln II. The absence of cysteine in rhesus monkey apoLP-gln II, called fraction IV by EDELSTEIN *et al.* [15], accounts for its inability to form dimers cross-linked by disulfied bonds, which occur in human apoLP-gln II [16].

The studies of SCANU *et al.* are important contributions in defining the structural relationships between human and rhesus monkey lipoproteins. More such information is needed for other classes of lipoproteins as well as for HDL, so that we can define the relationships between atherosclerosis and lipoprotein structure in monkey models for the human disease. For instance, the work of MORRIS and GREER [45] has shown a differential response in rhesus monkey HDL to cholesterol feeding. Perhaps complete structural analysis of HDL in animals with a high response to dietary cholesterol, compared to HDL of animals which respond only moderately, could demonstrate differences which would help explain the mechanism for the response.

The work of LEE and MORRIS [36] also is relevant to this discussion. The chemical structure of LDL in two spontaneous hyperlipoproteinemic rhesus monkeys, corresponding to the human type IIb phenotype as described by FREDRICKSON and LEVY [19] in human beings, was studied and compared with material obtained from normal monkeys fed the same monkey chow diet. The chemical composition of the LDL from the two type IIb animals was not significantly different than in the control animal LDL (table XI), although the data suggested that the cholesteryl ester percentage may have been higher in the LDL of type IIb animals. The classes of phospholipid present in LDL of the type IIb animals were present in the same distribution as for control LDL. Lecithin was about 70%, sphingomyelin about 10%, and lysolecithin, phosphatidylethanolamine, and phosphatidylserine + inositol were each about 5% of the total phospholipid. The distribution of fatty acids within triglycerides, cholesteryl esters, and phospholipids also did not differ significantly between control and type IIb monkey LDL. The molecular weights of the LDL, determined by sedimentation equilibrium analysis, were 3.3×10^6 daltons for both control and type IIb monkeys. The one area of some disagreement was in the amino acid composition of the delipidated apolipoprotein, although individual amino acids did not vary more than a few percent on the average. The authors concluded that the LDL of the type IIb hyperlipoproteinemic monkeys was structurally very similar and probably identical to normal LDL. Such a result would suggest that the primary defect in these animals was in the synthesis or catabolism of LDL *per se* rather than in the production of an abnormal lipoprotein which was metabolically distinct as a result of structural abnormalities.

RUDEL and CLARKSON [unpublished] compared the composition of the four lipoprotein fractions separated by agarose column chromatography and cholesterol-fed rhesus monkeys in controls (table XII). The compositions of LDL and HDL in the control animals appear to be almost identical to those found in human beings (table III). In this case, the percent of LDL present as cholesteryl ester and triglyceride for rhesus monkeys fed 45% of calories as fat are more comparable to the human LDL values than are those for rhesus monkeys fed monkey chow (table XI). It will be interesting to see if comparisons between larger groups of monkeys on different low cholesterol, high fat diets will confirm this difference in chemical composition of lipoproteins. Such findings in monkeys would carry important implications for human beings since diet modification is presently a recommended procedure for individuals known to be predisposed to atherosclerosis. Another point of interest is the similarity in composition between IDL and LDL. The only apparent difference is that IDL contain more triglyceride. It is unclear at this time to which analytical ultracentrifuge fraction IDL from the agarose column would correspond, although preliminary results in control animals indicate that much of the material is in the S_f 12–40 range.

The data in table XII show that LDL and IDL of cholesterol-fed rhesus monkeys – which eluted from the agarose chromatography column as larger particles than those of control monkeys (table X) – contained a higher percentage of cholesteryl ester than LDL and IDL of control animals. The results of LEE and MORRIS [37] for LDL isolated by preparative ultracentrifugation are in agreement with these findings. Table XII also shows that the VLDL from cholesterol-fed rhesus monkeys were quite different in composition than those of control animals in that they contain primarily cholesteryl ester and surprisingly little triglyceride. On the other hand VLDL from cholesterol-fed animals contained similar but not identical amounts of free cholesterol, phospholipid, and protein as VLDL from control animals. This may be significant if one considers these findings in terms of the 'core and coat' concept of very low density lipoprotein structure suggested for chylomicra by ZILVERSMIT [67]. The coat of the large lipoprotein particle is made up primarily of free cholesterol, phospholipid, and protein, and covers the surface of the particle. The large core inside this coat is primarily neutral lipid and in ZILVERSMIT's [67] study was principally triglyceride, although most of the cholesteryl ester and some free cholesterol of the original particle was also present in the core. It would appear possible that, in the cholesterol-fed rhesus monkey, the core of the VLDL is primarily cholesteryl ester. Since the percentage of phospholipid, free cholesterol, and protein is comparable in control and cholesterol-fed monkey VLDL, the amount of coat material may be similar, and the size of the particles may also be expected to be similar.

If anything, the data indicate that the amount of the surface material in the cholesterol-fed monkey VLDL is less than in control animals. Thus, the population of particles would be larger (less surface area per volume) if the core and coat concept can be applied in this case. Such findings lead to speculation about lipoprotein metabolism, which is presented here in the hope of stimulating further research into this area. SHORE and SHORE [60] have demonstrated that in human beings, subpopulations of VLDL can be isolated from plasma which contain different apolipoprotein components, i.e. the number and relative amounts of the nine different apolipoprotein bands identified by polyacrylamide gel electrophoresis varied between VLDL subfractions eluted from a concanavalin A-Sepharose affinity chromatography column. It is temping to postulate that the rhesus monkey fed large amounts of cholesterol produces a cholesteryl ester-rich VLDL which then accumulates in plasma since it cannot be cleared by the same mechanism as triglyceride-rich VLDL. In fact, it is possible that the metabolic products of such VLDL would also be different. This hypothesis might help to explain why the IDL and LDL of cholesterol-fed rhesus monkeys are abnormally large and contain high amounts of cholesteryl ester.

Another quite different situation has been described in rhesus monkeys in which the metabolism of lipoproteins is abnormal. KERR *et al.* [32] have reported a hyperchylomicronemia which was produced in infant and adolescent rhesus monkeys by feeding an excess of L-histidine in the formula diet. Pair-fed control animals did not receive excess histidine and were normal. The hyperchylomicronemia was shown to persist for several months. The β-lipoproteins, as identified by electrophoresis, were not elevated, and α-lipoproteins appeared to be decreased in concentration in the histidine fed animals. Chylomicra made up some 40–90% of the total plasma lipoproteins 4 h after feeding, and the range only decreased to 25–75% of the total plasma lipoproteins after a 24-hour fasting period. Triglyceride concentrations ranged from 550 to 3,400 mg/dl in the affected animals. Although the mechanism by which elevated dietary histidine caused this effect in rhesus monkeys is not clear, the authors speculated that there was probably a defect in the clearing of chylomicra from plasma. The syndrome, as it is described, has many of the same features of human type I hyperlipoproteinemia [21], and may be an important model for this disease.

3. Lipoprotein Distribution in Other Macaques

The lipoprotein distribution in the serum of young male pigtailed macaques *(M. nemestrina)* fed monkey chow has been examined by analytical ultracentrifugation. The results of this study are shown in table IX. The distribution appears to be similar to that of other monkey chow-fed macaques, with LDL containing primarily an S_f 0–12 component which represents about 30% of total plasma lipoprotein. HDL is thus about 70% of the total, and HDL_2 is present in about twice the concentration of HDL_3. As in the rhesus monkeys, the absolute amount of lipoprotein and the distribution among lipoproteins is different from that of human beings, but so is the diet.

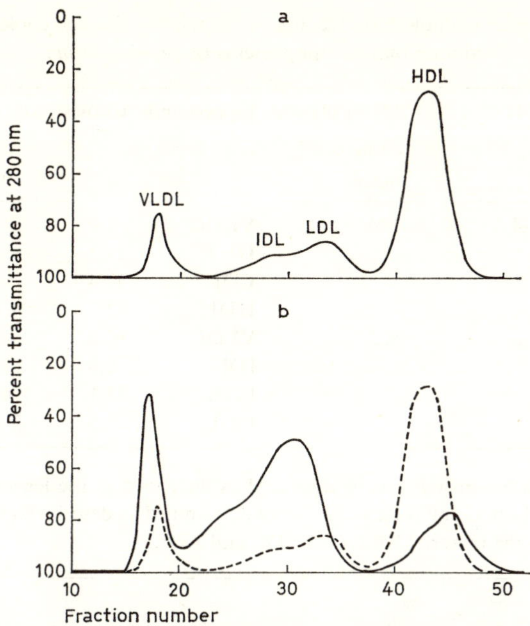

Fig. 4. Lipoprotein size distribution in *M. nemestrina* as determined by agarose column chromatography. Blood collection and lipoprotein separation were carried out as described in the legend of figure 1. In *a* is shown the separation for a representative young male pigtailed macaque with a plasma cholesterol concentration of 120 mg/dl, which was fed control diet A as described in table IV. In *b*, the solid line represents the separation of plasma lipoproteins of a pigtailed macaque (plasma cholesterol 750 mg/dl) fed test diet B as described in table IV. The dotted line in *b* represents the control animal pattern of *a* for comparison.

RUDEL *et al.* [unpublished] have studied the lipoprotein distribution in ten young male *M. nemestrina* fed a diet more like the normal human diet, in that it contained 45 % of the calories as fat. The lipoproteins from four animals on a control diet (0.05 mg cholesterol/cal), were separated using the agarose column chromatography method of RUDEL *et al.* [54]. Figure 4a shows the lipoprotein size distribution of a typical control animal. The distribution of cholesterol among lipoproteins in these animals (table XIII) was similar to that in rhesus monkeys on the same diet (table X), although slightly more VLDL and IDL cholesterol was found in *M. nemestrina*, and less was present in HDL. On the average, the agarose electrophoresis patterns of plasma lipoproteins of *M. nemestrina* also showed more pre-β material than for *M. mulatta*. When a group of six *M. nemestrina* were fed 1.0 mg cholesterol/ cal for 24 months, the mean plasma cholesterol was 822 mg/dl, compared to

Table XIII. The effect of cholesterol feeding on the relative size and cholesterol distribution among plasma lipoproteins of *M nemestrina*[1]

Diet	N	Mean plasma cholesterol mg/dl	Lipoprotein	Cholesterol %	mg/dl	Ve/V$_T$
Control	4	114	VLDL	3.2	4	0.350
			IDL[2]	21.3	25	0.483
			LDL	33.4	39	0.605
			HDL	42.1	49	0.773
Test	6	822	VLDL	10.6	87	0.349
			IDL[2]	8.8	72	–
			LDL	75.1	617	0.541
			HDL	5.5	45	0.780

[1] The plasma lipoproteins were separated as described in the legend of figure 1. The cholesterol distribution and relative size were determined as described in table X.

[2] Represents the material between VLDL and LDL.

117 mg/dl for the control animals. The lipoprotein size distribution was dramatically changed (fig. 4b) in a manner similar but not identical to what was seen in the rhesus monkey. The 'LDL' peak of cholesterol-fed animals was shifted so that it was intermediate in size (Ve/V$_T$ = 0.541) to the IDL (Ve/V$_T$ = 0.483) and LDL (Ve/V$_T$ = 0.605) of control animals (table XIII). No indication of separate peaks of IDL and LDL was seen in the cholesterol-fed pigtailed macaques, which was unlike the observations in rhesus monkeys. In cholesterol fed *M. nemestrina*, the HDL always eluted as a symmetrical peak which was not the case for cholesterol-fed *M. mulatta.*

The cholesterol-induced alterations in size and relative cholesterol distribution among lipoproteins of *M. nemestrina* were as pronounced as it was in *M. mulatta.* In addition, the composition of the lipoproteins of *M. nemestrina* also appeared to change in similar fashion [RUDEL and PAULA, unpublished]. Most of the material in the VLDL and LDL of cholesterol-fed *M. nemestrina* migrated in the β region on agarose electrophoresis. One exception present in some of the cholesterol-fed animals was a band which migrated between the origin and β region; it was isolated as part of the VLDL fraction from column chromatography. It may be worthy of note that some of the pigtailed macaques fed cholesterol developed a form of micronodular cirrhosis, and the severity of the cirrhosis seemed to correlate with the presence of the additional lipoprotein band seen on electrophoresis [9]. Presently, we are unsure of the relationship, if any, between the alteration of lipoprotein size and distribution and atherosclerosis in *M. nemestrina,* since none of the cholesterol-fed animals has yet been necropsied. It may be of interest that RAYMOND *et al.* [53] have done isotopic cholesterol balance studies in cholesterol-fed *M. nemestrina* and found that these animals failed to completely inhibit endogenous cholesterol synthesis. The *M. nemestrina* is a species for which much promise for future studies is held.

The distribution of lipoproteins in *M. arctoides* (stumptailed macaque) has been studied by analytical ultracentrifugation (table IX) and by dextran sulfate precipitation (table XIV). The number of animals studied was very small, however, and our present information must be considered only as preliminary. The lipoprotein distribution, determined on the analytical ultracentrifuge, is shown in table IX for a single control male. HDL was present in highest concentration, and was 65% of the total plasma lipoprotein. The lipoprotein distribution in serum pooled from six cholesterol-fed stumptailed macaques was also analyzed by analytical ultracentrifugation (table IX). The pattern of distribution is quite similar to the pattern seen in cholesterolfed rhesus monkeys although absolute concentrations were lower. The S_f 12–20 and 20–400 fractions were the lipoprotein classes highest in concentration and HDL was the lowest. The data shown in table XIV are in partial disagreement with the analytical ultracentrifuge data of table IX, and illustrate another case in which the results of dextran sulfate precipitation do not agree with those of centrifugation. The mean percent of serum cholesterol in HDL, as determined after dextran-sulfate precipitation for samples of five stumptailed macaques fed a control diet, was only 23%, whereas the total lipoprotein in HDL was 65% as shown in table IX. In most species of macaques on low cholesterol diets (table XIV) the percent of cholesterol in HDL has been close to 50%, and the analytical ultracentrifuge data (table IX) support this finding. This appears to be another instance in which dextran sulfate precipitation gave falsely low HDL cholesterol concentrations.

The other species of macaques for which we have some knowledge of lipoprotein distribution is the crab-eating macaque, *M. fascicularis.* The lipoprotein distributions – as shown in table IX for pooled plasma samples from young male *M. fascicularis* – are very similar to monkey chow-fed *M. mulatta* and *M. nemestrina.* No difference is apparent between Malaysian and Philippine animals. The relative amounts of HDL_2 and HDL_3 appear to vary somewhat between the different species of macaques, but at present there is not enough information on a large enough population to decide if

Table XIV. Distribution of serum cholesterol among lipoproteins of some Old World monkeys

Species	Sex	N	Diet[1]	Mean Serum cholesterol, mg/dl	Mean serum trigly-ceride, mg/dl	HDL cholesterol %	mg/dl
Cercopithecus							
aethiops	mixed[2]	8	m. chow	141	–	61.0	86
	male[3]	6	m. chow	125	50	36.1	45
	male[4]	5	A	132	–	28.0	37
	male[4]	5	B	378	–	15.0	57
talapoin	male[5]	8	m. chow	133	48	37.0	49
	male[5]	3	A	159	32	62.0	99
	male[5]	5	B	746	34	8.0	60
	female[5]	10	m. chow	124	35	41.0	51
	female[5]	4	A	164	32	55.0	90
	female[5]	6	B	1,105	80	6.0	66
Erythrocebus							
patas	mixed[2]	10	m. chow	96	–	43.0	41
Macaca							
arctoides	male[4]	5	A	167	–	23.0	38
	male[4]	5	B	774	–	3.0	23
fascicularis	female[5]	19	m. chow	128	68	46.5	60
	male[5]	20	m. chow	129	28	43.5	56
fascicularis	male[6]	12	m. chow	115	21	44.0	51
(Malaysian)	male[6]	6	A	164	31	33.0	54
	male[6]	6	B	947	25	1.0	13
fascicularis	male[6]	12	m. chow	121	28	38.0	45
(Philippine)	male[6]	5	A	157	30	45.0	71
	male[6]	6	B	827	29	2.0	18
nemestrina	male[6]	12	m. chow	126	63	49.0	62
	male[6]	6	A	236	71	39.0	91
	male[6]	6	B	1,007	92	5.0	47

[1] Diets A and B are as described in the legend to table IV.
[2] SRINIVASAN *et al.* [63].
[3] LOFLAND *et al.* [unpublished].
[4] BULLOCK *et al.* [8].
[5] HAMM and CLARKSON [unpublished].
[6] CLARKSON and LOFLAND [unpublished].

these relatively small differences will be consistently present. For instance, as discussed above, the two available reports disagree about the ratio of HDL_2 and HDL_3 in the monkey chow-fed rhesus monkeys.

The results of studying lipoprotein cholesterol distribution by heparin-manganese precipitation in monkey chow-fed male *M. fascicularis* (table XIV) are reasonably consistent with the results obtained by analytical ultra-centrifugation. The average of 43.5% of plasma cholesterol in HDL for 20 males was not significantly different from the average of 46.5% in HDL for females. However, the average of 28 mg/dl for plasma triglyceride in males was significantly lower than the average of 68 mg/dl for females at the $p < 0.01$ significance level. This finding suggests that there may be differences in lipoprotein distribution or in structure that are not detected by this technique. It is perhaps noteworthy that the female *M. fascicularis* reported here are adult and in most cases menstruating regularly.

The data in table XIV illustrate the response of lipoprotein cholesterol distribution of *M. fascicularis* to a diet change from monkey chow to low cholesterol (A) and high cholesterol (B) diets. In this experiment, the values were determined after the animals were on the diet four months. Both Malaysian and Philippine *M. fascicularis* responded to the low cholesterol diet with an increase in serum cholesterol. The Philippine animals had an increase in HDL cholesterol concentration, whereas the Malaysian animals maintained the same HDL levels. In animals fed high cholesterol diets, the Malaysian animals appeared to reach a higher serum cholesterol level than Philippine *M. fascicularis*. The HDL cholesterol levels were low in both population groups. In contrast, the *M. nemestrina* shown in table XIV, and treated in the same way as the *M. fascicularis*, increased serum cholesterol concentrations significantly on the low-cholesterol diet when compared to monkey chow diet. The HDL cholesterol concentration was significantly higher on the low cholesterol diet although the percent of total serum cholesterol in HDL was lower. The triglyceride concentrations in male *M. fascicularis* of both races were lower than those in *M. nemestrina*, suggesting further differences in lipoprotein distribution and structure between these two species. The results available to date suggest that although lipoprotein cholesterol distribution among macaque species fed monkey chow diet is similar, the response of different species to exogenous (dietary) cholesterol is not. The relationship of lipoprotein distribution to atherosclerosis in these populations is unknown and currently under investigation.

Another species of macaque, *M. nigra*, is unique in that some animals of the species develop diabetes spontaneously [27]. A primate experimental model for diabetes offers unusual opportunities for investigating lipoproteins under conditions of abnormal carbohydrate metabolism. Although detailed lipoprotein analyses were not reported by Howard [27], lipoproteins were separated by agarose electrophoresis and densitometry was then used to estimate relative distribution. Pre-β-lipoproteins (VLDL) as well as triglyceride concentrations appeared elevated among adult diabetic monkeys.

4. Lipoprotein Distribution of Cercopithecus

The distribution of plasma lipoproteins in a small group of African green monkeys, *Cercopithecus aethiops*, has been examined by several methods. Rudel *et al.* [unpublished] separated lipoproteins by agarose column chromatography and found that the lipoprotein cholesterol distribution in six male African green monkeys was about 2% in VLDL, 15% in IDL, 50% in LDL, and 33% in HDL. These animals were fed a semipurified diet which contained 40% of calories as fat (either butter or safflower oil) and 0.7 mg cholesterol per Cal. The distribution was the same in six animals fed the same diet but with 0.06 mg cholesterol per Cal. The mean plasma cholesterol concentration of animals on the former diet was 261 mg/dl and on the latter diet was 184 mg/dl.

In the analytical ultracentrifugation study of Lofland *et al.* [unpublished], shown in table IX, a pooled serum sample with a cholesterol concentration of 367 mg/dl from five African green males had a lipoprotein distribution of 10% in VLDL, 60% in LDL, and 30% in HDL, with a total lipoprotein concentration of 1,020 mg/dl. This is a very similar pattern of distribution to the column separation mentioned above, and to the pattern in normal human males shown in table II (17% in VLDL, 49% in LDL, 34% in HDL with a total lipoprotein concentration of 827 mg/dl.) The data of table XIV show that in African green monkeys feeding 1 mg cholesterol per Cal lowered the percent of cholesterol in HDL, while the serum cholesterol concentration increased from 132 to 378 mg/dl. However, the absolute concentration of plasma HDL cholesterol increased some 50%. It should be pointed out that the diet shown in table XIV caused much higher increases in plasma cholesterol concentration in three species of macaques even though the absolute concentrations of HDL was reduced. The plasma cholesterol concentrations and lipoprotein distributions seen in these studies on African green monkeys closely approximate those of humans. It should also be pointed out that the morphology of the atherosclerotic lesions produced in the cholesterol-fed African green monkeys [8], was that of lipid-containing intimal lesions with fibromuscular caps, similar to those seen in human beings.

Observations by HAMM and CLARKSON [unpublished] on *Cercopithecus talapoin* are also presented in table XIV. The percent of serum cholesterol in HDL of monkey chow-fed males and females was about 40%, which was also the value observed for monkey chow-fed male *C. aethiops* by LOFLAND *et al.* [unpublished]. In the talapoin monkeys after five months on low cholesterol (0.05 mg/Cal) semipurified diets, there was more cholesterol in HDL than in LDL + VLDL. This result occurred after a small but significant rise in serum cholesterol concentration. In animals placed for five months on the same diet, but with 1 mg cholesterol per Cal, the serum cholesterol rose to very high levels and the percent and absolute amount of cholesterol in HDL decreased. The serum cholesterol concentrations in talapoin monkeys fed diet A for five months were more than twice as high as for African green monkeys fed this diet for six months (table XIV).

5. Studies of Lipoproteins in Baboons

DE LA PENA *et al.* [49] have published values for the distribution of plasma lipids in the lipoproteins of 15 adult male and 15 adult female baboons maintained on control diets. The average plasma cholesterol in females was 129 mg/dl, and the percent distribution of cholesterol was 3% in VLDL (d < 1.006), 14% in LDL_1 (d 1.006–1.019), 27% in LDL_2 (d 1.019–1.063), and 56% in HDL (d 1.063–1.21), while the percentages for triglyceride, which averaged 29 mg/dl in females, were 6% in VLDL, 16% in LDL_1, 23% in LDL_2, and 55% in HDL. The average plasma cholesterol in males was 111 mg/dl, and the percent distribution was 3% in VLDL, 5% in LDL_1, 27% in LDL_2, and 65% in HDL. The average triglyceride concentration was 28 mg/dl in males, and was distributed as follows: 10% VLDL, 20% LDL_1, 24% LDL_2, and 46% HDL. Such a triglyceride distribution is most unlike that of human beings, and may reflect, in part, the low levels of triglyceride in the plasma. The higher amount of triglyceride in VLDL and LDL_2 in males was significantly different from that of females. The relatively higher amount of cholesterol in LDL_1 and LDL_2 in females than in males was also significantly different, as was the total plasma cholesterol concentration. SRINIVASAN *et al.* [63] found a mean plasma cholesterol concentration of 105 mg/dl in 12 baboons fed monkey chow, when lipoproteins were separated by flocculation with heparin-Ca^{++}, 32% of cholesterol was associated with LDL + VLDL, a result in very close agreement with those reported above.

BLATON *et al.* [4] have studied baboons fed cholesterol-containing diets. They found that both α- and β-lipoproteins, when separated by the rather unusual method of electrochromatography, were increased when the plasma cholesterol increased from 117 to 203 mg/dl. The β-lipoproteins contained about 50% of the plasma cholesterol on both control and high cholesterol diets. The absolute concentration of each lipoprotein group studied increased approximately twofold. The authors report compositional changes in the lipoprotein lipids. The total cholesterol-to-phospholipid ratio of the β-lipoproteins decreased in animals fed cholesterol compared to controls; the cholesterol-to-phospholipid ratio of the α-lipoproteins increased. The fatty acid patterns appeared to be similar between lipoproteins for each lipid class. Only small changes were apparent when animals were fed cholesterol-enriched diets. The observed change was probably due mostly to the rather large difference in fatty acid patterns between the two diets fed. The control diet was rat cake, while the cholesterol-enriched diet was rat cake plus 15% egg yolk and 15% butter.

SHAPIRO et al. [59] reported a cholesterol distribution among lipoproteins separated from serum of two monkey chow-fed male baboons by dextran sulfate-Ca^{++} precipitation of LDL + VLDL. The average serum cholesterol concentration was 125 mg/dl and about 50% of the lipoprotein cholesterol was in HDL and 50% in LDL + VLDL. The percent of total cholesterol esterified was about 80% in HDL and about 75% in LDL + VLDL. The pattern of appearance of label in HDL and in LDL + VLDL monitored at daily intervals was similar when exogenous cholesterol was radioactive endogenous cholesterol waslabeled with radioactive mevalonate. The rate appearance of labeled free cholesterol exceeded the rate of appearance of ester cholesterol in bothlipoprot eins when derived either from endogenous or exogenous cholesterol. The cholesteryl ester of the HDL fraction appeared to be morerapidly labeled than that of LDL + VLDL in the first hour immediately after radioactive cholesterol administration, suggesting that the mechanism of incorporation of cholesterol into individual lipoproteins may differ; at longer time intervals, however, equilibration seemed to occur.

KRITCHEVSKY et al. [33] fed baboons diets in which the type of carbohydrate was varied, while the amount and type of fat were held constant. The serum cholesterol levels of the baboons on the experimental diets averaged near 150 mg/dl and the serum triglycerides averaged near 110 mg/dl, both values being about 40 mg/dl higher than for control animals. About 64% of the serum cholesterol of animals on the special diets was present in LDL + VLDL (as determined by dextran sulfate-Ca^{++} precipitation), whereas 57% was present in this fraction in control animals.

Apolipoproteins present in the serum of baboons [22] appeared to be similar to those in human HDL, except for apoLP-gln-II, which contained no cysteine, and therefore did not form disulfide-linked dimers. This result for baboon HDL is similar to the result of EDELSTEIN *et al.* [15] for HDL of rhesus monkeys, discussed above.

IV. Concluding Comments

Although the role of circulating lipoproteins in the development of atherosclerosis remains uncertain, it seems likely that the absolute concentration of lipoproteins in plasma, their relative distribution and their macromolecular structure may all be factors which interact to determine the relative atherogenicity of a particular plasma. Available information in these areas has been presented for nonhuman primates. It is clear that we are only beginning to understand the mechanism involved in the control of the complex plasma lipoprotein system. However, by surveying the presently available information, we may be able to better evaluate the direction for future research.

From the authors' point of view, some aspects do emerge from this survey which bear consideration. Lipoprotein distribution and structure in nonhuman primates are measurably influenced by diet and other environmental factors. In attempts to use nonhuman primates to experimentally mimic the experience of human beings, the environmental situation as well as individual species variation become even larger considerations. Some species appear to respond to monkey chow (a diet very atypical to that of human beings) in a similar manner to a diet more typical of man (40–45% of calories as fat, moderate cholesterol concentration) with absolute and relative lipoprotein concentrations comparable to those found in man, e.g. *C. aethiops*. In other species, lipoprotein levels on any diet yet tested do not appear to mimic closely the human situation, e.g. *S. sciureus*.

Due to the wide variety of environmental influences on lipoprotein distribution and structure, nonhuman primates would appear to be an ideal group of animals in which to test the effects of specific variables. As opposed to human beings, these animals can be (and should be) maintained under conditions in which the effects of single variables can be monitored. The results presented here, demonstrating the effect of graded amounts of dietary cholesterol on lipoprotein distribution and structure, illustrate a case in point.

References

1 ARMSTRONG, M. L.; CONNOR, W. E., and WARNER, E. D.: Xanthomatosis in rhesus monkeys fed a hypercholesterolemic diet. Archs Path. *84:* 227–237 (1967).

2 BARCLAY, M.: Lipoprotein class distribution in normal and diseased states; in NELSON Blood lipids and lipoproteins: quantitation, composition, and metabolism, pp. 585–704 (Wiley-Interscience, New York 1972).

3 BARCLAY, M.; BARCLAY, R. K., and SKIPSKI, V. P. High density lipoprotein concentrations in men and women. Nature, Lond. *200:* 362–363 (1963).

4 BLATON, V.; HOWARD, A. N.; GRESHAM, G. A.; VANDAMME, D., and PEETERS, H.: Lipid changes in the plasma lipoproteins of baboons given an atherogenic diet. I. Atherosclerosis *11:* 497–507 (1970).

5 BREWER, H. B., jr.; LUX, S. E.; RONAN, R., and JOHN, K. M.: Amino acid sequence of human apoLp-Gln II (apoA-II), an apolipoprotein from the high density lipoprotein complex. Proc. natn. Acad. Sci. USA *69:* 1304–1308 (1972).

6 BREWER, H. B., jr.; SHULMAN, R.; HERBERT, P.; RONAN, R., and WEHRLY, K.: The complete amino acid sequence of an apolipoprotein obtained from human very low density lipoprotein (VLDL). Adv. exp. Med. Biol. *26:* 280 (1972).

7 BRUNNER, D.; MANELIS, G., and ALTMAN, S.: Physical activity lipoproteins and ischemic heart disease. Pathol. Microbiol. *30:* 648–652 (1967).

8 BULLOCK, B. C.; LEHNER, N. D. M.; CLARKSON, T. B.; FELDNER, M. A.; WAGNER, W. D., and LOFLAND, H. B.: Comparative primate atherosclerosis. I. Tissue cholesterol concentration and pathologic anatomy. Exp. molec. Pathl. *22:* 151–175 (1975).

9 BULLOCK, B. C.; PAULA, R., and RUDEL, L. L.: Dietary-induced cirrhosis in the pig-tailed macaque *(Macaca nemestrina)*. Fed. Proc. Fed. Am. Socs exp. Biol. *33:* 626 (1974).

10 BULLOCK, B. C.; CLARKSON, T. B.; LEHNER, N. D. M.; LOFLAND, H. B., and ST. CLAIR, R. W.: Atherosclerosis in *Cebus albifrons* monkeys. III. Clinical and pathologic studies. Expl molec. Path. *10:* 39–62 (1969).

11 BURSTEIN, M. et SAMAILLE, J.: Sur un dosage rapide du cholesterol lié aux α- et aux β-lipoproteines du serum. Clin. chim. Acta *5:* 609 (1960).

12 CAMEJO, G.; BOSCH, V.; ARREAZA, C., and MENDEZ, H. C.: Early changes in plasma lipoprotein structure and biosynthesis in cholesterol-fed rabbits. J. Lipid Res. *14:* 61–68 (1973).

13 CORNWELL, D. G. and KRUGER, F. A.: Molecular complexes in isolation and characterization of plasma lipoproteins. J. Lipid Res. *2:* 110–134 (1961).

14 DeLALLA, O. F.; ELLIOTT, H. A., and GOFMAN, J. W.: Ultracentrifugal studies of high density serum lipoproteins in clinically healthy adults. Am. J. Physiol. *170:* 333–337 (1954).

15 EDELSTEIN, C.; LIM, C. T., and SCANU, A. M.: The serum high density lipoproteins of Macacus rhesus. II. Isolation, purification, and characterization of their two major polypeptides. J. biol. Chem. *248:* 7653–7660 (1973).

16 EDELSTEIN, C.; LIM, C. T., and SCANU, A. M.: On the subunit structure of the protein of human serum high density lipoprotein. I. A study of its major polypeptide component (Sephadex fraction III). J. biol. Chem. *247:* 5842–5849 (1972).

17 FRASER, R.; DUBIEN, L.; MUSIL, F.; FOSSLIEN, E., and WISSLER, R. W.: Transport of cholesterol in thoracic duct lymph and serum of rhesus monkeys fed cholesterol with various food fats. Atherosclerosis *16:* 203–216 (1972).

18 FREDERICKSON, D. S.; GOTTO, A. M., and LEVY, R. I.: Familial lipoprotein deficiency; in STANBURY, *et al.* The metabolic basis of inherited disease, pp. 493–530 (McGraw-Hill, New York 1972).

19 FREDERICKSON, D. S. and LEVY, R. I.: Familial hyperlipoproteinemia; in STANBURY *et al.* The metabolic basis of inherited disease, pp. 545–614 (McGraw-Hill, New York 1972).

20 FREDERICKSON, D. S.; LUX, S. E., and HERBERT, P. N.: The apolipoproteins. Adv. exp. med. biol. *26:* 25–56 (1972).

21 FREDERICKSON, D. S.; LEVY, R. I., and LEES, R. S.: Fat transport in lipoproteins: an integrated approach to mechanisms and disorders. New Engl. J. Med. *276:* 32–44, 94–103, 148–156, 215–226, 273–281 (1967).

22 GOLDSWORTHY, P. D.; LIM, G.; GLOMSET, J. A., and VOLWILER, W.: Polypeptides of baboon serum high density lipoproteins. Fed. Proc. Fed. Am. Socs exp. Biol. *32:* 547A (1973).

23 GREENE, D. G.: Cholesterol metabolism and lipoprotein composition of hypo- and hyperresponding squirrel monkeys consuming saturated and unsaturated fat diets; M. S. thesis, Bowman Gray School of Medicine, Winston-Salem (1974).

24 GULBRANDSEN, C. L.; WILSON, R. B., and LEES, R. S.: Conversion of human plasma very low density to low density lipoproteins in the squirrel monkey. Circulation *44:* suppl. II, p. 10 (1971).

25 HAVEL, R. J.: Early effects of fat ingestion on lipids and lipoproteins of serum in man. J. clin. Invest. *37:* 848–854 (1957).

26 HAVEL, R. J.; EDER, H. A., and BRAGDON, J. H.: The distribution and chemical composition of ultracentrifugally separated lipoproteins in human serum. J. clin. Invest. *34:* 1345–1353 (1955).

27 HOWARD, C. F., jr.: Spontaneous diabetes in *Macaca nigra*. Diabetes *21:* 1077–1090 (1972).

28 HULLY, S. B.; COOK, S. G.; WILSON, W. S.; NICHAMAN, M. Z., and HATCH, F. T.: Quantitation of serum lipoproteins by electrophoresis on agarose gel: standardization in lipoprotein concentration units (mg/100 ml) by comparison with analytical ultracentrifugation. J. Lipid Res. *12:* 420–433 (1971).

29 ILLINGWORTH, D. R. and PORTMAN, O. W.: Secretion of newly synthesized apo low density lipoprotein (LDL) into the plasma of squirrel monkeys. Circulation *47–48:* suppl. IV, p. 111 (1973).

30 ILLINGWORTH, D. R. and PORTMAN, O. W.: Exchange of phospholipids between low and high density lipoproteins of squirrel monkeys. J. Lipid Res. *13:* 220–227 (1972).

31 ILLINGWORTH, D. R. and PORTMAN, O. W.: Independence of phospholipid and protein exchange between plasma lipoproteins *in vivo* and *in vitro*. Biochim. biophys. Acta *280:* 281–289 (1972).

32 KERR, G. R.; WOLF, R. C., and WAISMAN, H. A.: A disorder of lipid metabolism associated with experimental hyperhistidinemia in *Macaca mulatta*. Symp. zool. Soc. Lond. *17:* 371–392 (1966).

33 KRITCHEVSKY, D.; DAVIDSON, L. M.; SHAPIRO, I. L.; KIM, H. K.; KITAGAWA, M. MALHOTRA, S.; NAIR, P. P.; CLARKSON, T. B.; BERSOHN, I., and WINTER, P. A. D.: Lipid metabolism and experimental atherosclerosis in baboons: influence of cholesterol-free, semisynthetic diets. Am. J. clin. Nutr. *27:* 29–50 (1974).

34 KRITCHEVSKY, D.; TEPPER, S. A.; ALAUPOVIC, P., and FURMAN, R. H.: Cholesterol content of human serum lipoproteins obtained by dextran sulfate precipitation and by preparative ultracentrifugation. Proc. Soc. exp. Biol. Med. *112:* 259–262 (1963).

35 LASSER, N. L. and ALLEBAUGH, J. F.: Density distribution of plasma lipoproteins in rhesus monkeys: differential effect of an atherogenic diet in males and females; in SCHETTLER and SCHLIERF Atherosclerosis III, pp. 874 (Springer, Berlin 1974).

36 LEE. J. A. and MORRIS, M. D.: Characterization of the serum low density lipoproteins of normal and two spontaneous hyperbetalipoproteinemic rhesus monkeys. Biochem. Med. *10:* 245–257 (1974).

37 LEE, J. A. and MORRIS, M. D.: The effects of cholesterol feeding on the composition of rhesus monkey serum low density lipoprotein. Fed. Proc. Fed. Am. Socs exp. Biol. *31:* 701A (1972).

38 LEWIS, L. A.; OLMSTED, F.; PAGE, I. H.; LAWRY, E. Y.; MANN, G. V.; STARE, F. J.; HANIG, M.; LAUFFER, M.; GORDON, T., and MOORE, F. E.: Serum lipid levels in normal persons. Findings of a cooperative study of lipoproteins and atherosclerosis. Circulation *16:* 227–245 (1957).

39 LINDGREN, F. T. and JENSEN, L. C.: The isolation and quantitative analysis of serum lipoproteins; in NELSON Blood lipids and lipoproteins: quantitation, composition, and metabolism, pp. 181–274 (Wiley-Interscience, New York 1972).

40 LINDGREN, F. T.; ADAMSON, G. L.; JENSEN, L. C., and HATCH, F. T.: Quantification of plasma lipoproteins by agarose gel electrophoresis and comparison with analytic ultracentrifugation. Circulation *44:* suppl. I, pp. 11–58 (1971).

41 LOFLAND, H. B.; ST.CLAIR, R. W., and GREENE, D. G.: Certain aspects of lipid metabolism in the young squirrel monkey *(Saimiri sciureus)*; in SCHETTLER and SCHLIERF Atherosclerosis III, pp. 376–380 (Springer, Berlin 1974).

42 LOFLAND, H. B.; ST. CLAIR, R. W.; MACNINTCH, J. E., and PRICHARD, R. W.: Atherosclerosis in New World primates. Archs Path. *83:* 211–214 (1967).

43 MARGOLIS, S. and CAPUZZI, D.: Serum lipoprotein synthesis and metabolism; in NELSON Blood lipids and lipoproteins: quantitation, composition, and metabolism, pp. 825–880 (Wiley-Interscience, New York 1972).

44 MARSH, J. B. and WHEREAT, A. F.: The synthesis of plasma lipoprotein by rat liver. J. biol. Chem. *234:* 3196–3200 (1959).

45 MORRIS, M. D. and GREER, W. E.: Hyperalphalipoproteinemia in cholesterol-fed rhesus monkeys. Fed. Proc. Fed. Am. Socs exp. Biol. *31:* 727A (1972).

46 MORRIS, M. D. and FITCH, C. D.: Spontaneous hyperbetalipoproteinemia in the rhesus monkey. Biochem. Med. 2: 209–215 (1968).

47 NICHOLS, A. V.: Human serum lipoproteins and their interrelationships. Adv. biol. med. Phys. *11:* 109–158 (1967).

48 NICHOLS, A. V.; DOBBIN, V., and GOFMAN, J. W.: Influence of dietary factors upon human serum lipoprotein concentrations. Geriatrics *12:* 7–17 (1957).

49 PENA, A. DE LA; MATTHIJSSEN, C., and GOLDZIEHER, J. W.: Normal values for blood constituents of the baboon. II. Lab. Anim. Sci. *22:* 249–257 (1972).

50 PORTMAN, O. W. and ANDRUS, S. B.: Comparative evaluation of three species of New World monkeys for studies of dietary factors, tissue lipids, and atherogenesis. J. Nutr. *87:* 429–438 (1965).

51 PUCAK, G. J.; LEHNER, N. D. M.; CLARKSON, T. B.; BULLOCK, B. C., and LOFLAND, H. B.: Spider monkeys (*Ateles* sp.) as animal models for atherosclerosis research. Expl molec. Path. *18:* 32–49 (1973).

52 RACHMILEWITZ, D.; STEIN, O.; ROHEIM, P. S., and STEIN, Y.: Metabolism of iodinated high density lipoproteins in the rat. II. Autoradiographic localization in the liver. Biochim. biophys. Acta *270:* 414–425 (1972).

53 RAYMOND, T. L.; LOFLAND, H. B., and CLARKSON, T. B.: Lack of feedback control of cholesterol synthesis in pig-tail macaques. Circulation *47–48:* suppl. IV, p. 43 (1973).

54 RUDEL, L. L.; LEE, J. A.; MORRIS, M. D., and FELTS, J. M.: Characterization of plasma lipoproteins separated and purified by agarose-column chromatography. Biochem. J. *139:* 89–95 (1974).

55 SCANU, A. M.; EDELSTEIN, C.; VITELLO, L.; JONES, R., and WISSLER, R.: The serum high density lipoproteins of Macacus rhesus. I. Isolation, composition, and properties. J. biol. Chem. *248:* 7648–7652 (1973).

56 SCANU, A. M. and WISDOM, C.: Serum lipoproteins structure and function. Annu. Rev. Biochem. *41:* 703–730 (1972).

57 SCHONFELD, G.; GULBRANDSEN, C. L.; WILSON, R. B., and LEES, R. S.: Catabolism of human very low density lipoproteins in monkeys: the appearance of human very low density lipoprotein peptides in monkey high density lipoproteins. Biochim. biophys. Acta *270:* 426–432 (1972).

58 SCHUMAKER, V. N. and ADAMS, G. H.: Circulating lipoproteins. Annu. Rev. Biochem. *38:* 113–136 (1969).

59 SHAPIRO, I L.; JASTREMSKY, J. A.; EGGEN, D. A., and KRITCHEVSKY, D.: Cholesterol metabolism in the baboon. Lipids *3:* 136–142 (1968).

60 SHORE, V. G. and SHORE, B.: Heterogeneity of human plasma very low density lipoproteins. Separation of species differing in protein components. Biochemistry *12:* 502–507 (1973).

61 SHULMAN, R.; HERBERT, P.; WEHRLY, K.; CHESEBRO, B.; LEVY, R. I., and FREDRICKSON, D. S.: The complete amino acid sequence of apoLP-ser: an apolipoprotein obtained from human very low density lipoprotein. Circulation *46:* suppl II, pp. 11–246 (1972).

62 SKIPSKI, V. P.: Lipid composition of lipoproteins in normal and diseased states; in NELSON Blood lipids and lipoproteins: quantitation, composition and metabolism, pp. 471–583 (Wiley-Interscience, New York 1972).

63 SRINIVASAN, S. R.; McBRIDE, J. R.; RADHAKRISHNAMURTHY, B., and BERENSON, G. S.: Comparative studies on serum lipoprotein and lipid profiles in subhuman primates. Comp. biochem. Physiol. *47:* 711–716 (1974).

64 SRINIVASAN, S. R.; DALFERES, E. R., jr.; RUIZ, H.; PARGAONKAR, P. S.; RADHAKRISHNAMURTHY, B., and BERENSON, G. S.: Rapid serum lipoprotein changes in spider monkeys on short-term feeding of high cholesterol-high saturated fat diet. Proc. Soc. exp. Biol. Med. *141:* 154–160 (1972).

65 ST. CLAIR, R. W.; MACNINTCH, J. E.; MIDDLETON, C. C.; CLARKSON, T. B., and LOFLAND, H. B.: Changes in serum cholesterol levels of squirrel monkeys during importation and acclimation. Lab. Invest. *16:* 828–832 (1967).

66 WINDMUELLER, H. G. and LEVY, R. I.: Production of β-lipoprotein by intestine in the rat. J. biol. Chem. *243:* 4878–4884 (1968).

67 ZILVERSMIT, D. B.: The surface coat of chylomicrons: lipid chemistry. J. Lipid Res. *9:* 180–186 (1968).

Dr. LAWRENCE L. RUDEL and Dr. HUGH B. LOFLAND, Arteriosclerosis Research Center Departments of Comparative Medicine and Pathology, Bowman Gray School of Medicine, *Winston-Salem, NC 27103* (USA)

Prim. Med., vol. 9, pp. 267–299 (Karger, Basel 1976)

Cholesterol Metabolism in Nonhuman Primates[1]

DOUGLAS A. EGGEN

Departments of Pathology and Biometry,
Louisiana State University Medical Center, New Orleans, La.

Contents

I. Introduction

The metabolism of cholesterol has played a central role in studies of the pathogenesis and etiology of atherosclerosis since atherosclerotic lesions were first produced experimentally by feeding cholesterol to rabbits [4]. It has generally been presumed that cholesterol metabolism is related to atherosclerosis through its effect on the concentration of serum cholesterol or its lipoprotein carriers. Experimental atherosclerotic lesions in nonhuman pri-

[1] Supported by USPHS, NIH grant HL-08974.

mates have invariably been accompanied by hypercholesterolemia, and the concentration of cholesterol in serum has been reported in every study. In addition, many other aspects of cholesterol metabolism have also received attention. This chapter reviews current knowledge of whole-body cholesterol metabolism in nonhuman primates. The reader is referred to other chapters in this volume which cover the metabolism of cholesterol in the arterial wall [94] and the lipoproteins [105]. Chapters devoted to atherosclerosis in nonhuman primate species discuss the methods used to produce hypercholesterolemia [6, 21, 22, 82]. No attempt will be made to review exhaustively the literature on the effects of diet on the concentration of serum cholesterol in nonhuman primates. Published data on the concentration of serum cholesterol in nonhuman primates fed maintenance or basal diets have recently been tabulated [88].

Cholesterol metabolism is also related to the pathogenesis of clinically significant lesions other than atherosclerosis, namely gallstones and xanthomata. Recent articles have reviewed the use of nonhuman primates as models of lithogenesis [33, 58, 84, 90, 103, 107, 109]. Xanthomatosis has been observed incidentally when atherogenic diets are fed to nonhuman primates [79, 115, 123] and one paper has described the rhesus monkey as a model for xanthomatosis [8].

Aspects of whole-body cholesterol metabolism in nonhuman primates that may influence the serum cholesterol concentration include the following: (a) the size of the *body pool* of cholesterol that is exchangeable with that of the serum, i.e. excluding the central nervous system; (b) the *absorption* of cholesterol from the intestinal lumen whether it be of dietary or endogenous origins; (c) the *synthesis* of cholesterol in those tissues that equilibrate with serum cholesterol, i.e. excluding synthesis in nerve tissue; (d) the *catabolism* of cholesterol to other compounds, principally bile acids; (f) the *distribution* of cholesterol among the various body tissues, and (g) *excretion* of cholesterol and bile acids. Methods of investigating these parameters of cholesterol metabolism in nonhuman primates will be discussed and results obtained in several primate species will be reviewed.

II. Methods Applicable to Nonhuman Primates

Most methods being used to measure whole-body cholesterol metabolism were designed for studying humans in metabolic wards. Several of the methods, however, have been adapted for studying nonhuman primates and other animals.

A. Isotope Kinetics

One method that measures the decline of specific activity of serum cholesterol following an intravenous pulse dose of radio-labeled cholesterol has been referred to as the *isotope kinetic* method [51]. In early investigations using this method [18, 39, 45, 118] the serum-miscible body cholesterol was considered to exist as a single pool (one-pool model). The size of this pool was determined by isotope dilution using an estimate of specific activity obtained by extrapolating the log-linear phase of the decay curve to the time of injection. Since some of the activity was lost from the system during the equilibration period the value of pool size so obtained was in excess of the true value [51]. The terminal slope of the decay curve, however, gave valid estimates of the fractional turnover of the total miscible pool, provided data were obtained for a sufficiently long time after injection. The value of turnover rate obtained as the product of fractional turnover rate and pool size was, of course, also in excess of the true value [51].

GOODMAN and NOBLE [47] first showed that the decay of serum specific activity in man can be represented as the sum of two exponentials. This is the pattern of decay to be expected if the exchangeable cholesterol of the body consisted of two pools. One pool, which includes the serum, exchanges rapidly, and the other pool exchanges slowly with serum cholesterol. More recently, it has been reported that three exponentials are required to fit adequately data obtained for longer than 12 weeks following injection of the isotope [48]. The model is then extended to include a second peripheral pool exchanging with the rapidly miscible pool.

The author has applied the isotope kinetic method to an adult male rhesus monkey *(Macaca mulatta)* and an adult male baboon *(Papio cynocephalus)* in an unpublished pilot study of cholesterol metabolism. A tracer dose of [4-^{14}C]-cholesterol suspended in saline with Tween-80 (Sigma Chemical Co., St. Louis, Mo.) was given intravenously after the animals had been fed a commercial monkey food for 18 months. The specific activity of serum cholesterol was determined 15 times during the following 149 days. The diet was then changed to an atherogenic diet containing fat at 40% of calories and cholesterol at 1 mg/kcal [38]. Three weeks later a second tracer dose of [4^{14}C]-cholesterol was given, and again 15 determinations of serum specific activity were made during the following 105 days.

The logarithm of the serum cholesterol specific activity as a function of time after injection of radiocholesterol was fitted to the logarithm of the sum of two or three exponentials using an iteration procedure described by SNEDECOR and COCHRAN [111]. This procedure is analogous to the graphical 'peeling' procedure using semilog plots. Results obtained were nearly identical to results obtained when the data were analyzed using the weighted least squares method described by DELL *et al.* [28]. By either method the addition of the third pool to the model gave a highly significant (p < 0.01) reduction in the residual variance for all four animal-diet combinations. As was shown for man by GOODMAN *et al.* [48] the addition of a fourth term (pool) to the model in this pilot study did not result in a significant additional decrease in the residual variance.

The results of the analysis for the baboon on the basal diet are shown in figure 1 (two-pool) and figure 2 (three-pool). The improvement in fit obtained with addition of the third pool to the model is evident. A diagramatic representation of the model is also shown above each fitted curve. The values of parameters given in figures 1 and 2 have been

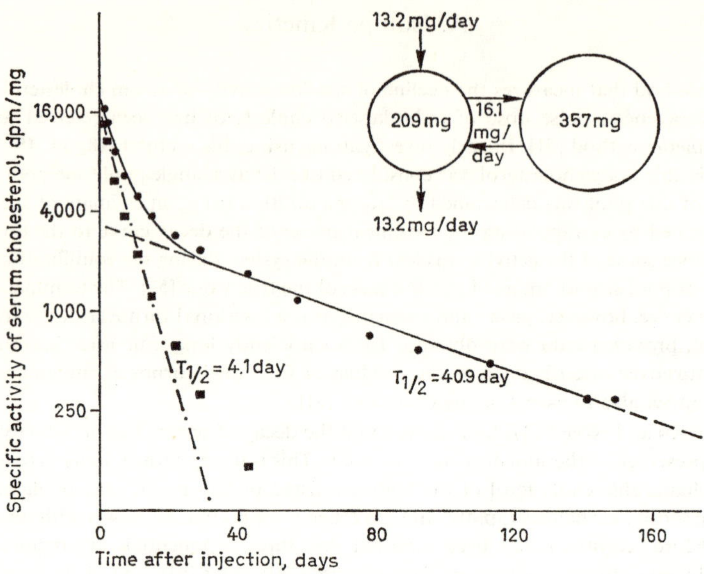

Fig. 1. Two-pool analysis of decay of serum cholesterol specific activity following intravenous administration of [4-¹⁴C]-cholesterol to a baboon. The young adult male baboon was fed a basal, cholesterol-free, commercial primate ration for 18 months prior to and during this decay period. The solid line shows the fit obtained as the sum of the two linear components shown as dashed lines. Measured values are shown as circles and the difference between measured values and linear components are shown as squares. Values derived from the analysis are given in the insert where data are expressed per kg of body weight. The areas of circles are proportional to the calculated mass of each pool.

normalized by dividing by body weight. The areas of circles representing the pools in these diagrams are proportional to the sizes of the pools (normalized). It has been assumed that entry of cholesterol into the system and exit from the system occur only by way of the central pool containing the serum. With this assumption, values derived for total pool size represent the minimum values [48]. WILSON [127] has shown that in the baboon this minimum value is approximately equal to the value of the exchangeable pool determined by direct carcass analysis.

Table I shows results obtained for the two-pool and three-pool analyses of data for each animal on each diet. Here, as in figures 1 and 2, data are expressed on the basis of 1 kg body weight. The first row gives the square root of the mean squared deviation about the fitted curve expressed as a fraction of the predicted value. Values for the approximate 95-percent confidence limits for each parameter are listed in parentheses. The changes in size of the pools, production rate for rapidly changing pool, or slope of the linear partion of the decay that result from inclusion of the third term in the model are similar to those reported for man by GOODMAN *et al.* [48].

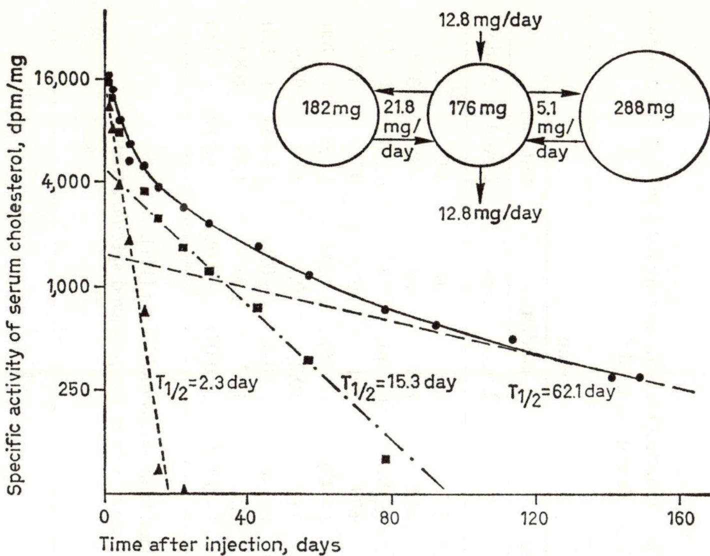

Fig. 2. Three-pool analysis of same data shown in figure 1. The solid line shows the fit obtained which is the sum of the three linear components shown as dashed lines. Difference values are shown as squares or triangles. Values derived from the analysis are given in the insert where data are expressed per kg of body weight. The areas of circles are proportional to the derived mass of each pool.

Subsequent to this pilot study, the addition of a third pool has also been required in the analysis of 22 of 23 experiments on rhesus monkeys in which the decay was followed for 128 or 225 days [40].

These results indicate that in the baboon and rhesus, as in man, the body cholesterol which exchanges with serum cholesterol can be considered to be contained in three pools or three groups of pools. In the pool which includes the serum cholesterol, the exchange between tissues is so rapid that equilibration is essentially complete within one to two days after administration of the labeled cholesterol. The other pools exchange cholesterol with the rapidly miscible pool only, and do not exchange with each other. They differ in the rate of exchange of cholesterol with the rapidly miscible pool. It is probable that if the conditions of the experiment could be controlled better or if more data points could be measured with greater precision and over longer periods of time, the model would have to be extended to include other pools or other pathways for interchange between pools.

Although the two- or three-pool models described above are obviously an oversimplification, they are useful because they allow determination of several parameters of whole-body cholesterol metabolism. Some of these parameters have been checked by independent methods. Thus, for man it has been shown [51] that measures of production rate derived by kinetic analysis are slightly greater than the measures of turnover rate obtained by the sterol balance method. Similar differences between the two measures have been obtained in rhesus monkeys [40]. MANNING *et al.* [81] reported somewhat greater differences in the

Table I. Parameters of cholesterol metabolism derived from analysis of decay of serum cholesterol specific activity for a baboon and a rhesus monkey on two diets. Comparison of two-pool and three-pool models

	Baboon				Rhesus			
	basal diet		test diet		basal diet		test diet	
	2-pool	3-pool	2-pool	3-pool	2-pool	3-pool	2-pool	3-pool
Residual deviation[1], %	11.4	5.2	4.5	2.7	4.2	2.6	4.0	1.9
Size of rapidly exchanging pool, mg/kg^2	209	176	269	212	188	173	408	370
	(±37)	(±22)	(±27)	(±46)	(±16)	(±15)	(±40)	(±34)
Size of total exchangeable pool, mg/kg^2	566	646	672	679	591	605	972	1278
	(±60)	(±120)	(±42)	(±42)	(±25)	(±21)	(±56)	(±700)
Production rate for rapidly exchanging pool	13.2	12.8	14.8	14.5	10.2	10.1	17.0	14.9
mg/day/kg^2	(±0.5)	(±0.8)	(±0.7)	(±0.6)	(±0.4)	(±0.4)	(±0.8)	(±3.8)
Slope of long-term decay of specific activity,	1.70	1.11	1.85	1.52	1.42	1.33	1.43	0.80
%/day	(±0.70)	(±0.50)	(±0.14)	(±0.14)	(±0.07)	(±0.08)	(±0.12)	(±0.8)

Values in parentheses are the approximate 95-percent confidence limits determined from gauss multipliers [111].

[1] Square root of the mean squared fractional deviation of the observed specific activity from the predicted specific activity.

[2] Data normalized by dividing by body weight in kg.

rhesus. Production rate would be expected to exceed fecal excretion since it is known that there is some loss of cholesterol through the skin [13] and there may be some flux into nerve tissue [19] or other tissues which equilibrate very slowly. WILSON [127] has determined the size of the total exchangeable pool in baboons both by kinetic analysis using the two-pool model and by total carcass analysis. He obtained good agreement between the two methods and concluded that this offered substantial proof of the validity of the two-pool method and of the assumption that essentially all of the newly synthesized cholesterol first enters the rapidly miscible pool.

B. Isotopic Steady State

This method involves labeling of the dietary cholesterol at a constant specific activity, then following the serum specific activity until it approximates an asymptotic level [87, 127]. The ratio of the asymptotic serum specific activity to the dietary specific activity is the fraction of the exchangeable body pool derived from the diet. Under steady-state conditions this ratio is also the fraction of the total body turnover derived from diet. ZILVERSMIT and WENTWORTH [132] have derived expressions relating the time course of buildup of serum specific activity to the parameters of the two-pool model and have shown that the time for attaining equilibrium levels of serum cholesterol specific activity can be reduced considerably by orally administering and appropriate single pulse dose of labeled cholesterol prior to beginning the continuous label. These expressions can also be used to obtain a correction for failure to achieve steady-state levels before discontinuing the dietary label [132].

C. Chromatographic Balance

The mass of cholesterol and products of the catabolism of cholesterol excreted in feces can be determined by combined gas-liquid chromatography and thin-layer chromatography [49, 86]. In applying these methods to determine the loss of body cholesterol in feces it is necessary to correct for variations in fecal flow [54]. In some human patients on liquid formula diets it is also necessary to correct for degradation of cholesterol to products no longer recognizable as steroids [1, 29, 50]. The balance methods developed in the laboratories of AHRENS permit a correction for these factors by use of phytosterols or chromic oxide as markers [27, 51, 54]. In addition to these problems encountered in studying human subjects the precise quantitation of the dietary intake of sterols and the quantitative collection of feces are the major difficulties in applying balance methods to other primates.

D. Isotopic Balance

When the serum miscible pool of cholesterol has been labeled, the endogenous[2] excretion of cholesterol or products of catabolism of cholesterol can be determined as the ratio

[2] Throughout this chapter the term endogenous excretion will refer to that fraction of the fecal excretion derived from sources in equilibrium with serum cholesterol pool, i.e. that secreted in the bile or sloughed off of intestinal mucosa. As measured isotopically this does not include any cholesterol synthesized in the intestines that may be excreted directly into the lumen.

of activity in fecal neutral or acidic steroids to the specific activity of serum cholesterol at the time of formation of the bile [51, 55]. This method assumes an equilibration of activity between biliary cholesterol or newly formed bile acids and serum cholesterol [11].

E. General Considerations

The concept of turnover, by definition, implies a dynamic steady-state system. There are few if any situations in which a true steady state for cholesterol metabolism exists; however, for most purposes a steady state can be considered to be attained if the criteria given by GRUNDY and AHRENS [51] are satisfied. These criteria are: constant plasma cholesterol concentration, unchanging fecal excretion of steroids, and constant body weight. To these might be added the stipulation that pool sizes do not change with time. The author [38] has recently presented evidence that pool size may continue to increase with time in some rhesus monkeys even after five to six months on a high cholesterol diet.

When a steady-state condition can be assumed, total body turnover of cholesterol is measured by determining the rate at which cholesterol either enters or leaves the system. Entry in the system occurs either by absorption of exogenous (dietary) cholesterol or by synthesis of cholesterol, primarily in liver and intestines [30]. Significant losses from the system occur through excretion of neutral steroids in feces, catabolism of cholesterol to acidic steroids or hormones, or losses through sebum, hair, or sloughed skin.

The indirect methods for determining the miscible pool size based on the isotope analyses discussed above require considerable time (four to eight weeks) for the administered activity to attain either a uniform distribution or a log-linear decay in specific activity. It is not possible to obtain serial measures of pool size taken days or weeks apart because of the length of time required by these measurements. Short-term changes in body pool size, therefore, must be based on balance considerations. The total body pool or the serum miscible pool of cholesterol can also be determined by direct analysis of mass of cholesterol or activity of cholesterol in the carcass [69, 127].

In considering intestinal absorption of cholesterol it is necessary to consider absorption of total luminal cholesterol which includes dietary cholesterol, cholesterol secreted in the bile, and cholesterol either secreted by the intestines or in cells sloughed from intestinal mucosa. It is generally assumed that all sources of luminal cholesterol are indistinguishable insofar as absorption is concerned [125]; however, DIETSCHY and WILSON [31] have presented considerations which cast doubt on this assumption. Absorption of cholesterol has been measured by a variety of methods [15, 51, 53, 98, 112, 113, 128]. QUINTAO [98] has compared four of these methods for measurements of absorption in man and has found that the most reliable appears to be 'method II' of GRUNDY and AHRENS [51] in which the fraction of labeled dietary cholesterol that is not absorbed is determined by a solution of simultaneous equations relating the mass and activity of cholesterol in diet, serum and feces. This method and methods I and III [51] can only be applied with diets containing cholesterol. Other methods, which measure the absorption of tracer amounts of labeled cholesterol administered orally, can be used with essentially cholesterol-free diets. One such method – introduced by BORGSTROM and co-workers [10,15] and evaluated by SODHI et al. [112, 113] and by QUINTAO et al. [98] – involves measuring the change in the ratio of labeled cholesterol to labeled β-sitosterol on passage through the intestines. This method has been applied to baboons and rhesus monkeys in the author's laboratory [40], and has been found to give results consistent with 'method I' of GRUNDY and AHRENS [51]. Another method, described by ZILVERSMIT [131], involves feeding a meal containing labeled cholesterol and simultaneously injecting a tracer dose of cholesterol labeled with a different isotope. Although the latter method is reliable for rats ingesting diets containing cholesterol [131] it has not yet been shown to be valid for use on cholesterol-free diets for use in primates.

Under steady-state conditions for cholesterol metabolism, total body synthesis of cholesterol that enters the exchangeable pool can be determined indirectly as the difference between total excretion and total ingestion. Relative rates of synthesis can be determined by measuring the rate of incorporation of labeled substrate (mevalonate or acetate) into cholesterol [30, 63, 65, 67].

The rate of catabolism of cholesterol to bile acids has been determined by measuring the rate of excretion of acidic steroids in feces under steady-state conditions [35]. The conversion of cholesterol to other steroids is quantitatively insignificant in studies of whole-body cholesterol metabolism.

By combining the isotopic procedures with the procedures for chromatographic analysis of neutral and acidic steroids, described by MIETTINEN [86] and GRUNDY [49], it is possible to achieve a fairly complete picture of whole-body cholesterol metabolism under steady-state conditions. Some of these procedures require a minimum of several months of experimental manipulation; therefore, it is not possible to repeat the total procedure to determine changes occurring during periods as short as several days or weeks. Some aspects of cholesterol metabolism, such as fecal excretion or absorption can, however, be determined on a daily or weekly basis.

The methods discussed above have been used by investigators to examine whole-body cholesterol metabolism in several species of nonhuman primates. In the remainder of this review, the results of published studies will be examined and some additional unpublished results from the author's laboratory will be presented.

III. Rhesus Monkey

In spite of early unsuccessful attempts to induce hypercholesteremia and atherosclerotic lesions in rhesus [57, 60], this species has become the most commonly used primate for experimental studies of atherosclerosis. The pioneering studies of TAYLOR and co-workers [122–124] first demonstrated its potential for such studies. They also demonstrated the extreme variability among individual monkeys in the case with which hypercholesteremia can be induced with high fat, high cholesterol diets [25]. This variability has also been shown in many subsequent investigations. It has restricted the interpretation of many studies in which homogeneity of response to the atherogenic diet is desirable. An understanding of the factors causing this variability could be of considerable value in designing studies using this model and may aid in developing modes of treatment of hypercholesterolemia.

The first measurements of cholesterol metabolism in the rhesus were reported by MANALO-ESTRELLA *et al.* [74]. They examined the relationship between the level of dietary cholesterol and the production of hypercholesterolemia and concluded that there is a hypercholesterolemic threshold for dietary cholesterol in the rhesus at about 50 mg/day or approximately 150–200 mg/kcal/day. By examining cholesterol synthesis in liver slices, these investigators also confirmed the earlier observation [26] that dietary cholesterol suppresses hepatic synthesis of cholesterol in the rhesus monkey. Extrapolating to the whole animal they suggested that the rate of hepatic synthesis subject to dietary inhibition is similar in magnitude to the rate of absorption expected with ingestion at the dietary threshold level. In light of the large variability among animals, the data presented might also have been interpreted as indicating that some animals will respond with detectable increases in serum cholesterol at lower levels of dietary cholesterol than other animals. Evidence in favor of a continuous response of serum cholesterol was recently presented by ARMSTRONG and MEGAN [9] who fed rhesus monkeys diets varying only in cholesterol content (range 0–389 mg/kcal) for 18 months and found that serum and carcass cholesterol concentrations increased with increasing diet cholesterol at all levels employed.

In two experiments, the author and his associates have fed an atherogenic diet [38] to groups of 36 and 37 young adult male rhesus monkeys for 12 weeks. Prior to the atherogenic diet period, and after feeding a basal commercial monkey food for periods of two weeks to eight months, the serum cholesterol concentrations of both groups were normally distributed with means and standard deviations of 129 $\pm$ 24 and 139 $\pm$ 22 mg/dl, respectively. The average serum cholesterol concentration for individual animals during the 12 weeks of feeding atherogenic diets were also normally distributed with mean and standard deviation of 431 $\pm$ 137 and 418 $\pm$ 124 mg/dl, respectively. The coefficients of variation of serum cholesterol concentration among animals ranged from 22 to 37% for the six samples taken during this 12-week atherogenic diet period. In the first of these studies [40] the six animals with highest and six with lowest concentration of serum cholesterol were fed the atherogenic diet an additional 38 weeks for a total of 50 weeks. In both high- and low-responding groups, the serum cholesterol concentration increased for the first 10–12 weeks then remained essentially constant for the remaining 38–40 weeks.

ARMSTRONG *et al.* [7] have examined the effect of an atherogenic diet on the total body content of cholesterol and on the distribution of cholesterol among the tissues in rhesus monkeys. They determined cholesterol in many tissues from a group of rhesus that had been fed a diet high in fat and cholesterol for 17 months and a second group that had been fed a low fat, cholesterol-free diet for a similar period. They found that the group fed high cholesterol diet had a fivefold greater mean plasma cholesterol concentration and twofold greater total body cholesterol content than the control group. Approximately half of the increase in total body cholesterol occurred in the blood and liver with the bulk of the remainder in tissues that are widely distributed such as skin and muscle. Although the increase in concentration of cholesterol in the accessible cardiovascular system ranged from threefold in peripheral arteries to sevenfold in the aorta, this increase contributed only a small fraction of the increase in total cholesterol content of the body.

MANNING *et al.* [81] have also studied whole body cholesterol metabolism in rhesus monkeys consuming an atherogenic diet. The rate of fecal excretion of endogenous neutral and acidic sterols was determined by the isotopic balance procedure. The two-pool model was used to analyze the decay of serum cholesterol specific activity during 58 days following intravenous administration of radiolabeled cholesterol. The values obtained for production rate exceeded the total fecal excretion of neutral sterols and bile acids by 28–53%. These investigators also applied the method of isotopic steady state and found that the fraction of the miscible cholesterol pool (or fraction of turnover) derived from an atherogenic diet containing 1 mg cholesterol/kcal ranged from 83 to 92% in six rhesus. From this they estimated that absorption of dietary cholesterol ranged from 130 to 190 mg/day or from 23 to 35% of 550 mg cholesterol ingested per day.

The author has recently reported studies of cholesterol metabolism in a group of eight rhesus monkeys while they were being fed a basal commercial monkey food, and again while they were fed the basal diet supplemented with saturated fat at 40% of calories and cholesterol at 1 mg/kcal [38]. Absorption of dietary cholesterol in these rhesus, calculated by method I of GRUNDY and AHRENS [51], was found to range from 44 to 54% of an average 320 mg/day ingested. The diet contributed an average of 69% of the miscible cholesterol pool in these rhesus. The half-time for decay of serum cholesterol specific activity during the log-linear phase following injection of radiocholesterol ranged from 34 to 51 days. This did not change significantly on transition to the atherogenic diet. There was evidence that the miscible pool of cholesterol continued to increase even though the animals had been on the atherogenic diet for five months prior to the balance measurements.

The author has also studied cholesterol metabolism in the high- and low-responding rhesus referred to above [40] while they were being fed the atherogenic diet and again after return to the basal monkey food. Production rates determined by analysis of isotope kinetic data using the three-pool model have also been found to exceed values for fecal excretion obtained by isotopic balance methods, but the difference was not as marked as for the rhesus studied by Manning *et al.* [81]. The percentage of dietary cholesterol absorbed by these groups as measured by method I of Grundy and Ahrens [51] was similar to that in the eight rhesus studied earlier [38]. Absorption was found to be greater in the high-responding than in the low-responding group. The fraction of the miscible pool derived from diet was also significantly greater in high- than low-responding animals while they were fed the atherogenic diet. Five months after return to the basal diet, both serum cholesterol concentration and the percentage of luminal cholesterol absorbed, measured by the method of Borgstrom [15], continued to be higher in the high-responding than in the low-responding group. During both diet periods, the rate of fecal excretion of either neutral or acidic sterols was similar for the two groups. These results are in striking contrast to the results reported by Lofland *et al.* [71] on hyper- and hyporesponding squirrel monkeys which will be discussed later.

Several other aspects of cholesterol metabolism have been studied in rhesus monkeys. Taylor *et al.* [123], Armstrong *et al.* [8], Mann and Andrus [79] and Stary [115] have reported that rhesus monkeys which develop severe hypercholesterolemia will, after several months, develop subcutaneous and tendinous xanthomata. The distribution of this cholesterol deposition differs in some respects from that in man; however, it has been suggested that the rhesus can provide a useful experimental model for 'studying the metabolism of primate tissues that predispose to xanthoma formation' [8]. Frazer *et al.* [42] have used the rhesus to examine the effect of different dietary fats on the mode of transport of cholesterol in thoracic duct lymph and in serum. They found that the ester/free ratio of cholesterol in the different lipoproteins of thoracic duct lymph varies with the nature of the dietary fat, and postulated that this may affect the subsequent metabolism of the absorbed cholesterol. This may explain the differences in atherogenicity observed for different dietary fats [43]. Pitkin *et al.* [92] have studied the interchange of cholesterol between maternal and fetal circulations of rhesus and have found that there is an appreciable net transfer from mother to fetus. At term, approximately 43% of fetal serum cholesterol was derived from the mother.

Dowling *et al.* [36] described an experimental model for studying biliary physiology in nonhuman primates. These and other investigators have applied this method to many aspects of the physiology and pathophysiology of bile acids and other constituents of bile in the rhesus monkey [16, 17, 34, 35, 37, 56, 101, 102, 104, 110, 116, 117].

IV. Baboon

The earliest published reports of studies of cholesterol metabolism in baboons are those of SAVAGE et al. [106] who demonstrated that in the baboon, cholesterol is synthesized from pyruvate. WERTHESSEN, as cited by KRITCHEVSKY [62], used radiolabeled acetate to study the time course of appearance of synthesized cholesterol in serum-free cholesterol. KRITCHEVSKY and co-workers [62, 66, 108] used radio-mevalonate and radio-cholesterol in baboons to study the rates of appearance of labeled cholesterol of exogenous or endogenous origins in the free and ester fractions of the α- and β-lipoproteins fractions. They concluded that in several respects their results were similar to observations reported for man. With values for intercept and half-time for decay obtained using the one-pool model, and with the procedure described by GIDEZ and EDER [44] for estimating the fraction of radio-mevalonate converted to cholesterol, these investigators estimated pool size and turnover or synthesis rates in baboons. These were found to be similar to values in man obtained using similar methods. The work of this group prior to 1970 has been summarized by KRITCHEVSKY [62]. In later studies KRITCHEVSKY and co-workers have examined the effects of dietary free fatty acids [64] and carbohydrates [63] on various aspects of cholesterol metabolism in baboons.

EGGEN and co-workers [39, 118] examined cholesterol metabolism in seven baboons used in a study of dietary factors in experimental atherosclerosis [119, 120]. Previously published results from these seven animals were based on the one-pool model for analysis of kinetic data. The data have since been analyzed using an iteration procedure analogous to that of DELL et al. [28]. Since the three-pool model did not provide a statistically significant improvement in fit to this data (there were fewer data points with greater residual error than for the unpublished pilot study referred to in section II. A), the two-pool model was used. Although the diet combinations for each of these seven animals were different, neither total miscible pool (range 0.6–1.2 mg/kg body weight), rapidly miscible pool (range 180–450 mg/kg body weight), nor production rate (range 11–25 mg/day/kg body weight) appeared to be correlated with serum cholesterol concentration, level of dietary cholesterol or protein, or degree of unsaturation of dietary fat. The fraction of serum cholesterol derived from diet for three baboons on high cholesterol diets ranged from 51 to 57%. This is about twice the contribution observed by KAPLAN *et al.* [59] for human subjects ingesting diets with similar cholesterol content.

In a subsequent study, EGGEN and STRONG [41] examined differences between young adult and old adult male baboons in the changes in cholesterol metabolism resulting from a transition from a basal to an atherogenic diet. Although the increase in serum lipids with transition to the atherogenic diet differed little between the two age groups, the contribution of dietary cholesterol to the miscible cholesterol pool was greater in the younger than in the older animals. It is probable that both an increased rate of absorption of dietary cholesterol and a greater inhibition of hepatic cholesterol synthesis in the younger animals contributed to the differences in fractional dietary contribution. The minor differences between the younger and older baboons in mean serum cholesterol and triglyceride concentrations on the atherogenic diet indicated that only a small part of the variability in response of serum lipids to dietary fat and cholesterol is likely to be a result of differences in age. Age could be a significant factor, however, in studies designed to examine other aspects of cholesterol metabolism.

In a later study involving eight young adult male baboons, the author [38] found that transition from a low fat, basal monkey food to a diet in which the basal food is supplemented with fat (40% of calories) and cholesterol (1 mg/cal) caused a 75-percent increase in the rate of fecal excretion of neutral sterols but did not change the rate of excretion of bile acids significantly. These baboons absorbed from 22 to 55% of an average of 780 mg cholesterol ingested per day, and this absorbed cholesterol contributed from 57 to 63% of the miscible cholesterol pool. There was a slight but statistically significant increase in fractional turnover rate (decrease in half-time) on transition from basal to atherogenic diet.

WILSON [127] studied baboons ingesting a diet high in cholesterol and fat to examine the validity of the two-pool model for analysis of isotopic cholesterol die-away curves. The isotopic balance technique and the isotope kinetic method were applied simultaneously. Measures of total miscible pool size obtained from the two-pool kinetic analysis and from direct analysis of activity in the total carcass were in good agreement. These findings support the validity of the two-pool method and the assertion that the only significant entry to the system and exit from the system occurs via the rapidly exchanging pool. These analyses also permitted assessment of the approximate distribution of the cholesterol from each tissue among the two miscible pools and the nonexchanging pool. WILSON concluded that although only a few tissues contribute the bulk of cholesterol in any one pool, cholesterol from all three pools is present in most if not all tissues.

In another study, WILSON [128] examined the relation between absorption and synthesis of cholesterol in the baboon. With low dietary cholesterol, hepatic synthesis accounts for as much as three-fourths of the daily production rate. Hepatic synthesis was completely inhibited only when the daily absorption approximated the amount that was synthesized by the liver under conditions of low dietary cholesterol. Synthesis in the intestines, the other major source for endogenous cholesterol, is not completely inhibited even under very high levels of dietary cholesterol. By studying other baboons that had been subjected to ileal diversion, WILSON showed that the action of the absorbed cholesterol cannot be the only mechanism by which dietary cholesterol inhibits hepatic synthesis. He suggested that bile acids may play a role in the inhibition of hepatic synthesis.

Stones have been observed incidentally in the gallbladder and bile duct of autopsied male baboons in several of the studies by STRONG and co-workers. Most of these animals had been fed high-fat diets; however, it is possible that in some animals the stones were present when captured in the wild. McGILL *et al.* [83] found no gallstones in 163 baboons autopsied after being trapped directly from the wild. GLENN and McSHERRY [46] and McSHERRY *et al.* [85] have reported the occurrence of cholesterol gallstones in 3 of 21 adult female baboons that had not been fed high-cholesterol diets. They suggested that the baboon would be an excellent model for studying the pathogenesis of cholesterol cholelithiasis. These investigators have performed several studies to examine various aspects of the biliary physiology of this species [84, 85] and have recently adapted the method of DOWLING *et al.* [36] for use in the baboon. They have reported [85] that the bile composition in the baboon is indistinguishable from that in man.

V. Squirrel Monkey

The first reports of studies of cholesterol metabolism in the squirrel monkey were those of DIETSCHY and WILSON [30, 31]. These investigators measured, *in vitro*, the incorporation of radioacetate into cholesterol in a variety of tissues from squirrel monkeys that had been fed diets with various levels of cholesterol. They showed that on low-cholesterol diets the liver and gastrointestinal tract accounted for 86 and 11% of the total cholesterol synthesis, respectively. In monkeys fed high levels of cholesterol, hepatic synthesis was reduced to 4% of its basal rate, whereas synthesis in the gastrointestinal tract was still about half of its basal rate so that the gastro-intestinal tract now became the major site of cholesterol synthesis. Within the gastrointestinal tract, the terminal ileum, esophagus and colon had the highest rates of synthesis. Fasting reduced hepatic synthesis by a factor of more than 10, whereas intestinal synthesis was decreased by a factor of only 2. Biliary diversion, however, had a major effect on synthesis in the small bowel (eightfold increase) and a somewhat lower effect on hepatic synthesis (twofold increase). Synthesis in the colon was not affected by biliary diversion. These results are in good agreement with those reported in the rat [32]; therefore, these authors concluded that the large differences that have been observed among species in the relative contribution of endo-genous and exogenous sources to the miscible body pool are not due to basic differences in the control of cholesterol synthesis, but must be due to differences in absorption, turnover, excretion or other parameters of choles-terol metabolism.

WILSON [126] has demonstrated directly that cholesterol synthesized in the intestines does enter the circulating pool of squirrel monkeys, and that, in cholesterol-fed monkeys, very little synthesized cholesterol enters the serum pool when the intestinal contribution is diverted. This result confirmed the earlier finding that hepatic synthesis is nearly completely inhibited by dietary cholesterol. By labeling serum and dietary cholesterol with different isotopes and following the time course of the specific activity in the intestinal lymph and in the serum, WILSON was also able to show unequivocally that the dilution of cholesterol radioactivity between intestinal lumen and thoracic duct lymph resulted from addition of nonradioactive cholesterol newly syn-thesized by the intestinal wall. He derived a lower limit of 2 mg/day for the rate of synthesis in the intestinal wall of the squirrel monkey and showed again that intestinal synthesis was not greatly diminished by cholesterol feeding. In this study, approximately 50–64% of the miscible cholesterol pool was derived from dietary cholesterol, in contrast to values of 85% observed for the rat [87] and 10–40% observed in man [59, 129].

LOFLAND et al. [69] have applied the isotope kinetic method to four groups of squirrel monkeys ingesting diets differing in source of fat and containing 0.5% cholesterol by weight. An additional group received a diet identical to one of the four high cholesterol diets but without added chole-sterol. The total miscible body pool of cholesterol in these monkeys was determined by carcass analysis after removing the central nervous system. The total miscible body pool of cholesterol in the four groups fed high cholesterol exceeded that in the group fed a low cholesterol diet. The fraction of this difference accounted for by the increase in size of the serum cholesterol pool differed from one dietary fat to another. The investigators concluded that the reduction in serum cholesterol concentration by unsaturated fats is mediated through a shift in distribution of cholesterol among body pools. In comparing the measured value of the total body pool with the values calculated from the two-pool model these investigators found that agreement was satisfactory only for the group on low-cholesterol diet. They concluded that the two-pool model was not valid for use in monkeys having expanded body pool. This conclusion might be questioned, since analysis of isotope kinetic data by methods which provide an assessment of confidence limits for derived parameters has shown that values for pool size of the slowly miscible pool can be determined only within very wide confidence limits, especially when based on relatively few data points taken over a short period of time. Also, it has been shown [127] that much of the nonexchangeable cholesterol of the baboon lies outside of the brain and spinal cord, whereas these were the only such tissues excluded in the carcass analyses of these squirrel monkeys.

LOFLAND *et al.* [71] have also reported studies of cholesterol metabolism in groups of squirrel monkeys that differed widely in the response of serum cholesterol concentration to dietary cholesterol. The combined methods of isotope kinetics, isotopic steady state, and isotopic and chromatographic balance were used to examine possible mechanisms by which some animals (hyporesponders) are better able to compensate for increased dietary cholesterol than others (hyperresponders). This study showed that the hypo-responding squirrel monkeys fed a liquid formula high cholesterol diet had a much greater increase in fecal excretion of bile acids than did the hyper-responding squirrel monkeys. The two groups were apparently similar in the rate of absorption of dietary cholesterol, as well as in the rate of synthesis and in the rate of fecal excretion of endogenous neutral sterols. The expansion of body pools was greater in the hyperresponders than in the hypo-responders.

In the same study [71], the transition from basal cholesterol-free diet to the high cholesterol diet occurred during the measurement of the isotope decay. Since the analysis of decay of serum cholesterol specific activity by the two-pool method has not been shown to be valid under such conditions, the measurements of the size of either the rapidly miscible or total cholesterol pool for the two groups might be in error, and this may have masked any difference in shift of cholesterol between body pools. The estimates of the difference in rate of synthesis might also have been subject to considerable error, since they were based on measurements obtained in two different dietary periods in which the steady state for cholesterol metabolism is likely not to have been attained. The methods used to measure the rate of absorption or rate of excretion of neutral and acidic sterols were not based on the assumption of steady state [71]. It would seem justified, therefore, to conclude that one factor that enables hyporesponding squirrel monkeys to maintain low serum cholesterol concentration when ingesting large amounts of cholesterol is their ability to increase markedly the rate of catabolism of cholesterol. These results differ from those abtained for groups of high- and low-responding rhesus monkeys [40] discussed above.

In a study of cholesterol metabolism in five squirrel monkeys, the author [38] reported that a mean of 42% of cholesterol was absorbed from a diet containing cholesterol at a concentration of 1 mg/kcal and saturated fat at 40% of calories. This result is lower than the means of 55 and 62% reported by LOFLAND *et al.* [71] for the groups of hypo- and hyperresponding squirrel monkeys, respectively. One explanation for this difference might be that in the latter study the diet was an essentially bulk-free, liquid formula diet with unsaturated fat, whereas in the former study the diet was a grain-based commercial chow with added saturated fat. The mean long-term half-time (15.5 days) of the decay of serum cholesterol specific activity in the five squirrel monkeys [38] was nearly identical to that calculated from the curve shown for the hyporesponding squirrel monkeys [71]. There was, however, a marked difference in the mean rates of excretion of bile acids in the two studies. Five months after beginning the atherogenic diet, EGGEN [38] found a mean and standard deviation for bile acid excretion of 17 ± 5 mg/day, whereas two and a half months after beginning the liquid formula diet, LOFLAND *et al.* [71] observed bile acid excretion of 25 and 95 mg/day for hyper- and hyporesponding monkeys, respectively (data read from graphs presented [71]). The differences between the two reports may have also resulted from the large differences in diets used.

VI. Cebus Monkey

The earliest report of a study of whole-body cholesterol metabolism in a primate was that of PORTMAN and SINISTERRA [95]. These investigators followed the time course of the specific activity of labeled free and ester cholesterol in the serum after intravenous administration to cebus monkeys consuming diets containing corn oil or lard as a source of fat. They also studied the rate of disappearance of activity from serum following the intragastric administration of radiocholesterol in single test diets to cebus monkeys being fed basal diets. The experiments performed during continuous ingestion of high cholesterol diets were continued for only 12 days after injection of isotope and therefore the decay half-times (6.6–8.8 days) do not represent the true terminal rate; however, the half-times (17–34 days) obtained after feeding various test meals were based on studies of longer duration and therefore probably do represent the terminal rate of decay for this species. Only one animal was studied on each test diet and since more recent studies indicate that there is considerable variability among animals on any one diet, the conclusion that the fat content of the single test meal affects the absorption or fractional turnover rate of cholesterol may be questioned.

Sugano and Portman [121] studied the esterification of cholesterol and the rates of formation of various fatty acid esters of cholesterol in cebus monkeys fed basal diets. They examined the specific activity of free cholesterol and of the individual cholesteryl esters in plasma following intravenous administration of labeled free cholesterol or mevalonate and intragastric administration of labeled cholesterol. From these *in vivo* studies and from *in vitro* studies of esterification of cholesterol in plasma they concluded that the plasma esterification activity was of considerable significance in maintenance of the level of the plasma cholesterol esters.

In 1953, Mann *et al.* [80] reported that hypercholesterolemia and atherosclerotic lesions could be produced in cebus monkeys by feeding a diet high in cholesterol and unsaturated fat but deficient in organic sulfur compounds. In subsequent studies, Mann [75–77] examined the mechanism by which sulfur deficiency disturbed sterol metabolism. In 1961, he reported [76] that addition of as little as 0.01% cholesterol to a sulfur-deficient diet caused a perceptible rise in serum cholesterol concentration. In 1966, he applied the methods of isotopic steady state and the one-pool isotope kinetic analysis to compare various aspects of cholesterol metabolism in cebus monkeys that were ingesting either complete or sulfur-deficient diets, with or without added cholesterol [77]. He found that dietary contribution to the miscible pool in cebus on high-cholesterol diets with complete protein was similar to that for cebus on diets that were prepared with sulfur-deficient protein (soy). He concluded that sulfur-deficient diets did not interfere with the negative feedback mechanism controlling cholesterol metabolism. One-pool analysis of decay of serum specific activity revealed that the fractional turnover rate was similar on soy protein and on casein feeding. The increase in pool size with addition of cholesterol to the diet was similar for animals receiving either protein. By examining the relationship between pool size and serum cholesterol concentration, MANN [77] concluded that there was a threshold pool size (8g) beyond which sulfur deficient animals are unable to maintain normal levels of serum cholesterol. Since body weight (which varied markedly among the animals of this study) was not considered, it might be questioned whether this conclusion was justified. If pool size per kilogram body weight is plotted against serum cholesterol concentration (using the data tabulated [77]), there appears to be no significant trend for either group. For a given pool size per kilogram body weight, however, the serum cholesterol does appear to be greater on sulfur-deficient than on normal diets. Thus sulfur-deficient diets apparently do alter the distribution of cholesterol between the serum and other tissues in the cebus monkey.

Cholesterol metabolism has been studied in cebus also by LOFLAND and co-workers [20, 70, 72]. In one study, cholesterol metabolism [72] was compared in juvenile and adult cebus while they were being fed basal and atherogenic diets. Adult cebus responded with greater increase in serum cholesterol concentration than juvenile cebus when a high-cholesterol, high-fat diet was substituted for the basal diet. Study of serum cholesterol specific activity following a single test meal containing radio-cholesterol revealed that a similar fraction of the labeled cholesterol appeared in the total serum pool in both age groups. The authors concluded that a difference in absorption of cholesterol between young and old cebus was not responsible for the difference in serum concentration. Estimates of pool size and fractional turnover rate were obtained by one-pool analysis of the decay of serum activity following a single intravenous injection of radio-cholesterol. Results were interpreted to indicate that turnover rate was greater in young than in adult cebus and that this was the only difference observed that might explain the differences in serum cholesterol concentration in the two age groups.

LOFLAND *et al.* [70] has reported additional results obtained in the juvenile and adult cebus monkeys used for the study discussed above [72]. The observed age difference in response of serum cholesterol concentration to an atherogenic diet occurred primarily in the males. Juvenile and adult females differed little in their response. Isotope kinetic data were analyzed using the two-pool method to determine size of rapidly miscible pool, production rate, half-time, and rate constant for transfer out of the rapidly miscible pool. An estimate of the total miscible body cholesterol pool was obtained by extrapolation from measured tissue concentration and organ weight. As might be expected, this method gave smaller pool size than was obtained earlier by one-pool kinetic analyses [72]. Fecal endogenous excretion of combined neutral and acidic steroids was measured on a subsample of these cebus monkeys by the isotopic balance method, and the fraction of the rapidly miscible pool than was excreted per day was calculated. The adult males which had the largest elevation of serum cholesterol concentration also had the largest expansion of the rapidly miscible and total body pools. The rate of excretion of cholesterol was also greatest in the adult males, but this represented a smaller fraction of the rapidly miscible pool. The activity per gram of tissue and specific activity of cholesterol in various tissues varied considerably from animal to animal. Confidence limits for parameters estimated by the two-pool model were not given; however, unless the residual error was less than 1 % it is likely that, with only 6 data points over 90 days following administration of isotope, the confidence limits for the derived parameters would be extremely wide, and conclusions based on the values derived might be questioned.

VII. Other Species

In a recent report, MANN [78] summarized results obtained in studies of cholesterol metabolism in three chimpanzees over a period of several years. He presented data on the size of the rapidly miscible pool and the long-term half-time for decay based on the two-pool analysis of the decline in specific activity following a single pulse dose given while the chimpanzees were being fed diets varying in source of protein and amount of added cholesterol. He determined sterol clearance by way of feces, sebum, urine and respiratory gases, and presented data on the effect of several hypocholesteremic agents on clearance by these various pathways.

RAYMOND *et al.* [100] have reported that studies using the two-pool isotopic kinetic analysis and the isotopic steady-state method have shown that *Macaca nemestrina* lacks the mechanism for negative feedback by diet cholesterol on hepatic cholesterol synthesis.

The response of serum lipids to atherogenic diets have been reported for a number of other species including *Macaca fascicularis* [61, 73, 96], *M. arctoides* [91], and *Ateles* sp. [24, 89, 97]; however, studies of other aspects of cholesterol metabolism have not been reported for these species.

VIII. Variations in Cholesterol Metabolism
within and among Species

It is apparent from results discussed above that there is a considerable difference in the way different animals and different species respond to perturbations in the steady state for cholesterol metabolism. An understanding of these differences could be valuable in design of experimental studies of atherosclerosis or lipid metabolism and may provide insight into mechanisms of regulation of cholesterol metabolism in man.

Most studies of whole-body cholesterol metabolism reported to date have involved limited numbers of a single species, and since experimental conditions and methods have varied considerably, it is difficult to obtain meaningful comparisons. Several investigators, however, have examined two or more species using similar methods, and measures of whole body cholesterol metabolism have been obtained in some of these comparative studies.

PORTMAN and ANDRUS [93] found little difference among three New World species (squirrel, woolly and cebus monkeys) in their response to an atherogenic diet. LEHNER *et al.* [68] fed an atherogenic diet to four species and found that response increased in the order: woolly monkey < squirrel monkey < African green monkey < stumptailed macaque. In a recent paper, COREY *et al.* [24] reported that although the relative responses of four species depended on the nature of the carbohydrate and lipid component of the diet, the effect on serum cholesterol concentration of adding cholesterol to the diet increased in the order: spider monkey < cebus < squirrel monkey < cynomolgus monkey. ANDRUS *et al.* [3] and BLATON *et al.* [14] have reported on the exaggerated response of the chimpanzee to atherogenic diets.

In most studies in man, the response of serum cholesterol to change in diet was relatively small, seldom exceeding 100 mg/dl, so that man appears to fall at the low end of the scale for sensitivity to changes in serum cholesterol concentration by manipulation of diet. Thus, with respect to the effect of atherogenic diet on serum cholesterol concentration, man, the baboon, and some New World primates (i.e. *Ateles*) are relatively refractile, while the macaques and the chimpanzee appear to have an exaggerated response, similar to the rabbit. Other species, such as most New World species, have an intermediate response.

In a study involving eight animals of each of three species [38], the mean concentration of cholesterol in the serum during 18 months on an atherogenic diet was greatest in rhesus, least in the baboon and at an intermediate level in the squirrel monkey. All differences were statistically significant. Because of the large difference in body size of these species, it is difficult to make comparisons of parameters, such as turnover or excretion rates which are directly related to body weight. However, examination of changes in rate of excretion of neutral and acidic steroids on transition from a basal chow to the atherogenic diet indicated that the relative ability to increase excretion of one or both of these fractions did not explain the differences in serum cholesterol concentration among these species. There were differences in the fraction of total body pool derived from the atherogenic diet and in the percentage of diet cholesterol absorbed that paralleled the differences in serum cholesterol concentration among these species. Thus, relative ability to absorb dietary cholesterol and possibly relative ability to inhibit synthesis are the most likely factors responsible for differences in response of serum lipids to atherogenic diet in the three species. There were also large differences in fractional turnover rate (slope of long-term decay of serum specific activity). It was severalfold greater in the squirrel monkey than in either the baboon or the rhesus monkey. The rate of decay for the rhesus was only slightly greater than that for the baboon. It is tempting to speculate that this large difference in fractional turnover rate between squirrel monkey and baboon or rhesus reflects a difference between New World and Old World monkeys, especially since the results of LEHNER et al. [68], LOFLAND et al. [70] and MANN [77] for cebus indicate a large fractional turnover rate, similar to those for the squirrel monkey. It is possible, however, that these differences in fractional turnover rate only reflect the differences in metabolic rate of animals of different size. The fractional turnover rates that have been reported for man, an Old World primate, are similar to those observed for the baboon and rhesus, so that in this respect the cholesterol metabolism of these Old World species more nearly approximates that of man than does that of the smaller New World species.

The number of animals in most studies did not contain enough individuals at the extremes of response to provide valid indications of the correlation among the various parameters of cholesterol metabolism. For this reason, LOFLAND et al. [71] and the author [40] have examined cholesterol metabolism in groups of animals selected from those at the extremes of response of serum cholesterol to atherogenic diets. As indicated in previous discussion of these studies, it appears that there is a major difference between squirrel and rhesus monkeys in the factors that might explain the differences in response of serum cholesterol among individual animals. In the squirrel monkey [71], the increase in the rate of excretion of acidic steroids on transition to atherogenic diet was much greater in the hyporesponding than in the hyperresponding group. These groups of squirrel monkeys did not differ either in absorption of dietary cholesterol or in the fraction of the miscible cholesterol pool derived from diet. In the rhesus monkey, however, the author [40] has found that the high- and low-responding animals differ significantly in the absorption of cholesterol either on atherogenic diets or on basal, cholesterol-free diets, and also differ in the fraction of the miscible body pool of cholesterol derived from atherogenic diets. These groups of rhesus also differed in the rate of excretion of acidic steroids; however, on both diets the high-responding group excreted more acidic steroids than the low-responding group. The rate of neutral steroid excretion was similar in the two groups for both rhesus [40] and squirrel monkeys [71]. In neither species did the change in the relative distribution of cholesterol among the body pools appear to explain the large difference in serum cholesterol concentrations.

It is likely that some of the apparent differences between rhesus and squirrel monkeys in the factors mediating the variability in response of individual animals is a result of the large difference in the nature of the diets used in these two studies. The study in squirrel monkeys [71] used a bulk-free liquid formula diet, whereas the study in rhesus [40] used a diet based on whole grains with considerable natural bulk. Speculation on the significance of the apparent difference between these species is not justified until the role of this and other differences in experimental procedure can be assessed.

A number of studies have been reported in which the effect of dietary factors on whole body metabolism in man have been examined [2, 11, 23, 52, 99, 114, 130]. Most of these studies have been conducted in metabolic wards and many of the patients have had abnormal serum lipids. The majority of such studies have used semisynthetic liquid formula diets. The emphasis has been on the effect of substituting unsaturated for saturated fat on the metabolism of cholesterol [2, 11, 23, 52, 114, 130]; this was found either to increase excretion of neutral sterols or bile acids, or to reduce the rate of absorption of dietary cholesterol, or to cause a shift of cholesterol from serum to other body pools, or to have no effect on these parameters. Within any study there was generally much variability from one subject to another. Thus, there appears to be no consistent finding regarding the mechanism causing lowering of serum cholesterol concentration on substitution of unsaturated fat for saturated fat.

In two reports [52, 99] the effect of adding cholesterol to a diet containing unsaturated fat was examined. Both studies indicated that adding cholesterol to the diet increased excretion of endogenous steroids but caused no change in excretion of bile acids. Both studies found that in some patients there is a decrease in cholesterol synthesis when cholesterol was added to the diet. Again, the variability among patients was large and the number of patients was too small to permit evaluation of the correlation between response of serum cholesterol concentration and other parameters of cholesterol metabolism.

IX. Concluding Remarks

One objective of studies of cholesterol metabolism in nonhuman primates is to obtain a comparison between these species and man in the response of various parameters of cholesterol metabolism to dietary or other factors. Before such comparison can be made, it is necessary that there be a known average response of 'normal' man to the factor under consideration. It is also necessary that the different species be studied under conditions as similar as possible to those used in eliciting the response in man or, at least that possible effects of deviations in procedure for different species are known. From the foregoing discussions it is clear that these conditions have not been fulfilled in most studies and therefore it is probably not valid at this time to state that any species is 'similar' to man or is more like man than any other with regard to most parameters. The results to date have shown that the methods applicable to man can also be used in other species of primates. Further effort should be made to define more precisely the validity of the nonhuman primate model for the study of dietary or other factors in cholesterol metabolism of man.

References

1 AHRENS. E. H., jr.: A review of the evidence that dependable sterol balance studies require a correction for the losses of neutral sterols that occur during intestinal transit; in JONES Proc. 2nd Int. Symp. Atherosclerosis, pp. 248–252 (Springer, Berlin 1970).

2 ALI, S. S.; KUKSIS, A., and BEVERIDGE, J. M. R.: Excretion of bile acids by three men on corn oil and butterfat diets. Can. J. Biochem. *44:* 1377–1388 (1966).

3 ANDRUS, S. B.; PORTMAN, O. W., and RIOPELLE, A. J.: Comparative studies of spontaneous and experimental atherosclerosis in primates. II. Lesions in chimpanzees including myocardial infarction and cerebral aneurysms. Prog. biochem. Pharmacol., vol. 4, pp. 393–419 (Karger, Basel 1968).

4 ANITSCHKOV, N. N. und CHALATOV, S.: Über experimentelle Cholinesterin-steatose. Zentbl. allg. Path. path. Anat. *24:* 1 (1913).

5 ANTONIS, A. and BERSOHN, I.: The influence of diet on fecal lipids in South African white and Bantu prisoners. Am. J. clin. Nutr. *11:* 142–155 (1962).

6 ARMSTRONG, M. L.: Atherosclerosis in rhesus and cynomolgus monkeys; in STRONG Atherosclerosis in primates (this volume).

7 ARMSTRONG, M. L.; CONNOR, W. E., and WARNER, E. D.: Tissue cholesterol concentration in the hypercholesterolemic rhesus monkey. Archs Path. *87:* 87–92 (1969).

8 ARMSTRONG, M. L.; CONNOR, W. E., and WARNER, E. D.: Xanthomatosis in rhesus monkeys fed a hypercholesterolemic diet. Archs Path. *84:* 227–237 (1967).

9 ARMSTRONG, M. L. and MEGAN, M.: Plasma and carcass cholesterol in rhesus monkeys after low and intermediate levels of dietary cholesterol. Circulation *43–44:* suppl. II, p. 3 (1971).

10 ARNESJO, B.; NILSSON, A.; BARROWMAN, J., and BORGSTROM, B.: Intestinal digestion and absorption of cholesterol and lecithin in the human. Scand. J. Gastroent. *4:* 653–665 (1969).

11 AVIGAN, J. and STEINBERG, D.: Sterol and bile acid excretion in man and the effects of dietary fat. J. clin. Invest. *44:* 1845–1856 (1965).

12 BERGER, J.; REDINGER, R. N., and SMALL, D. M.: Instrument for sampling and measuring bile flow. Med. Biol. Engng *8:* 19–24 (1970).

13 BHATTACHARYYA, A. K.; CONNOR, W. E., and SPECTOR, A. A.: Excretion of sterols from the skin of normal and hypercholesterolemic humans; implications for sterol balance studies. J. clin. Invest. *51:* 2060–2070 (1972).

14 BLATON, V.; VANDAMME, D., and PEETERS, H.: Chimpanzee and baboon as biochemical models for human atherosclerosis; in GOLDSMITH and MOOR-JANKOWSKI Medical primatology 1972, part III, pp. 306–312 (Karger, Basel 1972).

15 BORGSTROM, B.: Quantification of cholesterol absorption in man by fecal analysis after the feeding of a single isotope-labeled meal. J. Lipid Res. *10:* 331–337 (1969).

16 CAMPBELL, C. B.; COWLEY, D. J., and DOWLING, R. H.: Dietary factors affecting biliary lipid secretion in the rhesus monkey. Eur. J. clin. Invest. *2:* 332–341 (1971–72).

17 CAMPBELL, C. B. and DOWLING, R. H.: Effect of diet on bile volume and total bile salt secretion in rhesus monkeys. Aust. Ann. Med. *19:* 422 (1970).

18 CHOBANIAN, A. V.; BURROW, B. A., and HOLLANDER, W.: Body cholesterol metabolism in man. II. Measurement of the body cholesterol miscible pool and turnover rate. J. clin. Invest. *41:* 1738–1744 (1962).

19 CHOBANIAN, A. V. and HOLLANDER, W.: Body cholesterol metabolism in man. I. The equilibration of serum and tissue cholesterol. J. clin. Invest. *41:* 1732–1737 (1962).

20 CLARKSON, T. B.; BULLOCK, B. C., and LEHNER, N. D. M.: Pathologic characteristics of atherosclerosis in new world monkeys. Prog. biochem. Pharmacol., vol. 4, 420–428 (Karger, Basel 1968).

21 CLARKSON, T. B.; HAMM, T. E.; BULLOCK, B. C., and LEHNER, N. D. M.: Old World monkeys in atherosclerosis research; in STRONG Atherosclerosis in primates (this volume).

22 CLARKSON, T. B.; LEHNER, N. D. M.; BULLOCK, B. C.; LOFLAND, H. B., and WAGNER, W. D.: Atherosclerosis of New World monkeys; in STRONG Atherosclerosis of primates (this volume).

23 CONNOR, W. E.; WITIAK, D. T.; STONE, D. B., and ARMSTRONG, M. L.: Cholesterol balance and fecal neutral steroid and bile acid excretion in normal men fed dietary fats of different fatty acid composition. J. clin. Invest. *48:* 1363–1375 (1969).

24 COREY, J. E.; HAYES, K. C.; DORR, B., and HEGSTED, D. M.: Comparative lipid response of four primates species to dietary changes in fat and carbohydrate. Atherosclerosis *19:* 119–134 (1974).

25 COX, G. E.; TAYLOR, C. B.; COX, L. G., and COUNTS, M. A.: Atherosclerosis in rhesus monkeys. I. Hypercholesterolemia induced by dietary fat and cholesterol. Archs Path. *66:* 32–52 (1958).

26 COX, G. E., NELSON, L. G.; WOOD, W. B., and TAYLOR, C. B.: Effect of dietary cholesterol on cholesterol synthesis in monkeys tissues *in vitro.* Fed. Proc. Fed Am. Socs exp. Biol. *13:* 31 (1954).

27 DAVIGNON, J.; SIMMONDS, W. J., and AHRENS, E. H., jr.: Usefulness of chromic oxide as an internal standard for balance studies in formula-fed patients and for assessment of colonic function. J. clin. Invest. *47:* 127–138 (1968).

28 DELL, R. B.; SCIACCA, R.; LIEBERMAN, K.; CASE, D. B., and CANNON, P. J.: A weighted least squares technique for the analysis of kinetic data and its application to the study of renal ^{133}xenon washout in dogs and man. Circulation Res. *32:* 71–84 (1973).

29 DENBESTEN, L.; CONNOR, W. E.; KENT, T. H., and LIN, D.: Effect of cellulose in the diet on the recovery of dietary plant sterols from the feces. J. Lipid Res. *11:* 341–345 (1970).

30 DIETSCHY, J. M. and WILSON, J. D.: Cholesterol synthesis in the squirrel monkey. Relative rates of synthesis in various tissues and mechanisms of control. J. clin. Invest. *47:* 166–174 (1968).

31 DIETSCHY, J. M. and WILSON, J. D.: Regulation of cholesterol metabolism (3 parts). New Engl. J. Med. *282:* 1128–1138, 1179–1183, 1241–1249 (1970).

32 DIETSCHY, J. M. and SIPERSTEIN, M. D.: Effect of cholesterol feeding and fasting on sterol synthesis in seventeen tissues of the rat. J. Lipid Res. *8:* 97–104 (1967).

33 DOWLING, R. H.: The enterohepatic circulation. Gastroenterology *62:* 122–140 (1972).

34 DOWLING, R. H.; MACK, E., and SMALL, D. M.: Biliary lipid secretion and bile composition after acute and chronic interruption of the enterohepatic circulation in the rhesus monkey. IV. Primate biliary physiology. J. clin. Invest. *50:* 1917–1926 (1971).

35 DOWLING, R. H.; MACK, E., and SMALL, D. M.: Effects of controlled interruption of the enterohepatic circulation of bile acids by biliary diversion and by ileal resection on bile salt secretion synthesis and pool size in the rhesus monkey. J. clin. Invest. *49:* 19–24 (1970).

36 DOWLING, R. H.; MACK, E.; PICOTT, J.; BERGER, J., and SMALL, D. M.: Experimental model for the study of the enterohepatic circulation of bile in rhesus monkeys. J. Lab. clin. Med. *72:* 169–176 (1968).

37 DOWLING, R. H.; WHITE, J., and LOWLEY, D., *et al.*: The effect of dietary fat on bile composition in monkeys with intact EHCs. Eur. J. clin. Invest. *1:* 39a (1971).

38 EGGEN, D. A.: Cholesterol metabolism in the rhesus monkey, squirrel monkey and baboon. J. Lipid Res. *15:* 139–145 (1974).

39 EGGEN, D. A.; NEWMAN, W. P., and STRONG, J. P.: Absorption and turnover of cholesterol in the baboon. The baboon in medical research. Proc. 2nd Int. Symp., vol. 2, pp. 559–569 (Univ. of Texas Press, Austin 1967).

40 EGGEN, D. A. and STRONG, J. P.: Cholesterol metabolism in rhesus: extremes of response of serum cholesterol to atherogenic diet. Circulation *45–46:* suppl. II, p. 250 1972).

41 EGGEN, D. A. and STRONG, J. P.: Diet and cholesterol metabolism in young and old baboons. Atherosclerosis *12:* 359–369 (1970).

42 FRASER, R.; DUBIEN, L.; MUSIL, F.; FOSSLIEN, E., and WISSLER, R. W.: Transport of cholesterol in thoracic duct lymph and serum of rhesus monkeys fed cholesterol with various food fats. Atherosclerosis *16:* 203–216 (1972).

43 GETZ, G. S.; VESSELINOVITCH, D., and WISSLER, R. W.: A dynamic pathology of atherosclerosis. Am. J. Med. *46:* 657–674 (1969).

44 GIDEZ, L. I. and EDER, H. A.: Cholesterol turnover in man; in HORNING Effects of drugs on synthesis and mobilization of lipids, pp. 67–75 (Pergamon Press, Oxford 1963).

45 GIDEZ, L. I. and EDER, H. A.: Cholesterol turnover in man. Proc. 1st Int. Pharmacological Meet. Effects of drugs on synthesis and mobilization of lipids, vol. 2, pp. 67–76 (Pergamon Press, Oxford 1963).

46 GLENN, F. and McSHERRY, C. K.: The baboon and experimental cholelithiasis. Archs Surg., Chicago *100:* 105–108 (1970).

47 GOODMAN, D. S. and NOBLE, R. P.: Turnover of plasma cholesterol in man. J. clin. Invest. *47:* 231–240 (1968).

48 GOODMAN, D. S.; NOBLE, R. P., and DELL, R. B.: Three-pool model of the long-term turnover of plasma cholesterol in man. J. Lipid Res. *14:* 178–188 (1973).

49 GRUNDY, S. M.: Quantitative isolation and gas-liquid chromatographic analysis of total fecal bile acids. J. Lipid Res. *6:* 397–410 (1965).

50 GRUNDY, S. M. and AHRENS, E. H., jr.: An evaluation of the relative merits of two methods for measuring the balance of sterols in man: isotopic balance versus chromatographic analysis. J. clin. Invest. *45:* 1503–1515 (1966).

51 GRUNDY, S. M. and AHRENS, E. H., jr.: Measurements of cholesterol turnover, synthesis and absorption in man, carried out by isotope kinetic and sterol balance methods. J. Lipid Res. *10:* 91–107 (1969).

52 GRUNDY, S. M. and AHRENS, E. H., jr.: The effects of unsaturated dietary fats on absorption, excretion, synthesis and distribution of cholesterol in man. J. clin. Invest. *49:* 1135–1152 (1970).

53 GRUNDY, S. M.; AHRENS, E. H., jr., and DAVIGNON, J.: The interaction of cholesterol absorption and cholesterol synthesis in man. J. Lipid Res. *10:* 304–314 (1969).

54 GRUNDY, S. M.; AHRENS, E. H., jr., and SALEN, G.: Dietary β-sitosterol as an internal standard to correct for cholesterol losses in sterol balance studies. J. Lipid Res. *9:* 374–387 (1968).

55 HELLMAN, L.; ROSENFELD, R. S.; INSULL, W., jr., and AHRENS, E. H., jr.: Intestinal excretion of cholesterol: a mechanism for regulating plasma levels. J. clin. Invest. *36:* 898 (1957).

56 HERMAN, A. H.; REDINGER, R. N., and SMALL, D. M., *et al.*: The effects of biliary tract pressure on bile flow, bile salt secretion and bile salt synthesis in the primate. Surgery, St Louis *70:* 140–146 (1971).

57 HUEPER, W. C.: Experimental atherosclerosis in macacus rhesus monkeys. Am. J. Path. *22:* 1287–1289 (1946).

58 JAVITT, N. B. and MCSHERRY, C.: Pathogenesis of cholesterol gallstones. Hosp. Pract. *8:* 39–48 (1973).

59 KAPLAN, J. A.; COX, G. E., and TAYLOR, C. B.: Cholesterol metabolism in man. Archs Path. *76:* 359–368 (1963).

60 KAWAMURA, R.: Neue Beiträge zur Morphologie und Physiologie der Cholinesterin-steatose, p. 159 (G. Fischer, Jena 1927); cited by ARMSTRONG and WARNER Morphology and distribution of diet-induced atherosclerosis in rhesus monkeys. Archs Path. *92:* 395–401 (1971).

61 KRAMSCH, D. M. and HOLLANDER, W.: Occlusive atherosclerotic disease of the coronary arteries in monkey *(Macaca irus)* induced by diet. Expl molec. Path. *9:* 1–22 (1968).

62 KRITCHEVKY, D.: Cholesterol metabolism in the baboon. Trans. N.Y. Acad. Sci. *32:* 821–831 (1970).

63 KRITCHEVSKY, D.; DAVIDSON, L. M.; SHAPIRO, I. L.; KIN, H. K.; KITAGAWA, M.; MALHOTRA, S.; NAIR, P. P.; CLARKSON, T. B.; BERSOHN, I., and WINTER, P. A. D.: Lipid metabolism and experimental atherosclerosis in baboons: influence of cholesterol-free, semi-synthetic diets. Am. J. clin. Nutr. *27:* 29–50 (1974).

64 KRITCHEVSKY, D.; JASTREMSKY, J. A.; SHAPIRO, I. L.; KRITCHEVSKY, E. S., and NAIR, P. P.: Effects of free fatty acids on cholesterol metabolism in the baboon. The baboon in medical research, vol. 2, pp. 547–557 (Univ. of Texas Press, Austin 1967).

65 KRITCHEVSKY, D.; KRITCHEVSKY, E.; NAIR, P. P.; JASTREMSKY, J., and SHAPIRO, I. L.: Effect of free fatty acids on cholesterol metabolism in the baboon. Nutr. Diet. *9:* 283–299 (1967).

66 KRITCHEVSKY, D.; SHAPIRO, L., and WERTHESSEN, N. T.: Biosynthesis of cholesterol in the baboon. Biochem. biophys. Acta *65:* 556–557 (1962).

67 LEHNER, N. D. M.; CLARKSON, T. B.; BELL, F. P.; ST. CLAIR, R. W., and LOFLAND, H. B.: Effects of insulin deficiency, hypothyrodism and hypertension on cholesterol metabolism in the squirrel monkey. Expl molec. Path. *16:* 109–123 (1972).

68 LEHNER, N. D. M.; CLARKSON, T. B.; BULLOCK, B. C.; LOFLAND, H. B.; ST. CLAIR, R. W., and PRICHARD, R. W.: Studies on atherosclerosis of some nonhuman primates; in GOLDSMITH and MOOR-JANKOWSKI Medical primatology 1970, pp. 873–885 (Karger, Basel 1971).

69 LOFLAND, H. B., jr.; CLARKSON, T. B., and BULLOCK, B. C.: Whole body sterol metabolism in squirrel monkeys *(Saimiri sciureous)*. Expl molec. Path *13:* 1–11 (1970).

70 LOFLAND, H. B., jr.; CLARKSON, T. B.; ST. CLAIR, R. W.; LEHNER, N. D. M., and BULLOCK, B. C.: Atherosclerosis in *Cebus albifrons* monkeys. I. Sterol metabolism. Expl molec. Path. *8:* 302–313 (1968).

71 LOFLAND, H. B., jr.; CLARKSON, T. B.; ST. CLAIR, R. W., and LEHNER, N. D. M.: Studies on the regulation of plasma cholesterol levels in squirrel monkeys of two genotypes. J. Lipid Res. *13:* 39–47 (1972).

72 MacNintch, J. E.; St. Clair, R. W.; Lehner, N. D. M.; Clarkson, T. B., and Lofland, H. B.: Cholesterol metabolism and atherosclerosis in cebus monkeys in relation to age. Lab. Invest. *16:* 444–452 (1967).

73 Malmaos, H.; Wigand, G., and Kockum, I.: Experimental hypercholesterolemia and hypertriglyceridemia in cynomolgus monkeys fed saturated fat and cholesterol. J. Atheroscler. Res. *5:* 474–482 (1965).

74 Manalo-Estrella, P.; Cox, G. E., and Taylor, C. B.: Atherosclerosis in rhesus monkeys. VII. Mechanism of hypercholesteremia: hepatic cholesterologenesis and the hypercholesteremic threshold of dietary control. Archs Path. *76:* 413–423 (1963).

75 Mann, G. V.: Effects of sulfur compounds on hypercholesteremia and growth in cysteine-deficient monkeys. Am. J. clin. Nutr. *8:* 491–497 (1960).

76 Mann, G. V.: Sterol metabolism in monkeys; dietary cholesterol and protein in maintenance of sterol compensation. Circulation Res. *9:* 838–844 (1961).

77 Mann, G. V.: Cystine deficiency and cholesterol metabolism in primates. Circulation Res. *18:* 205–212 (1966).

78 Mann, G. V.: Sterol metabolism in the chimpanzee; in Goldsmith and Moor-Jankowski Medical primatology 1972, part III, pp. 324–335 (Karger, Basel 1972).

79 Mann, G. V. and Andrus, S. B.: Xanthomatosis and atherosclerosis produced by diet in an adult rhesus monkey. J. Lab. clin. Med. *48:* 533–559 (1956).

80 Mann, G. V.; Andrus, S. B.; McNally, A., and Stare, F. J.: Experimental atherosclerosis in cebus monkey. J. exp. Med. *98:* 195–218 (1953).

81 Manning, P. J.; Clarkson, T. B., and Lofland, H. B.: Cholesterol absorption, turnover, and excretion rates in hypercholesterolemic rhesus monkeys. Expl molec. Path. *14:* 75–89 (1971).

82 McGill, H. C., jr.; Mott, G. E., and Bramblett, C. A.: Experimental atherosclerosis in the baboon; in Strong Atherosclerosis in primates (this volume).

83 McGill, H. C., jr.; Strong, J. P.; Holman, R. L., and Werthessen, N. T.: Arterial lesions in the Kenya baboon. Circulation Res. *8:* 670–679 83.

84 McSherry, C. K. and Glenn, F.: The baboon and experimental cholelithiasis. Archs Surg., Chicago *100:* 105–108 (1970).

85 McSherry, C. K.; Javitt, N. B., and DeCarvalho, J., *et al.:* Cholesterol gallstones and the chemical composition of bile in baboons. Ann. Surg. *173:* 569–577 (1971).

86 Miettinen, T. A.: Quantitative isolation and gas-liquid chromatographic analysis of total dietary and fecal neutral steroids. J. Lipid Res. *6:* 411–424 (1965).

87 Morris, M. D.; Chaikoff, I. L.; Felts, J. M.; Abraham, S., and Fansah, N. O.: The origin of serum cholesterol in the rat: diet versus synthesis. J. biol. Chem. *224:* 1039–1045 (1957).

88 Morrow, A. C. and Terry, M. W.: Cholesterol and other plasma lipids in nonhuman primates: a tabulation from the literature. Primate Information Letter, Regional Primate Research Center at the University of Washington, Seattle (1972).

89 Newman, W. P., III; Eggen, D. A., and Strong, J. P.: Comparison of arterial lesions and serum lipids in spider and rhesus monkeys on an egg and butter diet. Atherosclerosis 19: 75–86 (1974).

90 Osuga, T. and Portman, O. W.: Experimental formation of gallstones in the squirrel monkey. Gastroent. jap. *7:* 323 (1972).

91 PICK, R. and KATZ, L. N.: Cholesterol-fat induced atherosclerosis in the stumptail macaque *(Macaca speciosa)*. Circulation *39:* suppl. III, p. 20 (1969).

92 PITKIN, R. M.; CONNOR, W. E., and LIN, D. S.: Cholesterol metabolism and placental transfer in the pregnant rhesus monkey. J. clin. Invest. *51:* 2584–2592 (1972).

93 PORTMAN, O. W. and ANDRUS, S. B.: Comparative evaluation of three species of New World monkeys for studies of dietary factors, tissue lipids, and atherogenesis. J. Nutr. *87:* 429–438 (1965).

94 PORTMAN, O. and ILLINGWORTH, D. R.: Arterial metabolism in primates; in STRONG Atherosclerosis in primates (this volume).

95 PORTMAN, O. W. and SINISTERRA, L.: Dietary fat and hypercholesteremia in the cebus monkey. II. Esterification and disappearance of cholesterol-4-C^{14}. J. exp. Med. *106:* 727–742 (1957).

96 PRATHAP, K. and LAU, K. S.: Spontaneous and experimental arterial lesions in the malaysian long-tailed monkey; in GOLDSMITH and MOOR-JANKOWSKI Medical primatology 1972, part III, pp. 343–349 (Karger, Basel 1972).

97 PUCAK, G. J.; LEHNER, N. D. M.; CLARKSON, T. B.; BULLOCK, B. C., and LOFLAND, H. B.: Spider monkeys (*Ateles* sp.) as animal models for atherosclerosis research. Expl. molec. Path. *18:* 32–49 (1973).

98 QUINTAO, E.; GRUNDY, S. M., and AHRENS, E. H., jr.: An evaluation of four methods for measuring cholesterol absorption by the intestine in man. J. Lipid Res. *12:* 221–232 (1971).

99 QUINTAO, E.; GRUNDY, S. M., and AHRENS, E. H., jr.: Effects of dietary cholesterol on the regulation of total cholesterol in man. J. Lipid Res. *12:* 233–247 (1971).

100 RAYMOND, T. L.; LOFLAND, H. B., and CLARKSON, T. B.: Lack of feedback control of cholesterol synthesis in pigtail macaques. Circulation *68:* suppl. IV, p. 43 (1973).

101 REDINGER, R. N.; HERMAN, A. H., and SMALL, D. M.: Effects of biliary tract obstruction and illness on total secretion and relative composition of biliary bile salt, phospholipid and cholesterol in the rhesus monkey. Gastroenterology *58:* 1058 (1970).

102 REDINGER, R. N.; HERMAN, A. H., and SMALL, D. M.: Effect of diet on bile composition in the rhesus monkey. Gastroenterology *60:* 198 (1971).

103 REDINGER, R. N. and SMALL, D. M.: Bile composition, bile salt metabolism and gallstones. Archs intern. Med. *130:* 618–630 (1972).

104 REDINGER, R. N. and SMALL, D. M.: Primate biliary physiology. VIII. The effect of phenobarbital upon bile salt synthesis and pool size, biliary lipid secretion, and bile composition. J. clin. Invest. *52:* 161–172 (1973).

105 RUDEL, L. L. and LOFLAND, H. B.: Lipoproteins in nonhuman primates; in STRONG Atherosclerosis in primates (this volume).

106 SAVAGE, N.; GILLMAN, J., and GILBERT, C.: Unimpaired synthesis of cholesterol from 2-^{14}C-sodium pyruvate in diabetic baboon *(Papio ursinus)*. S. Afr. J. med. Sci. *25:* 71–175 (1960).

107 SCHOENFIELD, L. J.: Animal models of gallstone formation. Gastroenterology *63:* 189–191 (1972).

108 SHAPIRO, I. L.; JASTREMSKY, J. A.; EGGEN, D. A., and KRITCHEVSKY, D.: Cholesterol metabolism in the baboon. Lipids *3:* 136–142 (1968).

109 SMALL, D. M.; DOWLING, R. H., and REDINGER, R. N.: The enterohepatic circulation of bile salts. Archs intern. Med. *130:* 552–573 (1972).

110 SMALL, D. M.; REDINGER, R. N., and HERMAN, A. H.: Primate biliary physiology. X. Effects of diet and fasting on biliary lipid secretion and relative composition and bile salt metabolism in the rhesus monkey. Gastroenterology *64:* 610–620 (1973).

111 SNEDECOR, G. W. and COCHRAN, W. G.: Statistical methods; 6th ed., p. 465 (Iowa State Univ. Press, Ames 1967).

112 SODHI, H. S.; HORLICK, L.; NAZIR, D. J., and KUDCHODKAR, B. J.: A simple method for calculating absorption of dietary cholesterol in man. Proc. Soc. exp. Biol. Med. *137:* 277–279 (1971).

113 SODHI, H. S.; KUDCHODKAR, B. J.: VARUGHESE, P., and DUNCAN, D.: Validation of the ratio method for calculating absorption of dietary cholesterol in man. Proc. Soc. exp. Biol. Med. *145:* 107–111 (1974).

114 SPRITZ, N.; AHRENS, E. H., jr., and GRUNDY, S.: Sterol balance in man as plasma cholesterol concentrations are altered by exchanges of dietary fats. J. clin. Invest. *44:* 1482–1493 (1969).

115 STARY, H. C.: Progression and regression of experimental atherosclerosis; in GOLD- SMITH and MOOR-JANKOWSKI Medical primatology 1972, part III, pp. 356–367(Karger, Basel 1972).

116 STRASBERG, S. M.; DORN, B. C.; REDINGER, R. N.; SMALL, D. M., and EGDAHL, R. H.: Effects of alteration of biliary pressure on bile composition – a method for study. V. Primate biliary physiology. Gastroenterology *61:* 357–362 (1971).

117 STRASBERG, S. M.; DORN, B. C.; SMALL, D. M., and EGDAHL, R. H.: The effects of biliary tract pressure on bile flow, bile salt secretion and bile salt synthesis in the primate. Surgery, St. Louis *70:* 140–146 (1971).

118 STRONG, J. P.; EGGEN, D. A.; NEWMAN, W. P., III, and MARTINEZ, R. D.: Naturally occurring and experimental atherosclerosis in primates. Ann. N.Y. Acad. Sci. *149:* 882–894 (1968).

119 STRONG, J. P. and MCGILL, H. C., jr.: Diet and experimental atherosclerosis in baboons. Am. J. Path. *50:* 669–690 (1967).

120 STRONG, J. P.; ROSAL, J.; DEUPREE, R. H., and MCGILL, H. C., jr.: Diet and serum cholesterol levels in baboons. Expl molec. Path. *5:* 82–91 (1966).

121 SUGANO, M. and PORTMAN, O. W.: Fatty acid specificities and rates of cholesterol esterification *in vivo* and *in vitro*. Archs Biochem. Biophys. *107:* 341–351 (1964).

122 TAYLOR, C. B.; COX, G. E.; MANALO-ESTRELLA, P., and SOUTHWORTH, J.: Atherosclerosis in rhesus monkeys. II. Arterial lesions associated with hypercholesteremia induced by dietary fat and cholesterol. Archs Path. *74:* 16–34 (1962).

123 TAYLOR, C. B.; MANALO-ESTRELLA, P., and COX, G. E.: Atherosclerosis in rhesus monkeys. V. Marked diet-induced hypercholesteremia with xanthomatosis and severe atherosclerosis. Archs Path. *76:* 239–249 (1963).

124 TAYLOR, C. B.; PATTON, D. E., and COX, G. E.: Atherosclerosis in rhesus monkeys. VI. Fatal myocardial infarction in a monkey fed fat and cholesterol. Archs Path. *76:* 404–412 (1963).

125 TREADWELL, C. R. and VEHOUNEY, G. V.: Cholesterol absorption; in CODE Handbook of physiology. A critical, comprehensive presentation of physiological know-

ledge and concept, vol. 3, sect. 6, pp. 1407–1438 (Am. Physiological Society, Washington 1968).

126 WILSON, J. D.: Biosynthetic origin of serum cholesterol in the squirrel monkey: evidence for a contribution by the intestinal wall. J. clin. Invest. *47:* 175–187 (1968).

127 WILSON, J. D.: The measurement of the exchangeable pools of cholesterol in the baboon. J. clin. Invest. *49:* 655–665 (1970).

128 WILSON, J. D.: The relation between cholesterol absorption and cholesterol synthesis in the baboon. J. clin. Invest. *51:* 1450–1458 (1972).

129 WILSON, J. D. and LINDSEY, C. A., jr.: Studies on the influence of dietary cholesterol on cholesterol metabolism in the isotopic steady state in man. J. clin. Invest. *44:* 1805–1814 (1965).

130 WOOD, P. D. S.; SHIODA, R., and KINSELL, L. W.: Dietary regulation of cholesterol metabolism. Lancet *1966:* 604–607.

131 ZILVERSMIT, D. B.: A single blood sample dual isotope method for the measurement of cholesterol absorption in rats. Proc. Soc. exp. Med. *140:* 862–865 (1972).

132 ZILVERSMIT, D. B. and WENTWORTH, R. A.: Determination of the optimal priming dose for achieving an isotopic steady state in a two-pool system: application to the study of cholesterol metabolism. J. Lipid Res. *11:* 551–557 (1970).

Dr. D. A. EGGEN, Department Pathology, Louisiana State University Medical Center, 1542 Tulane Avenue, *New Orleans, LA 70112* (USA)

Prim. Med., vol. 9, pp. 300–320 (Karger, Basel 1976)

Reversibility of Fatty Streaks in Rhesus Monkeys[1]

JACK P. STRONG, DOUGLAS A. EGGEN and H. C. STARY

Department of Pathology,
Louisiana State University Medical Center, New Orleans, La.

Contents

[1] Supported by USPHS, NIH Grant HLO8974.

I. Introduction

The regression of experimentally induced atherosclerotic lesions in primates has been the subject of intensive investigation in the past five years. Armstrong [2] and Strong [19] in this volume have reviewed the reports published on these investigations. This chapter describes a recently completed experiment on the regression of diet-induced fatty streaks in rhesus monkeys. Lesions in animals fed an atherogenic diet for twelve weeks, and in animals fed the same diet for twelve weeks and then returned to a control diet were evaluated by a combination of gross morphologic techniques, histological and histochemical methods, electron microscopy, biochemical analyses, and radioautography. The gross and histological changes [9, 16], biochemical changes [13], ultrastructural features [6, 17], and cell proliferation [17] of regressing atherosclerotic lesions are or will be published separately. This chapter includes the salient findings from these reports and integrates them in terms of significance.

II. Materials and Methods

A. Design

The details of design and methodology are given in the papers referred to above and will only be summarized.

Forty-four male rhesus monkeys were conditioned by an animal importer for a minimum of 30 days on a commercial monkey food supplemented with fruits. After arrival in our laboratory the animals were placed on a basal diet consisting of a low-fat, low-cholesterol commercial primate food. Two groups of three monkeys (group IV and V) and one group of two monkeys (group VII) were selected randomly to be used as controls. The controls were fed only the commercial primate food throughout.

The remaining animals were placed on a high-fat, high-cholesterol atherogenic diet (table I) formulated by adding fat and cholesterol to ground commercial primate ration. One animal refused the diet and was removed from the study. Total serum cholesterol concentration was determined at 1, 2, 3, 4, 6, 8, 10 and 12 weeks after beginning the atherogenic diet. The animals were ranked on mean serum cholesterol concentration during the 12 weeks, and six animals with highest and six with lowest concentration were removed for a separate project to study response of cholesterol metabolism to cholesterol-supplemented diets [7, 8]. The remaining 24 animals had mean serum cholesterol concentrations ranging from 294 to 555 mg/dl. These animals were assigned to four groups in a randomized block design so that each group of six animals had similar mean cholesterol levels during the 12 weeks on the atherogenic diet.

After 12 weeks on the atherogenic diet the four groups of six animals each were returned to the basal diet. Subsequent treatment of these groups and the groups of control animals

Table I. Composition of atherogenic diet

Ingredient	g/kg
Monkey chow[1]	525
Casein[2]	85
Salt mixture (Hegsted IV)	11
Vitamin mixture[3]	5
Butter (unsalted)	150
Beef tallow	45
Cholesterol (USP)[2]	3.5
Water	175.5

[1] D & G baked primate ration, Price Wilhoite Co., Frederick, Md.
[2] Nutritional Biochemical Corp., Cleveland, Ohio.
[3] General Biochemical Inc., Chagrin Falls, Ohio.

Table II. Design of experiment and concentration (mg/dl) of total cholesterol in serum for 7 groups of rhesus

Group		N	Initial serum cholesterol	Diet period			
				atherogenic		basal	
				duration, weeks	mean serum cholesterol	duration, weeks	terminal serum cholesterol
I	Progression	6	120	12	444	2	226
II	Regression 32	6	129	12	443	32	109
III	Regression 64	6	147	12	450	64	122
IV	Control 14	3	153	0	–	14	142
V	Control 76	3	127	0	–	76	91
Proliferation subgroup							
VI {	Progression	2	170	12	420	3	173
	Regression 16	2	128	12	435	16	146
	Regression 40	2	122	12	450	40	133
VII	Control 52	2	117	0	–	52	120

is given in table II. One group of six animals (group I) was autopsied two weeks after return to basal diet; at that time mean serum cholesterol had dropped to 226 mg/dl or about twice the prediet basal concentration. This group reflects the extent and type of lesions existing just before regression was expected to begin. Group II and III fed the basal diet for 32 and 64 weeks, respectively, after 12 weeks on the atherogenic diet provided an indication of regression of lesions from the level attained in group I animals. One of the two groups of three control monkeys (group IV), was autopsied at the same time as group I. A second control group (group V) was autopsied at the same time as group III.

The last of the four groups of six animals (group VI) was used for studies on cell proliferation and fine structure [17]. Subgroups of two animals each were killed 3, 16, and 40 weeks after return to the regression diet. A group of two control animals (group VII) was killed with the subgroup on regression diet for 40 weeks.

All animals were caged individually, offered food once a day, and provided water *ad libitum*. They were in excellent health throughout the experiment.

B. Autopsy Procedure

Under anesthesia, a terminal blood sample was obtained, the monkeys were killed by exsanguination and complete autopsy was performed with special emphasis on the arterial system.

For groups I, II, III, IV, and V the aorta was removed from the body, bisected longitudinally, and blocks of intima-media were taken for electron microscopy. The left half of the aorta was frozen and stored at $-20°C$ until analyzed chemically. The right half was mounted on chipboard, fixed in buffered formalin, stained with Sudan IV, and sealed in a plastic bag for subsequent visual and histologic examination.

The heart was removed and weighed. Blocks of tissue from the proximal left coronary artery were prepared for electron microscopic examination. The other major vessels were removed and prepared for gross and histological examination in a manner similar to that described for the aorta. These vessels included the brachial, iliac-femoral and renal arteries. The common carotid arteries were obtained from most of the animals.

The percentage of the intimal surface with grossly detectable sudanophilic lesions was evaluated visually on two separate occasions using blindly coded specimens and the methods described for the International Atherosclerosis Project [11].

Animals used for cell proliferation studies (groups VI, VII) received intravenous injections of [3H]-thymidine 8 and 1 h before killing. Blocks of tissue for electron microscopy were taken from one half of the longitudinally bisected aorta and the remainder was stained with Sudan IV for gross evaluation. The other half of the aorta was rolled into two coils, fixed in Carnoy's solution and processed for histologic sections and radioautographs.

C. Serum and Tissue Lipids

Blood samples were obtained after on overnight fast, allowed to clot for 2 h at room temperature, and centrifuged for 15 min at 2,000 rpm to separate the serum. The serum was frozen until it was analyzed.

Serum total cholesterol and triglyceride concentrations were determined on blind duplicate samples by the methods of ABELL *et al.* [1] and VAN HANDEL and ZILVERSMIT [12], respectively. The intima was stripped from the left half of the aorta by fine dissection under magnification. This procedure resulted in stripping a portion of the media along with the intima. The dissected intima-media was minced and extracted by the Folch procedure [10]. Total cholesterol (after saponification) and free cholesterol were determined by gas-liquid chromatography. The cholesterol esters were separated by thin layer chromatography, and the fatty acids esterified to cholesterol were determined by gas-liquid chromatography.

D. Histologic Evaluation

1. Aorta

For groups I to V blocks of tissue were taken from three standard regions of the right half of the aorta. Each block was bisected, one half was mounted in paraffin and the other in gelatin, so that sections for histologic evaluation could be obtained from opposing surfaces. Sections cut from the paraffin block were stained with hematoxylin and eosin (HE), and van GIESON elastin stain (EVG). Frozen sections cut from the gelatin-embedded block were stained with Oil-Red-O (ORO).

These 72 sections were identified by randomly assigned two-digit numbers, and four different observers examined each slide and assigned a numeric grade for: (a) *intimal thickness* (measured by estimating the number of cell layers at the thickest part of the intima); (b) *intimal collagen* (grade of one to three recorded); (c) *intimal lipid* (a grade of one to five was assigned by comparing the ORO section with a standard series of four color photographs arranged in order of increasing amounts of intimal lipid), and (d) *medial lipid* (assessed by assigning a grade of one to three).

A consensus grade was determined for each measure on each block. Analysis of variance was used to determine the statistical significance of the differences in mean grade for the three blocks from each of the groups of animals.

2. Coronary Arteries

Blocks of tissue including the coronary artery and subjacent myocardium were taken from the formalin-fixed heart from two sites within the proximal 2 cm of the right coronary artery and two sites near the bifurcation of the left coronary artery. Sections were prepared from frozen and paraffin-embedded blocks as described for aorta. Coded ORO-stained frozen sections taken from the four sites were examined by three investigators and the extent of intimal lipid was recorded on a scale of 0 to 3.

E. Electron Microscopy

Blocks of tissue taken from standard sites in the aorta were fixed in buffered osmium tetroxide, dehydrated in graded alcohols and embedded in Maraglas. Semithin sections were cut from the Maraglas block with a glass knife, mounted on slides, stained with Paragon multiple stain (Paragon C. & C. Co., Bronx, N.Y.) and examined by light microscopy. Areas of special interest were fine sectioned with a diamond knife, stained with lead citrate and uranyl acetate, and examined and photographed in an RCA EMU 3F electron microscope.

F. Radioautography

Animals in groups VI and VII were injected with [³H]-thymidine in a dose of 0.5 μCi/g of body weight. Two i.v. injections were given to each monkey, one 8 h, and another 1 h before killing. Paraffin-embedded sections (5 μm thick) of the entire length of the aorta, one half of which had been rolled into a thoracic and an abdominal coil, were dipped in nuclear track emulsion. All slides were developed for seven days and stained in HE. For each animal, all labeled cells in intima and media were counted in at least one radioautograph of the entire length of the aorta.

III. Results

A. Serum Lipids

The serum cholesterol concentration of all animals fed the atherogenic diet increased rapidly at first, then more slowly, and attained a steady level by the 12th week. The mean concentration increased from a basal level of 128 to 536 mg/dl at 12 weeks (table II). The serum cholesterol concentration dropped rapidly after return to the basal diet, and attained basal level after one to four weeks. During the 12-week period on atherogenic diet, the mean serum triglyceride concentration increased from a basal level of 32 to 43 mg/dl, a small but statistically significant increase. The changes in triglyceride concentration following return to basal diet varied considerably from animal to animal, and the mean terminal serum triglyceride concentrations were not significantly different from those during the atherogenic diet period.

B. Aortic Intimal Lipid

Results of chemical analysis of the aortic intima plus inner media for groups I to V are presented in table III. The total cholesterol content of the aortic intima-media preparation increased more than twofold during the hypercholesterolemic period (compare groups I and IV). The relative increase in the free cholesterol fraction was less than that of the ester cholesterol fraction.

Most of the increase in free and esterified cholesterol in the intima disappeared during the first 32 weeks of regression. During the second 32-week regression period there was a further reduction of cholesterol in the intima,

Table III. Number of animals, mean content of total, free and ester cholesterol and ratio of free to ester cholesterol in aortic intima-media preparation, by group

Group	N	Weeks on diet: atherogenic/basal	Cholesterol, mg/g dry weight[1]			Free/ester ratio
			total	free	ester	
I	6	12/2	16.3	9.6	6.7	1.6
			(2.3)	(0.9)	(1.4)	(0.2)
II	6	12/32	9.0	6.9	2.1	3.5
			(0.4)	(0.3)	(0.2)	(0.4)
III	6	12/64	8.3	6.4	1.9	3.8
			(0.8)	(0.4)	(0.4)	(0.5)
IV	3	0/14	7.0	5.6	1.4	4.5
			(0.8)	(0.6)	(0.4)	(1.2)
V	3	0/76	7.5	6.1	1.3	4.7
			(0.4)	(0.3)	(0.2)	(0.5)

[1] Values in parentheses are standard error of mean.

so that the concentration of both free and ester fractions approximated that in the control animals.

The mass of individual cholesteryl esters per mg of dry weight of aortic intima-media was calculated from the percentage contribution of the individual ester and the total cholesterol ester [table IV]. There was an increase in all major fatty acid fractions during the 12 weeks on the atherogenic diet. The largest proportional increase was in stearic and oleic esters and the least proportional increase was in linoleic and arachidonic esters.

While there was a decrease in the cholesteryl esters of all fatty acids after a 32-week period on the basal diet, the effect of this diet on the different cholesteryl esters varied. During this period there was a loss of more than 85% of the 'induced' cholesteryl ester of the saturated and mono-unsaturated fatty acids: palmitic (16:0); palmitoleic (16:1); stearic (18:0), and oleic (18:1). The induced linoleate decreased by two thirds and arachidonate by a half during 32 weeks on the regression diet. There was little change in the cholesteryl ester fatty acids during the additional 32 weeks, except for a further small reduction in the esters of linoleic and arachidonic acids. Cholesteryl stearate and arachidonate returned to approximately baseline levels after 32 weeks on the basal diet. After 64 weeks on the regression diet, the concentration of four of the six major cholesteryl esters remained higher than that of the control animals. These were palmitic, palmitoleic, oleic and linoleic acid. Cholesteryl linoleate made up a somewhat higher proportion

Table IV. Mean cholesteryl ester fatty acids (μg/g dry weight)[1] in intima-media preparations of rhesus aortas

Group	Weeks on diet: atherogenic/basal	Number of animals	Cholesteryl ester[2]					
			16:0	16:1	18:0	18:1	18:2	20:4
I	12/2	6	1,193	472	663	2.576	1.443	85
			(244)	(99)	(163)	(652)	(236)	(23)
II	12/32	6	361	111	101	482	729	64
			(36)	(17)	(14)	(66)	(129)	(7)
III	12/64	6	372	130	103	429	644	48
			(76)	(29)	(8)	(93)	(180)	(10)
IV	0/14	3	230	83	111	298	316	54
			(60)	(26)	(19)	(91)	(111)	(32)
V	0/76	3	222	76	62	264	420	32
			(24)	(9)	(11)	(55)	(72)	(8)

[1] Figures represent mean; numbers in parentheses indicate standard error of the mean.
[2] Number of carbon atoms:number of double bonds.

of the total cholesteryl esters in the regression animals than in the control groups or the group killed after inducing lesions (group I).

C. Arterial Lesions

After 12 weeks on the atherogenic diet there were grossly visible sudanophilic lesions in the thoracic and abdominal aorta, the brachial arteries, the common carotid arteries and in the iliac-femoral arteries. Figure 1 shows the Sudan IV-stained aortas, brachial arteries and common carotid arteries for each group. Only a few scattered sudanophilic intimal lesions were present in the initial controls (group IV).

The intimal sudanophilia in aortas of group I was extensive. After the first 32 weeks of regression, sudanophilia decreased markedly. A further reduction occurred after the additional 32-week period of regression; however, even after 64 weeks of regression, the monkeys had slightly more aortic intimal sudanophilia than the control group fed basal diet throughout (group V).

Lesions which formed in the brachial and common carotid arteries regressed essentially to control levels during the subsequent basal diet period. Although they were not graded visual inspection indicated that the extent of lesions in the iliac-femoral arteries was less than in any of the other

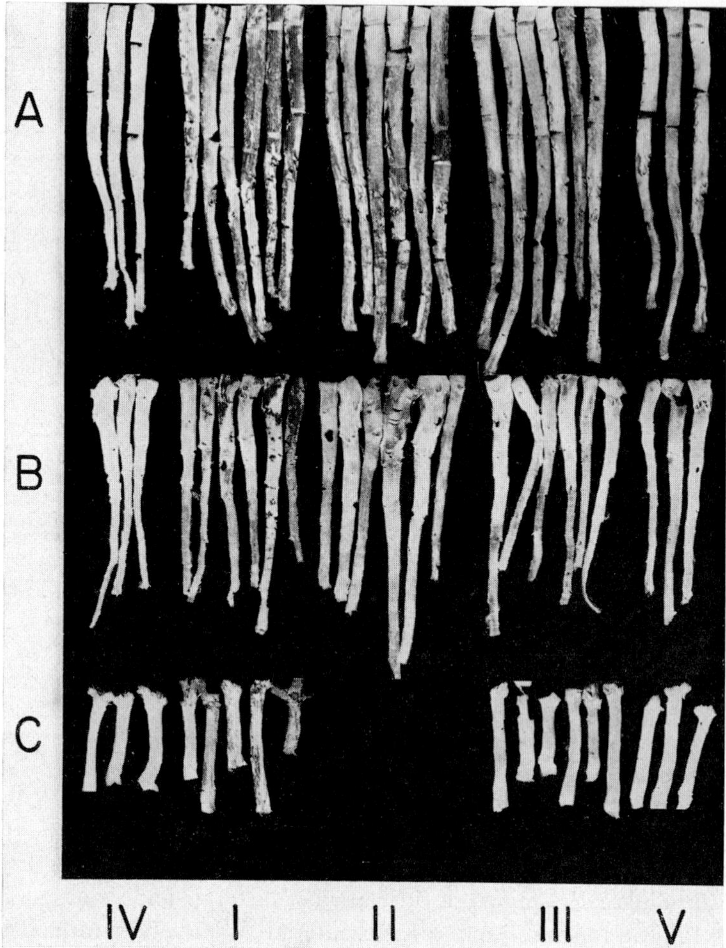

Fig. 1. Photographs of arteries showing reversibility of lesions. *A* Photograph of Sudan IV-stained left half of aorta from 24 rhesus monkeys. Within groups I, II and III, aortas are arranged from left to right in order of increasing mean serum cholesterol concentration during 12-week lesion induction period. *B* Photograph of Sudan IV-stained right brachial artery from 24 rhesus monkeys. Within groups I ,II and III aortas are arranged from left to right in order of increasing mean serum cholesterol concentration during 12-week lesion induction period. *C* Photograph of Sudan IV-stained left common carotid artery from rhesus monkeys of groups I, III, IV, and V. Within groups I and III the arteries are arranged from left to right in order of increasing mean serum cholesterol concentration during the 12-week lesion induction period.

Table V. Number of animals and mean percent of surface area with sudanophilic lesions in thoracic and abdominal aorta, right brachial artery and right common carotid artery, by group

Group	N	Weeks on diet: atherogenic/basal	aorta (left half)		brachial	carotid
			thoracic	abdominal		
I	6	12/2	56.7 (11.1)	28.8 (4.9)	15.0 (5.1)	15.2 (4.0)
II	6	12/32	20.8 (4.2)	15.3 (2.6)	6.3 (2.9)	–
III	6	12/64	15.3 (5.8)	8.5 (3.4)	1.7 (0.9)	0.7 (0.4)
IV	3	0/14	3.3 (3.3)	0.7 (0.7)	1.0 (1.0)	0.7 (0.7)
V	3	0/76	4.7 (2.9)	5.7 (1.2)	0.7 (0.7)	0.7 (0.3)

Values in parentheses are standard error of the mean.
Carotid arteries not available for group II.

arteries examined and that lesions of the iliac-femoral artery were more extensive in group I than in any other group.

The results of the quantitative evaluations of the aorta, brachial arteries and carotid arteries are given in table V. The mean percent of surface area involved with lesions was greater in group I than in group IV for all four arterial segments. The mean percent of surface involved with lesions was lower in group II than in group I for thoracic and abdominal aorta and for the brachial artery. For all arterial segments the mean percent of surface involvement after 64 weeks of regression was lower than in group I.

For all but the carotid artery, mean surface involvement was greater in the control group killed after 76 weeks on the basal diet (group V) than in the control group killed after 14 weeks on basal diet (group IV); however, this difference was statistically significant only for the abdominal aorta ($p < 0.025$).

D. Histologic Evaluation

1. Aorta

Histologic examination showed that the 12-week atherogenic diet period produced lesions characterized by accumulation of stainable lipid in a slightly thickened intima (fig. 2) with a small but variable amount of lipid sometimes occurring in the underlying media. One micron sections of Maraglas-embedded tissue examined by light microscopy, and fine sections examined by electron microscopy showed that the lesions typically consisted of foam cells below the endothelium, lipid droplets in intimal smooth muscle cells beneath this layer, and moderate amounts of extracellular lipid.

After 32 weeks of regression the total amount of intimal lipid was greatly decreased and foam cells were reduced in number. After 64 weeks of regression (fig. 3) foam cells had almost completely disappeared, lipid droplets were much less numerous in smooth muscle cells and the residual lipid was almost entirely extracellular. This remaining extracellular lipid was also decreased when compared to the other groups. Lipid in residual lesions frequently extended into the media (fig. 3).

The results of quantitative microscopic grading of the aortic sections are given in table VI. The mean grade for intimal lipid increased from 1.2 to 3.0 during the 12 weeks on the atherogenic diet, then decreased to 2.1 after 32 and to 1.6 after 64 weeks on regression diet. Analysis of variance indicated that these differences were statistically significant ($p < 0.01$). Medial lipid also increased during the 12 weeks on atherogenic diet; however, there was no apparent net loss of this lipid during 32 weeks on the regression diet. The mean grade for collagen did not increase during the atherogenic diet period but did increase significantly ($p < 0.05$) during the first regression period.

The effect of diet on thickness of the intima (table VII) was not consistent among the three anatomic sites. In the lower thoracic region (T_8) the intimal thickness paralleled the intimal lipid, i.e. it increased during progression and decreased again during regression. In the other two sites (T_2 or mid-abdominal), there was no increase in thickness during the progression period, but there was an increase in thickness during the regression periods.

2. Coronary Arteries

Only 1 of 24 sections from the six animals in the combined control groups IV and V showed lipid in the intima of the coronary arteries and this lipid was minimal in extent. Comparison of group I with groups IV and V indicated that lesions were induced in the coronary arteries during 12 weeks on the atherogenic diet. Little, if any, regression of atherosclerotic lesions of the coronary arteries was observed in sections from groups II or III. Findings in the coronary arteries should be interpreted with caution, however,

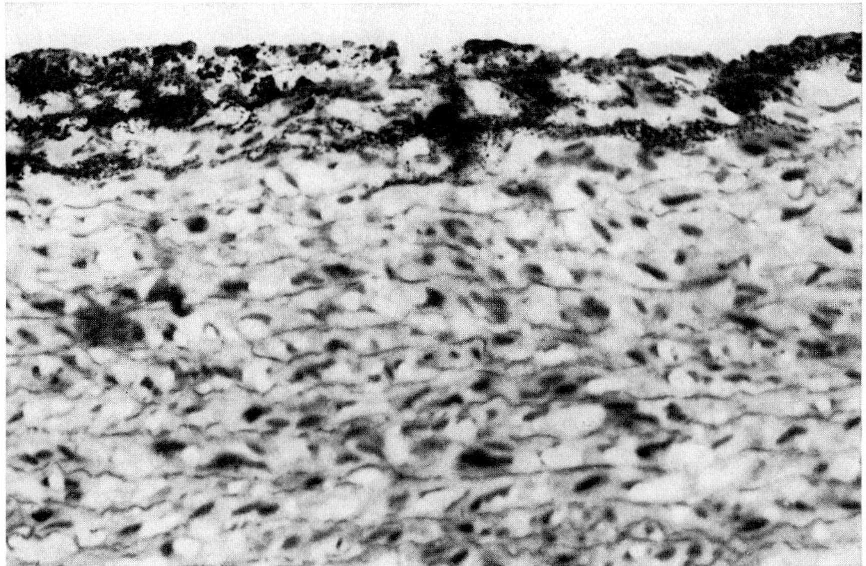

Fig. 2. Photomicrograph of lesion in thoracic aorta of rhesus from group I killed two weeks after return to basal diet. Intracellular and extracellular lipid is present in a slightly thickened intima. ORO, approximately $\times$ 420.

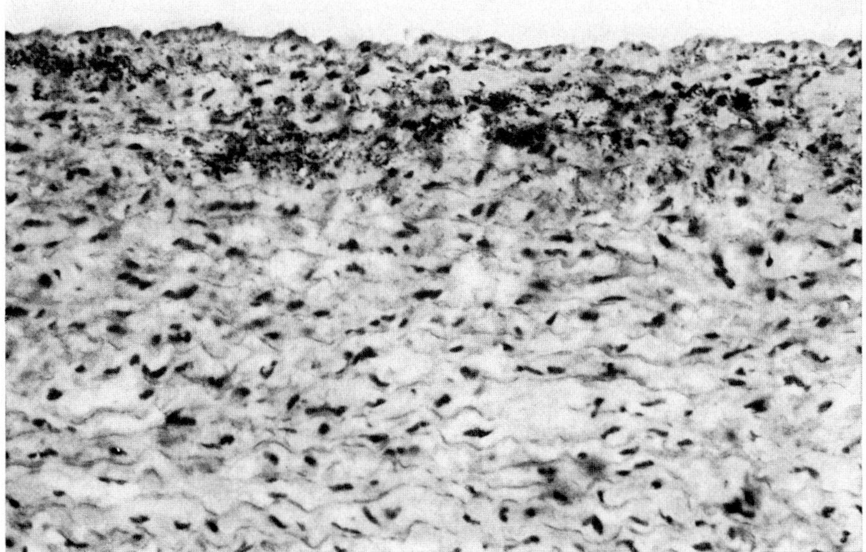

Fig. 3. Photomicrograph of residual lesion in thoracic aorta of rhesus killed 64 weeks after return to basal diet. The lipid which remains is predominantly extracellular and extends into the media. This selected section contains much more lipid than the typical section from animals in this group. ORO, approximately $\times$ 420.

Table VI. Number of sections graded and mean grade for lipid and collagen at three anatomic sites in aortas from four experimental groups. Means for group are for all sites combined and means for site are for all groups combined

| | N | Grade for lipid (ORO) | | Mean grade for intimal |
		intima	media	collagen (EVG)
Group				
I	18	3.0[1]	1.5[1]	1.2[2]
II	18	2.1	1.6	1.6
III	18	1.6	1.5	1.6
IV plus V[3]	18	1.2	1.0	1.2
Site				
Thoracic, T_2	24	1.9[1]	1.4[1]	1.5[1]
Thoracic, T_8	24	2.2	1.7	1.1
Abdominal	24	1.8	1.1	1.6

[1] Analysis of variance indicates that differences among means for group or for site are significant, $p < 0.01$. There was no interaction between group and site.

[2] Analysis of variance indicates that differences among means for group or for site are significant, $p < 0.05$.

[3] Combined control groups.

Table VII. Number of sections graded and mean grade for intimal thickness (cell layers) by treatment group and anatomic site

| Group | N | Site | | | |
| | | thoracic aorta | | abdominal | |
		T_2	T_8	aorta	Mean[1]
I Progression	6	2.7	2.4	2.5	2.5
II Regression 32	6	3.3	2.1	4.3	3.2
III Regression 64	6	3.0	1.6	4.2	2.9
IV plus V Control	6	2.8	1.6	2.5	2.3
Mean[1]	24	2.9	1.9	3.4	2.7

[1] Analysis of variance indicates that differences among sites or groups and interaction between site and group are statistically significant, $p < 0.01$.

because the extent of lesions was small and the histologic sampling was limited.

E. Cell Proliferation

Two control animals on the basal diet had no intimal lesions, and endothelial cells and intimal smooth muscle cells were labeled infrequently. Two animals given the atherogenic diet for 12 weeks and killed three weeks later had many intimal lesions consisting mainly of foam cells and intimal smooth muscle cells with lipid. There was frequent labeling of both foam cells and smooth muscle cells in the intimal lesions. Endothelial cells on the surface of lesions also showed more frequent labeling than endothelial cells in control animals given the basal diet only. Two animals given the atherogenic diet for 12 weeks and killed 16 weeks later still had numerous intimal lesions. These showed decreased cellularity due to loss of the majority of foam cells. Some of the residual foam cells were still labeled, although there appeared to be an overall reduction in the labeling frequency of all cell types in the intimal lesions. The two animals which received the atherogenic diet for 12 weeks and were killed 40 weeks later had a few small intimal lesions consisting of extracellular debris, some intimal smooth muscle cells with lipid, and rare solitary foam cells. None of the foam cells were labeled. The labeling frequency of smooth muscle cells and endothelial cells was similar to that of the control monkeys given the basal diet only.

F. Ultrastructure

Electron microscopic examination of the aorta revealed that the lipid of the developing fatty streak (group I) was both intracellular and extracellular. Intracellular lipid was in foam cells and in intimal smooth muscle cells (fig. 4). Foam cell lipid was in the form of vacuoles, cholesterol clefts, and residual bodies. Intimal smooth muscle cells contained lipid inclusions which were mainly of the electron-dense type. Intimal smooth muscle cells were usually below the foam cell layer and extended as far as the internal elastic lamina.

Some of the foam cells and a smaller number of smooth muscle cells were necrotic. Necrotic cell fragments and lipid inclusions were released and accumulated in the extracellular space. Additionally, the extracellular space contained elastic fibrils of varying size and appearance, collagen fibers, and

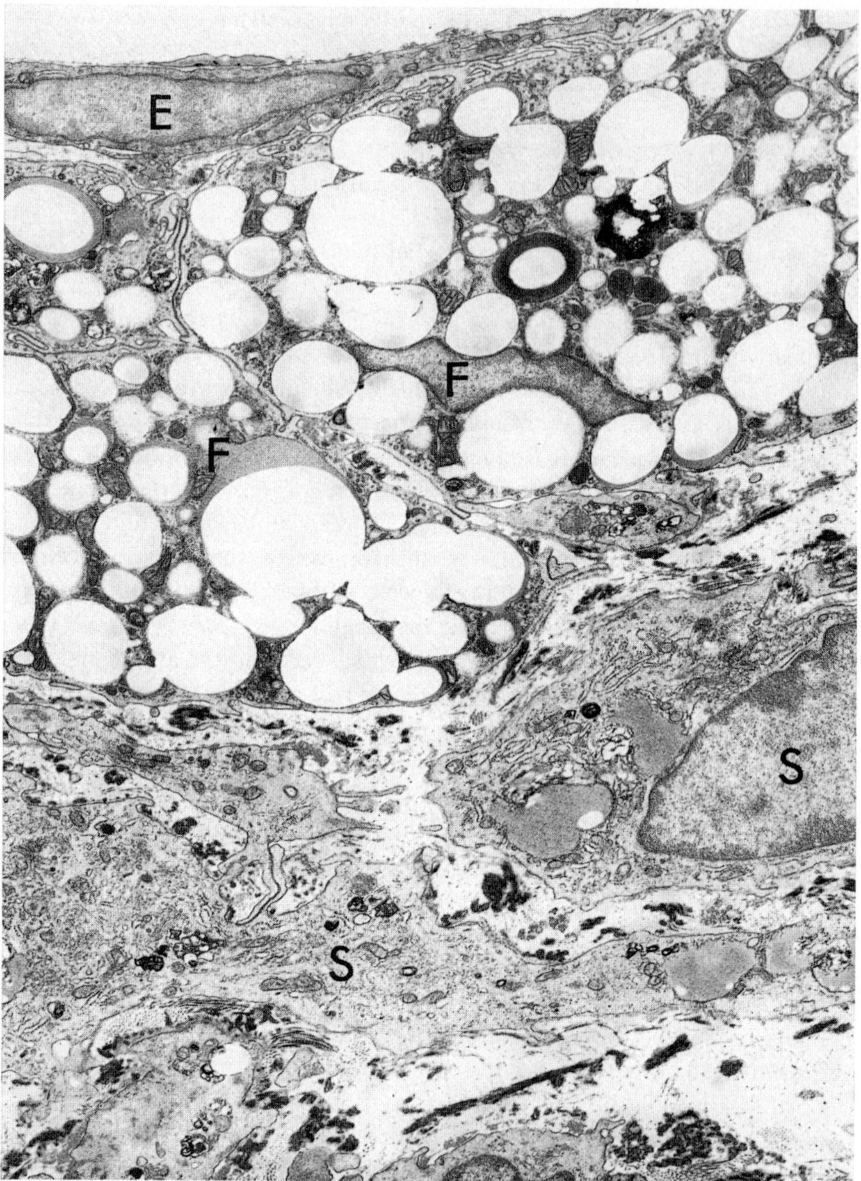

Fig. 4. Fatty streak in the descending thoracic aorta. Foam cells (F) are crowded in the upper intima and have predominantly electron-lucent lipid vacuoles. Reticulum-rich smooth muscle cells (S) contain homogeneous, electron-dense lipid inclusions. No extracellular lipid is visible. Endothelial cell (E). Electron micrograph from an adult monkey on a cholesterol and butter supplemented diet for 12 weeks, followed by unsupplemented food for 2 weeks. Mean serum cholesterol level during the cholesterol and butter period was 490 mg/dl. Rhesus No. 55. × 7,700.

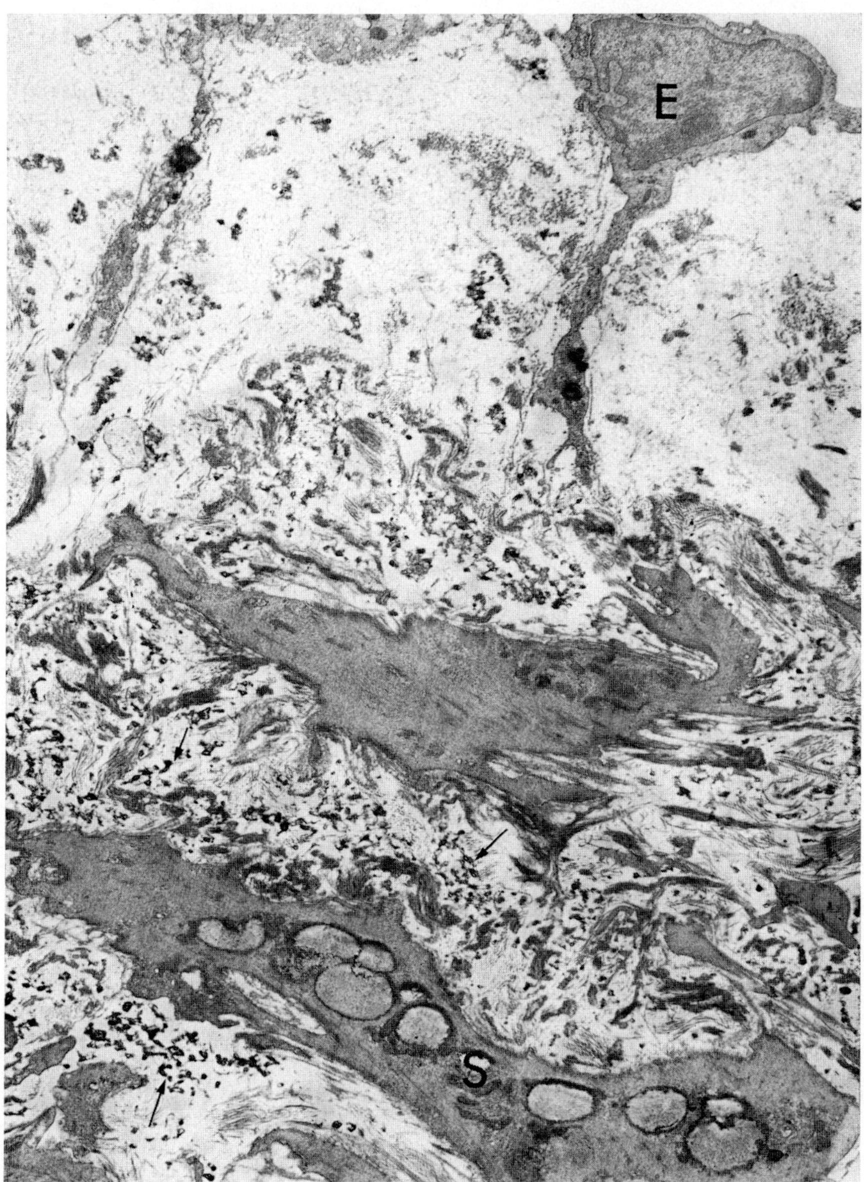

Fig. 5. Resolving intimal fatty streak in the abdominal aorta. Very electron-dense particles (arrows) in the extracellular space of the intima are abundant and represent lipid and cell debris derived from necrotic cells. A myofilament-rich smooth muscle cell (S) has lipid inclusions. There are no foam cells. Endothelial cell (E). Electron micrograph from an adult monkey maintained initially on a cholesterol and butter-supplemented diet for 12 weeks, followed by unsupplemented food for 40 weeks. Mean serum cholesterol level during the cholesterol and butter period was 430 mg/dl. Rhesus No. 62. $\times$ 6,300.

finely reticulated ground substance. Necrotic cell debris and lipid occasionally accumulated in extensive acellular pools between the foam cell layer and the internal elastic lamina.

Sixteen weeks after reversal of the diet there was a pronounced decrease in the number of foam cells. In those which remained, lipid inclusions were associated with peripheral tubular structures of lysosomal nature. The number of intimal smooth muscle cells with lipid did not seem to be greatly decreased at that time, and the nature of the intracytoplasmic lipid did not differ from that in group I. Extracellular lipid was increased in amount and the amount of necrotic cell fragments was similar to that seen in group I.

Thirty-two weeks after diet change there was a further decrease in the number of foam cells. Intact foam cells were rare. Persisting intracellular lipid was now mainly within intimal smooth muscle cells, although the overall number of lipid-containing smooth muscle cells had also decreased. After 40 weeks of regression (fig. 5) the changes were similar to those at 32 weeks. After 64 weeks of regression, no foam cells and few lipid-containing smooth muscle cells were observed. Extracellular lipid remained the predominant form of visible lipid, but it was greatly reduced as compared to the earlier periods. Collagen fibres appeared to be more numerous than in the intima of controls or of group I animals, and newly formed elastica appeared to be present in discontinuous layers.

IV. Discussion

A. Gross Lesions

The findings presented here clearly indicate regression of diet-induced fatty streaks in the aorta, carotid arteries, brachial arteries, and femoral arteries. Grossly visible sudanophilic lesions in the aorta regressed to nearly initial levels after 64 weeks on a basal diet. Previous studies in our laboratory indicate that early aortic lesions (fatty streaks) in the rhesus monkey require more than four months at low serum lipid concentrations before their gross extent is reduced appreciably.

The disappearance of grossly detectable lesions during the basal diet period was more nearly complete in the brachial, carotid and iliac-femoral arteries than in the thoracic or abdominal aorta. It is possible that the lesions induced in these arteries were not as far advanced as those in the more susceptible aorta, and therefore were more easily reversed. This finding might also reflect either a basic difference in the structure of these peripheral arteries or a difference in the hemodynamic conditions to which they are subjected.

B. Aortic Lipid

The selective accumulation and depletion of certain cholesteryl esters in the aorta demonstrates the dynamic nature of arterial cholesterol ester metabolism in this experimental model. Both types of dietary fat and metabolic activity of the aortic wall undoubtedly influence accumulation and efflux of cholesterol and its esters.

C. Histologic Evaluation

The histologic characteristics of the lesions changed dramatically during regression. A net loss of stainable lipid occurred in the intima confirming the findings on gross and chemical evaluation. After 64 weeks of regression, essentially all foam cells and most of the lipid inclusions in smooth muscle disappeared, and extracellular lipid in the intima decreased markedly.

We did not determine the mechanism by which the intimal lipid was removed. At least two possibilities exist. The lipid may have been metabolized *in situ* into a product no longer recognizable as lipid. It is more likely, however, that the lipid was transported out of the artery wall either through the endothelium or the adventitia via the vasa vasorum.

Histologic evaluation indicates that the decrease in lipid content of early atherosclerotic lesions in this study was accompanied by an increase in fibrous tissue. ARMSTRONG and co-workers [3, 5] have reported that regression of more advanced coronary lesions in rhesus results in a loss of lipid with increased density of the fibrous component; however, they did not determine whether this increased density was merely a concentration of preexisting collagen acquired in the progressive phase or an absolute increase in the amount of collagen. In a subsequent study in *Macaca fascicularis* [4] they concluded that there may be a net loss of alkali-soluble or acid-soluble collagen in the aortic intima-media during regression of lesions induced by feeding atherogenic diets for 18 months. In this latter study, however, collagen appeared to have increased in coronary arteries during the regression. Studies with longer follow-up of regressing lesions may clarify the role of collagen.

D. Cell Proliferation

The number of animals used in the radioautographic part of this experiment was small, and the results on cell proliferation must be considered as preliminary. Nevertheless, the experiment indicates that increased intimal cell proliferation induced by hypercholesterolemia can return to normal levels. This change occurred without special experimental manipulation or medication, and is attributed solely to the effect of removing the stimulus of high serum cholesterol levels. It thus seems unlikely that a permanent proliferative impulse is induced in arterial cells by a short period of hypercholesterolemia. Whether or not increased cell proliferation in advanced lesions such as fibrous or complicated plaques can be reduced with equal ease remains to be determined.

Hypercholesterolemia is accompanied by necrosis of some intimal cells. After the serum cholesterol was lowered, necrosis of intimal cells continued as the excess of newly proliferated cells was gradually cleared away. Cell proliferation returned to normal even though some cells were still undergoing necrosis and before cell debris and lipid released from necrotic cells disappeared from the intima.

V. Conclusion

In this study we have shown that, as measured by lipid accumulation, there is a regression of early diet-induced atherosclerotic lesions in rhesus monkeys when the dietary stimulus is removed and the serum cholesterol is allowed to return to basal levels of approximately 100–150 mg/dl. We have not shown a complete regression, since certain residual changes remain for as long as 64 weeks after removal of dietary stimulus.

The role of the fatty streak in the pathogenesis of more clinically significant atherosclerotic lesions is a subject of controversy. Evidence that the fatty streak is an essential precursor to the more advanced atherosclerotic lesion (fibrous plaque) is indirect and is based largely on similarities in topographic distribution of lesions [15, 18]. It has been shown, however, that the relative differences in extent of aortic fatty streaks among various human population groups are much less than the differences in fibrous plaques and may occasionally be of opposite direction [14]. One possible explanation for this difference is that some lesions classified grossly as fatty streaks may not progress to more advanced stages but remain instead as fatty streaks or regress toward normal, and that this propensity to regress varies in different populations. In the present study we have shown that, as assessed by visual, ultrastructural, radioautographic and chemical examination of the artery, the diet-induced experimental fatty streak in one primate species, the rhesus monkey, can regress.

Several important questions remain. First, the long-term fate of these regressed lesions will have to be evaluated. Will fibrosis continue to develop to a point where it either impinges on the lumen or forms the substrate for more advanced complicated lesions? Would the effect of a second pulse of hypercholesteremia on these 'regressed' lesions be similar to that of the initial stimulus or would it produce a more severe or less severe lipid deposition? What would be the result of several such episodes of progression or regression of lesions? Until answers to these questions are obtained, the full significance of these findings cannot be assessed.

Finally, in our study we have employed conditions both for progression and regression that are somewhat more extreme than those encountered in normal human experience. The question remains whether a significant regression of these early lesions will occur with reduction of serum lipid concentrations to levels more likely to be attainable in man. Answers to many of these questions may be obtained by further studies using an experimental model similar to the one described in this chapter.

References

1 ABELL, L. L.; LEVY, B. B.; BRODIE, B. B., and KENDALL, F. E.: A simplified method for the experimentation of total cholesterol in serum and demonstration of its specificity. J. biol. Chem. *195:* 357–366 (1952).

2 ARMSTRONG, M. L.: Atherosclerosis in rhesus and cynomolgus monkeys (this volume).

3 ARMSTRONG. M. L. and MEGAN, M. B.: Lipid depletion in atheromatous coronary arteries in rhesus monkeys after regression diets. Circulation Res. *30:* 675–680 (1972).

4 ARMSTRONG, M. L. and MEGAN, M. B.: Arterial fibrous proteins in cynomolgus monkeys after atherogenic and regression regimens. Circulation *43:* suppl. IV, p. 41 (1973).

5 ARMSTRONG, M. L.; WARNER, E. D., and CONNER, W. E.: Regression of coronary atheromatosis in rhesus monkeys. Circulation Res. *57:* 59–67 (1970).

6 CATSULIS, C.; Strong, J. P.; EGGEN, D. A., and TUCKER, C.: Unpublished.

7 EGGEN, D. A.: Cholesterol metabolism in nonhuman primates (this volume).

8 EGGEN, D. A. and STRONG, J. P.: Cholesterol metabolism in rhesus: extremes of response of serum cholesterol to atherogenic diet. Circulation *46* · suppl. II, p. 250 (1972).

9 EGGEN, D. A.; STRONG, J. P.; NEWMAN, W. P.; CATSULIS, C.; MALCOM, G. T., and KOKATNUR, M. A.: Regression of diet-induced fatty streaks in rhesus monkeys. Lab. Invest. *31(3):* 294–301 (1974).

10 Folch, J.; Ascoli, I.; Lees, M.; Meath, J. A., and Baron, F. N. Le: Preparation of lipid extracts from brain tissue. J. biol. Chem. *191:* 833–841 (1951).

11 Guzmán, M. A.; McMahan, C. A.; McGill, H. C., jr.; Strong, J. P.; Tejada, C.; Restrepo, C.; Eggen, D. A.; Robertson, W. B., and Solberg, L. A.: Selected methodologic aspects of the international atherosclerosis project. Lab. Invest. *10:* 479–497 (1968).

12 Handel, E. van and Zilversmit, D. B.: Micromethod for the direct determination of triglycerides. J. Lab. clin. Med. *50:* 152–157 (1957).

13 Kokatnur, M. G.; Malcom, G. T.; Eggen, D. A., and Strong, J. P.: Depletion of aortic free and ester cholesterol by dietary means in rhesus monkeys with fatty streaks. Atherosclerosis. *21:* 195–203 (1974).

14 McGill, H. C., jr.: Fatty streaks in the coronary arteries and aorta. Lab. Invest. *18:* 560–564 (1968).

15 Montenegro, M. H. and Eggen, D. A.: Topography of atherosclerosis in the coronary arteries. Lab. Invest. *18:* 586–593 (1968).

16 Stary, H. C.: Progression and regression of experimental atherosclerosis in rhesus monkeys; in Goldsmith and Moor-Jankowski Medical primatology 1972, part III, pp. 356–367 (Karger, Basel 1972).

17 Stary, H. C.: Cell proliferation and ultrastructural changes in regressing atherosclerotic lesions after reduction of serum cholesterol; in Schettler and Weizel Atherosclerosis III, pp. 187–190 (Springer, Berlin, 1974).

18 Solberg, L. A. and Eggen, D. A.: Localization and sequency of development of atherosclerotic lesions in the carotid and vertebral arteries. Circulation *43:* 711–724 (1971).

19 Strong, J. P.: Introduction and overview (this volume).

Jack P. Strong, M. D., Douglas A. Eggen, Ph. D., and H. C. Stary, M.D., Louisiana State University Medical Center, 1542 Tulane Avenue, *New Orleans, LA 70112* (USA)

Prim. Med., vol. 9, pp. 321–358 (Karger, Basel 1976)

Coronary Artery Fine Structure in Rhesus Monkeys: Nonatherosclerotic Intimal Thickening

H. C. STARY and JACK P. STRONG

Department of Pathology,
Louisiana State University Medical Center, New Orleans, La.

Contents

I. Introduction

The existence of nonatherosclerotic intimal thickening at arterial forks and around the ostia of branch vessels in the human aorta was described as early as 1883 by THOMA [39] who assumed it to be a universal feature in human arterial development. Such intimal thickening was also observed by

WOLKOFF in her studies of the coronary arteries of infants, children, and adults [43] and of a small sample of several animal species [44]. Many other authors have described similar intimal thickening in human coronary arteries [3–5, 8, 9, 11, 12, 19, 20, 22, 25, 28]. MCGILL *et al.* [17] found intimal thickening in the coronary arteries of normocholesterolemic baboons, MALINOW and STORVICK [15] observed similar changes in the coronary arteries of howler monkeys, and VLODAVER *et al.* [41] described intimal thickening in the coronary intima of rhesus and vervet monkeys.

Human coronary artery atherosclerosis is most severe near the main bifurcation of the left coronary artery [33], an area that normally has non-atherosclerotic intimal thickening. Since the topographic distribution of non-atherosclerotic thickening and atherosclerosis is similar not only in the coronary but also in other arteries [23, 36, 37, 42] it has been widely speculated that a close relationship exists between the two conditions. The nature of the relationship is controversial. Opinions range from the concept that nonatherosclerotic thickening is an essentially 'normal' anatomic structure in which atherosclerosis sometimes develops, to the concept that such intimal thickening is an integral part of atherosclerosis.

The present study supplements our knowledge of the structure of coronary intimal thickening by examination with the electron microscope, and provides an anatomic basis for a companion study on the fine structure of coronary atherosclerosis experimentally induced in rhesus monkeys [33].

II. Definition of Terms

A. Nonatherosclerotic Intimal Thickening

We use 'nonatherosclerotic intimal thickening' or 'intimal thickening' to refer to the ubiquitous thickening of arteries which consists of a variable quantity of smooth muscle cells and ground substance between the endothelium and the internal elastic lamina. In contrast to atherosclerosis, intimal thickening has no extracellular lipid and, when present at all, only infrequent intracellular lipid inclusions. We have distinguished two morphological types, the intimal cushion and diffuse intimal thickening. Although there are distinct ultrastructural differences between intimal cushions and diffuse intimal thickening, the two structures often are contiguous and one cannot be clearly delineated from the other.

Most authors have not distinguished between cushions and diffuse thickening; thus, the specific terms used by them can be assumed to apply to both structures. Such general terms for intimal thickening include 'musculo-elastic intimal lesion' [2], 'musculo-elastic intimal thickening' [11], 'musculo-elastic layering' [13], and 'fibromuscular intimal thickening' [16].

B. Intimal Cushion

'Intimal cushion' denotes the focal intimal thickening at arterial forks and at the ostia of branch vessels. The segment of intima involved is circumscribed, and the degree of thickening is greater than in diffuse thickening. Several authors have used this term [4, 23, 25]. Other terms used for this structure are 'spindle cell cushion' [34], 'intimal pad' [35–38], 'musculo-elastic plaque' [17], 'localized fibrous plaque' [20] and 'mucoid fibromuscular plaque' [18].

C. Diffuse Intimal Thickening

In diffuse intimal thickening, extensive segments of the arterial intima are involved rather uniformly with little variation in the degree of thickness. The degree of thickening is less than in intimal cushions. The term has been used by several authors [10, 21, 42]. Another term used is 'diffuse intimal fibrosis' [20].

III. Methods

A. Experimental Animals and Diet

Eight normal, male rhesus monkeys *(Macaca mulatta)* used as controls in three studies were used. Two animals had served as controls for an experiment on diet-induced atherosclerosis and its progression [30, 33], two in a study on regression of diet-induced atherosclerosis [30, 32] and four in a comparative study on atherosclerosis in three species of monkeys [6, 7]. All eight monkeys were purchased from Shamrock Farms, Middleton, N. Y., where they had been conditioned for six to eight weeks after arrival from India. During this period they were dewormed, tested for tuberculosis, and given a low fat, low cholesterol diet. After arrival at the laboratory and during the remainder of their lifetime, which varied from 8 to 102 weeks, the animals received D & G research animal laboratory diet (Price Wilhoite Co., Frederick, Md.) which has a low content of fat and cholesterol. The animals were young adult and adult; ages were estimated by body weight (table I) and tooth development.

Table I. Body weight at time of death and mean value of all serum total cholesterol determinations made in each animal

	Rhesus monkey No.							
	13	14	20	26	45	46	68	77
Weight at end of experiment, kg	7.4	7.8	7.3	7.4	4.5	4.7	8.7	7.7
Mean serum total cholesterol, mg/100 ml	131	177	120	162	130	180	110	130

B. Cholesterol Determination

Serum total cholesterol was determined on all animals by the method of ABELL *et al.* [1]. Serum samples were taken at one-week intervals initially, and less frequently later, from the time of the animals' arrival at the laboratory until their death. The mean value of all serum total cholesterol determinations for each animal was calculated (table I).

C. Electron Microscopy

The proximal extramyocardial part of the left coronary artery was dissected from the heart. Tissue taken for electron microscopy consisted of consecutive, complete cross-sections of the main stem of the left coronary artery, the bifurcation of the main stem into anterior descending and circumflex branches, and the adjacent left anterior descending branch. The tissue was fixed in two changes of buffered osmium tetroxide, dehydrated in graded alcohols, and embedded in Maraglas (29). The complete cross-sections of the artery were thick sectioned with a glass knife in an ultramicrotome, mounted on slides, stained with Paragon multiple stain (Paragon C. & C. Co., Bronx, N.Y.), and examined with the light microscope. Areas of special interest were selected, fine-sectioned with a diamond knife, stained with lead citrate and uranyl acetate, and examined and photographed on an RCA EMU 3F electron microscope.

IV. Results

A. Anatomy of the Left Coronary Artery

The left coronary artery originated from a single ostium and usually began as a distinct, single, main stem. The length of the main stem varied, and the initial bifurcation into anterior descending and circumflex branches was located 1.0–4.0 mm from the left coronary ostium. In a few cases there was no distinct main stem, and the anterior descending and the circumflex branches both originated at the left coronary ostium.

The main stem and the proximal segments of the anterior descending branch were embedded in subepicardial fat tissue. Four to eight millimeters beyond its origin the anterior descending branch penetrated the myocardium. Before becoming intramyocardial, the anterior descending gave off one or more smaller branches. The largest of these usually formed the intermediate descending branch. The anterior descending was occasionally somewhat larger in diameter than the circumflex; however, the two branches usually had about the same diameter. When the circumflex was larger than the anterior descending, it gave origin to the intermediate descending branch.

The coronary artery segment that was to be removed was measured before and after removal from the heart. When attached, even after diastolic collapse of the heart, the artery was stretched. When removed from the heart, the artery contracted about one third in length.

B. Distribution of Intimal Thickening

The following observations were made on semithin step-sections of Maraglas-embedded cross-sections of the proximal left coronary artery. All animals had intimal thickening of some degree. Thickening occurred in the form of either intimal cushions, or diffuse intimal thickening. The nature of the thickening varied according to location within the artery.

The highest degree of intimal thickening was a saddle-shaped cushion at the apex of the bifurcation of the anterior descending and circumflex branches, which projected in the direction of the lumen of the main stem (fig. 1, 2). The saddle-shaped cushion began gradually as two longitudinal cushions on opposite sides of the proximal segment of the main stem. The two cushions fused and reached a thickness of up to six cells at the apex of the bifurcation. Expanding distally and laterally into the two branch vessels, the saddle-shaped cushion decreased in thickness and became transformed into diffuse intimal thickening. The somewhat complicated architecture will be more easily understood by consulting figures 1 and 2, and by referring to the excellent description of the anatomic structure of arterial bifurcations by THOMA [40].

Smaller cushions were at the lips of branch vessel orifices. Away from arterial forks and away from branch vessel orifices, intimal thickening was diffuse. In some segments of the main stem and of the proximal branches of the left coronary artery, intimal thickening was absent entirely.

A compact internal elastic lamina was absent in the proximal portion of the main branch. Elastica was in the form of numerous delicate fibrils between muscle cells, and the intima could not be precisely distinguished from the media. Whereas a distinct internal elastic lamina appeared proximal to the bifurcation, it was frequently incomplete at the base of cushions.

C. Fine Structure of Cushions and Diffuse Thickening

Intimal cushions differed from diffuse intimal thickening not only in the degree of intimal thickness and the number of cell layers, but also in the nature of the cells and in the nature of the intercellular matrix. Cushions generally had an upper subendothelial layer with few cells and a loose intercellular matrix in which cells were widely separated and irregularly distributed. The deep layers bordering the internal elastic lamina had less intercellular matrix and a higher cell density, and the cells were stratified or clumped in groups (fig. 3–5).

Cushions had two main cell types. Myofilament-rich smooth muscle cells, closely resembling medial smooth muscle cells and occasionally smaller and more varied in shape, were the most common cell type. They were scattered throughout the cushions but tended to be the sole cell type at the base of cushions. The second cell type was a smooth muscle cell which was very rich in rough-surfaced endoplasmic reticulum and contained a variable, but always smaller, number of myofilaments than myofilament-rich smooth muscle cells. Bundles of myofilaments had 'dense bodies', and were frequently limited to the cell periphery where they were arranged parallel to the long axis of the cell. Serial thin sections of reticulum-rich smooth muscle cells showed that some cell segments lack myofibrils altogether. The rough-surfaced endoplasmic reticulum often formed prominent cisternae (fig. 11–13). These cells were surrounded by basement membrane which, however, was usually less prominent than in myofibril-rich smooth muscle cells. The cytoplasm also contained pinocytotic vesicles, a prominent Golgi apparatus, and mitochondria. Sections of cells completely lacking myofilaments resembled fibroblasts, but differed by the absence of the very long, slender cell projections which are prominent in fibroblasts. Reticulum-rich smooth muscle cells were scattered throughout cushions and tended to be more frequent in the layer near the lumen of the artery.

The intracellular matrix of cushions, especially in the upper layer, was loose, and consisted of abundant and finely reticulated material. The nature of the finely reticulated material has not been established; it most likely represents mucopolysaccharide ground substance (fig. 10–13). Collagen fibers and elastica were also present. Elastic fibrils were delicate and short, or fragmentary and granular. They were compact and pronounced in the deep layer of cushions where, occasionally, they seemed to be derived from a splintered internal elastic lamina. There was no extracellular lipid.

Diffuse intimal thickening was generally more homogeneous in character than intimal cushions, and similar in composition to the deeper layer of cushions. The cell density was greater than that of cushions (fig. 6). Intimal cells were almost entirely myofilament-rich smooth muscle cells (fig. 7, 8, 14) although an isolated reticulum-rich smooth muscle cell was present now and then. The intercellular matrix did not have as much of the loose, finely reticulated ground substance so abundant in cushions. Fibrils of collagen and elastica were present.

Degenerative changes in the cytoplasm of smooth muscle cells were found in both cushions and in diffuse thickening (fig. 9, 14). They were more common in cushions. Degenerative changes occurred mainly in the peripheral, isolated cell processes of intimal smooth muscle cells. Many of these degenerating cell extensions appeared to be separated from the adjacent, morphologically normal, main body of the cell. They had no basement membrane, and as isolated structures they were roughly spherical. The cytoplasm was edematous and less dense than normal cytoplasm. Although there sometimes were a few swollen microvesicles, there usually was complete dissolution of cytoplasmic structure with loss of vesicles, filaments, and mitochondria. Rupture of the cell membrane and complete dissolution of isolated cell processes was common.

Intracytoplasmic lipid was rare. The distribution of intracytoplasmic lipid in intimal cushions and in diffuse thickening paralleled that of degenerative cell change although the two were not necessarily associated and lipid inclusions were less common than degenerative change. Lipid was more often seen in reticulum-rich than in myofilament-rich smooth muscle cells; it occurred as moderately electron-dense, single or multiple inclusions (fig. 10, 13). Foam cells, that is the large, round cells without myofibrils or basement membrane and containing many lipid inclusions, were very rare. In every instance they were isolated and solitary and located in the portion of the intima just beneath the endothelium (fig. 9). Foam cells showed either advanced degenerative change or necrosis.

Endothelial cells on the surface of both cushions and diffuse thickening sometimes were larger than endothelial cells in nonthickened intima, and the cytoplasm was richer in cell organelles. Vesicles of rough-surfaced endoplasmic reticulum sometimes were dilated and increased in number. One or more small lipid inclusions were present irregularly (fig. 8, 11). Pinocytotic vesicles were abundant at the luminal surface and at the base of cells. Thin and thick filaments were prominent and sometimes formed filament bundles. Other cytoplasmic constituents such as mitochondria, Golgi apparatus, and Palade bodies were usually not prominent. A basement membrane was sometimes prominent and multilayered (fig. 15, 16). Endothelial cells on the surface of cushions containing a large amount of ground substance often did not have a basement membrane.

Separating cushions from the media, the internal elastic lamina was either a single, prominent, though often incomplete membrane or a multi-layered structure composed of delicate fibrils. Gaps were of varying size (fig. 3–5), and sometimes the ends of the internal elastic lamina overlapped (fig. 3). The internal elastic lamina that separated diffuse intimal thickening from the media was more often a single compact membrane. It was wavy in cross-sections of the artery as the result of postmortem contraction. Although there were gaps in this membrane, these were not as common nor as extensive as at the base of cushions.

In some arterial segments with gaps in the internal elastic lamina, medial smooth muscle cells seemed to be trapped in the gaps. Although the perpendicular arrangement of smooth muscle cells in such gaps of the internal elastic lamina suggests cell migration, entrapment might represent an artifact caused by contraction of the vessel.

The media was composed of medial smooth muscle cells which could be distinguished from intimal smooth muscle cells by their large size, regular arrangement, and greater prominence of the basement membrane. Distinction between intimal and medial smooth muscle cells was usually possible on this basis in spite of the frequent absence of the internal elastic lamina at the base of intimal cushions. Endoplasmic reticulum was scanty and limited to the perinuclear region. Intracytoplasmic lipid was seen infrequently; when present, it was different in nature from that in intimal cells. It usually formed single small droplets or grape-like clusters of small droplets, more dense than the lipid inclusions of intimal smooth muscle. The droplets were close to the organelles at the center of the cells and probably represent lipofuscin.

Adjacent to cushions and diffuse thickening, the space between internal elastic lamina and the first layer of medial smooth muscle cells sometimes contained swollen cell processes with degenerative change. These were identical to the degenerative changes described in isolated cell processes of intimal smooth muscle cells. Many of these degenerated cell portions were pressed against the internal elastic lamina where they accumulated (fig. 15–17). In figures 16 and 17 a degenerated cell process shows connection with the smooth muscle cell cytoplasm and is probably being freed from it. Rarely, degenerative change would involve an entire medial smooth muscle cell rather than only a cell projection. Such a cell with advanced and probably irreversible cytoplasmic damage indicating imminent cell lysis is shown in figures 18 and 19. There is formation of striking, intensely osmiophilic, concentric lamellae or membranes which often are collapsed and vacuolated. Some are complex structures containing degenerating mytochondria, lipid droplets, and disintegrating filaments. Completely collapsed multi-lamellated structures became transformed into residual bodies. Adjacent cytoplasm has an increased number of vesicles of smooth-surfaced endoplasmic reticulum, indicating increased lysosomal activity. Nuclear alterations were not observed in degenerating smooth muscle cells; they appear to be a late feature in the degenerative process.

V. Discussion

A. General Observations

Our findings indicate that nonatherosclerotic intimal thickening constitutes part of the coronary artery intima of normal rhesus monkeys and support the view that it may be universal in many species. Although we have found intimal thickening in every animal, there was some variation in degree. There also was some variation in the serum cholesterol levels within the normal low range in this group of eight monkeys. We found no relationship between the serum cholesterol level and the degree of intimal thickening. Dock [4] found variation of coronary intimal thickness in human fetuses of similar weight; he described a higher degree of intimal thickening in the arteries of males than females and these data are supported by the observations of Wilens [42]. Our own observations were entirely of male animals, and while we cannot comment on variation between sexes, we too have observed that the degree of intimal thickening may vary in animals of similar weight and similar estimated age.

B. Cellular Composition of Intimal Thickening

We found two main cell types in coronary intimal thickening. The predominant cell type was a myofilament-rich smooth muscle cell. The precise time at which this cell appears in the intima has not been determined, but smooth muscle cells were described by THOMA [39] in the aortic intima of two human fetuses at 30–33 weeks of gestation. WOLKOFF [43] who studied the coronary arteries of nine human subjects ranging in age from 8 months to 50 years, also recognized the cells in intimal thickening to be smooth muscle cells. Other authors who have studied intimal thickening in human coronary arteries [4, 8, 9, 19, 28] have not been specific as to the type of cell observed. WOLKOFF [44], who also studied the coronary arteries of one young and one old baboon, found splitting of the internal elastic lamina in the old animal. The nature of the cells seen in between the lamellae was not specified.

The second type of cell in the coronary intimal thickening of our monkeys was a reticulum-rich smooth muscle cell. Under this term we have included all smooth muscle cells with prominent rough-surfaced endoplasmic reticulum, not differentiating between those with many and others with few myofilaments. Occasionally, we saw spindle-shaped reticulum-rich cells lacking myofibrils altogether, and strongly resembling fibroblasts. We grouped these cells with the reticulum-rich smooth muscle cells since we had, in serial sectioning a few cells, occasionally found myofilaments in one part of the cytoplasm but not in another. We assumed that reticulum-rich cells lacking myofilaments might have some myofilaments in segments of the cell that we had not sectioned.

The incidence of reticulum-rich smooth muscle cells was highest in the subendothelial portions of large intimal cushions, a region particularly rich in the finely reticulated intercellular material which we take to be mucopolysaccharide ground substance. This close association indicates that reticulum-rich smooth muscle cells might produce ground substance, and supports evidence that smooth muscle cells can synthesize a wide variety of intercellular components in addition to maintaining a contractile function. RYAN *et al.* [26, 27] found similar cells with both contractile and protein-synthesizing elements in granulation tissue, and coined the descriptive and euphonic term 'myofibroblasts'.

STEHBENS and LUDATSCHER [38] studied the ultrastructure of the rabbit renal artery bifurcation and found smooth muscle cells to be the predominant cell type of intimal thickening. Undifferentiated cells were less common, but relatively more frequent in subendothelial than in deep portions of intimal cushions. Some of these cells resembled fibroblasts and may be identical to the reticulum-rich smooth muscle cells seen in a similar distribution in the coronary artery cushions of our monkeys. REALE and LUCIANO [24] who studied the fine structure of intimal cushions in small arterial forks near the thyroid and submaxillary glands and at the origin of an intercostal artery in rats, found intimal smooth muscle cells to be without cytological pecularities, not different from smooth muscle cells elsewhere in the arterial wall.

Reticulum-rich smooth muscle cells are selectively increased in the intima in atherosclerosis and probably also in a number of other pathological conditions of the arterial intima. In atherosclerotic lesions they have been widely noticed and described under various names by many authors. This subject is discussed more fully in the context of coronary atherosclerosis, elsewhere in this volume [33].

In 1866, LANGHANS [14] described spindle-shaped cells in the aortic intima of an infant dead four days after birth. LANGHANS then studied these cells in subjects up to an age of 79 years by specially processing the thick sections and separating individual cells by plucking tissue and cells apart mechanically. He found that, with advancing age, the spindle cells of childhood increased their size and the number of their cell processes. The intimal cells described by LANGHANS probably represent smooth muscle cells. This question of maturation or change with age in cells found in normal intima has, so far, not been restudied with the electron microscope. We do not know whether or not one of our two types of smooth muscle cells corresponds to the intimal cells seen by LANGHANS in infancy, and the other to the cells found by him to predominate in the normal intimal segments of old subjects.

A progressive increase in the degree of intimal thickening occurs in the postfetal period [8, 14, 28, 43]. It is unknown whether this is due to intramural division of cells or to migration of medial smooth muscle cells into the intima. Active cell proliferation observed in intimal cushions in radioautographic studies with tritiated thymidine [31, 34] suggests that an increase in the number of cells occurs through mitosis of existing intimal cells. On the other hand, migration of cells from the media into the intima is suggested by the observation of medial smooth muscle cells in gaps of the internal elastic lamina. We could not determine whether this phenomenon represents transmigration of medial smooth muscle cells into the intima or an artifact caused by contraction of the vessel.

C. Degenerative Cell Change and Necrosis

Evidence of cell damage was frequent in arterial segments with intimal thickening. Degenerative change occurred in the cytoplasm of both intimal and medial smooth muscle cells. There basically were two different types of degenerative change. The more common type occurred in projecting cell processes of intimal and medial smooth muscle cells and involved swelling and dissolution of these cell portions. The stages in this phenomenon are thought to consist of degeneration of the cytoplasm in a peripheral portion of the cell, sequestration, subsequent self-amputation of nonviable cell processes from the main body of the cell, and finally, lysis of free cell portions. We do not know if at an early stage this process can be reversed.

Degenerative change involving the entire cell and leading to necrosis and dissolution of the cell was an uncommon finding in this group of normo-cholesterolemic animals. Most susceptible to necrosis were the foam cells, although these were a rare component of the intima under normal conditions. Necrosis of whole smooth muscle cells was not seen in the intima, although this is a common feature of the atherosclerotic lesion [33]. In the media, severe degenerative change, presumably leading to necrosis, was rare and seen in only one medial smooth muscle cell.

Cellular debris has been reported in the intimal cushions and in the media of the renal artery bifurcation of normal rabbits [38]. It appears that arterial segments with cushions are subject to increased wear and tear, and the increased rate of cell proliferation observed in intimal cushions by tritiated thymidine radioautography [31, 34] might also be explained as the mechanism by which occasional necrotic cells are replaced. An equilibrium between cell death and cell replacement might exist in normal animals.

D. Absence of Extracellular Lipid

A few smooth muscle cells in the thickened intima contained usually small amounts of intracellular lipid. In addition, there were rare, isolated foam cells completely filled with lipid, most of them necrotic. In spite of the release of intracellular lipid into the extracellular space by necrotic foam cells, there was no accumulation of extracellular lipid in the intima. It is probable that small amounts of extracellular lipid and debris from necrotic cells are rapidly cleared from the intima. In the atherosclerotic lesions of hypercholesterolemic animals, on the other hand, the equilibrium between accumulation of lipid and debris, and removal appears to be unbalanced because of an oversupply of cell debris and lipid released from many necrotic cells, leading to the development of lesions with extracellular lipid.

VI. Conclusions

Rhesus monkeys with low serum cholesterol levels have nonatherosclerotic intimal thickening in the proximal left coronary artery. Intimal thickening is in the form of cushions, that is, thick elevations of the intima at the bifurcation of the artery and at the orifices of branch vessels, and in the form of less prominent but more extensive diffuse intimal thickening in arterial segments away from bifurcations and branch orifices. The boundary between the two forms is not always clear-cut since confluence is usual, and diffuse intimal thickening often becomes the slender extension of a cushion.

The fine structure of intimal cushions differs from that of diffuse thickening. Cushions have abundant ground substance and cells are loosely arranged, consisting of a mixture of myofilament-rich and reticulum-rich smooth muscle cells. Reticulum-rich cells are smooth muscle cells with a variable but generally small number of myofilament bundles and with abundant cisternae of rough-surfaced endoplasmic reticulum. Diffuse intimal thickening has little ground substance and consists of densely packed myofilament-rich intimal smooth muscle cells. Reticulum-rich cells are less frequent than in cushions.

Intracellular lipid occurs in small quantities in a fraction of the smooth muscle cells of intimal thickening. Foam cells are extremely rare and always isolated. Extracellular lipid is absent. Degenerative cell changes occur in some intimal and medial smooth muscle cells in segments of the coronary artery with intimal thickening.

Acknowledgments

The authors are indebted to Mrs. KATHY CONSENTINO and Mrs. CATHERINE VIAL for skillful technical assistance and to Dr. BARBARA LYNCH for criticism of the manuscript.

This work was supported by a grant-in-aid of research by the Louisiana Heart Association and by USPHS NIH, grant HL–08974.

Fig. 1. Schematic view of the arterial lumen at the bifurcation of the main stem of the left coronary artery. Two longitudinal intimal cushions (pale shading) in the main stem extend distally to form a thick saddle-shaped cushion on the apex of the bifurcation. Diminishing in degree, the intimal thickening continues distally to involve the entire circumference of the anterior descending and circumflex branches as diffuse intimal thickening.

Fig. 2. Schematic drawing of the artery segment shown in figure 1. On the left, view of the lumen; on the right, the corresponding arterial cross-sections. Intimal thickening is indicated by stippling. *Level I:* The main stem of the artery proximal to the bifurcation has paired, longitudinal intimal cushions. The location of the cushions corresponds to two grooves on the outside of the vessel. *Level II:* A thick saddle-shaped cushion forms at the apex of the bifurcation. Diffuse intimal thickening inconsistently extends to involve the circumference of the bifurcation. *Level III:* At the origin of the anterior descending and the circumflex branches, intimal thickening tends to involve the entire circumference of each branch diffusely. *Level IV:* Distal to the bifurcation intimal thickening is progressively less extensive.

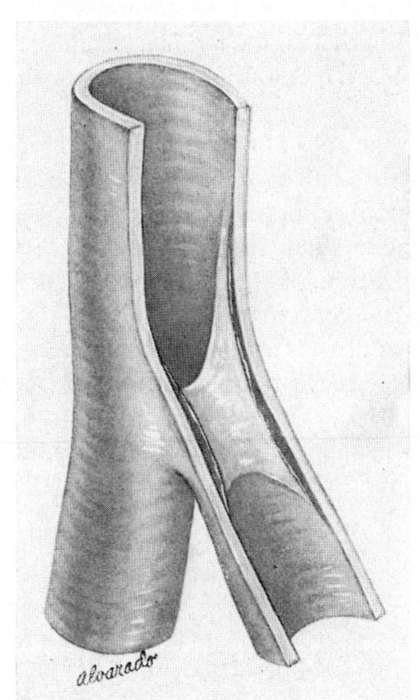

1

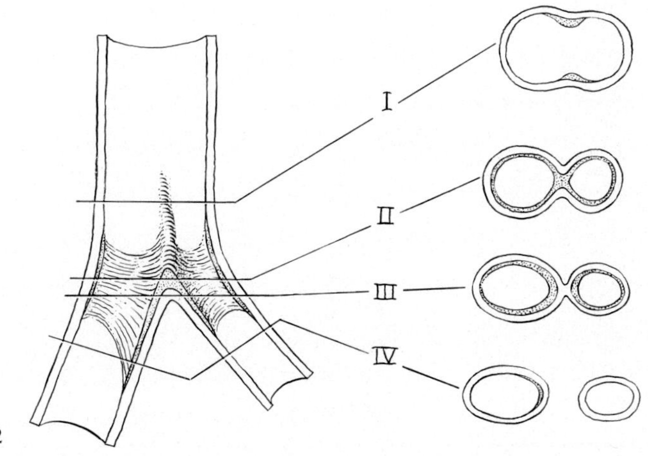

2

Fig. 3. Intimal cushion in the main stem at about level I in figure 2. There are several layers of smooth muscle cells, arranged less densely towards the surface of the thickening. Two ends of the internal elastic lamina overlap but do not join, creating a gap. Rhesus No. 45. Semithin Maraglas-embedded section. Paragon stain. Approximately × 740.

Fig. 4. Intimal cushion at the apex of the bifurcation at about level II in figure 2. The intercellular matrix is loose and abundant. Reticulum-rich smooth muscle cells are frequent in this type of cushion. The internal elastic lamina, granular and fibrilar rather than compact, is incomplete. Rhesus No. 47. Semithin Maraglas-embedded section. Paragon stain. Approximately × 740.

Fig. 5. Intimal cushion at the apex of the bifurcation at level III in figure 2. Large cushions tend to be less cellular than small cushions, especially near the lumen of the vessel. The internal elastic lamina is prominent but has one short gap. Rhesus No. 26. Semithin Maraglas-embedded section. Paragon stain. Approximately × 800.

Fig. 6. Diffuse intimal thickening in the lateral wall of the proximal anterior descending branch. The thickening is a continuation of the cushion seen in figure 5. The intima consists of one to two layers of densely arranged smooth muscle cells over a compact internal elastic lamina. Rhesus No. 26. Semithin Maraglas-embedded section. Paragon stain. Approximately × 800.

Fig. 7. Diffuse intimal thickening. The intima has two strata of myofilament-rich smooth muscle cells (S) over a continuous internal elastic lamina (IE) of uneven thickness. Endothelial cell (E). Rhesus No. 20. × 7,200.

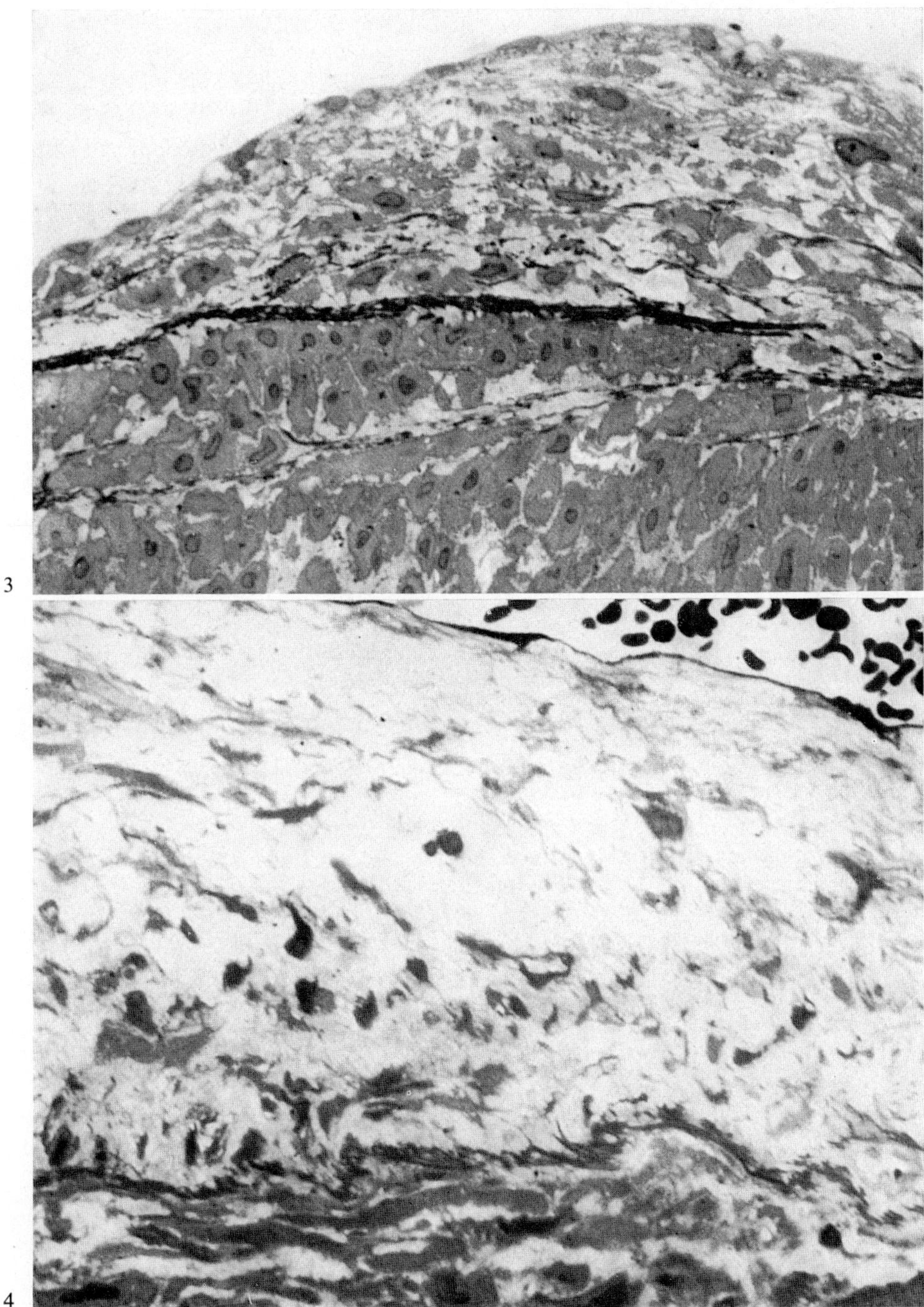

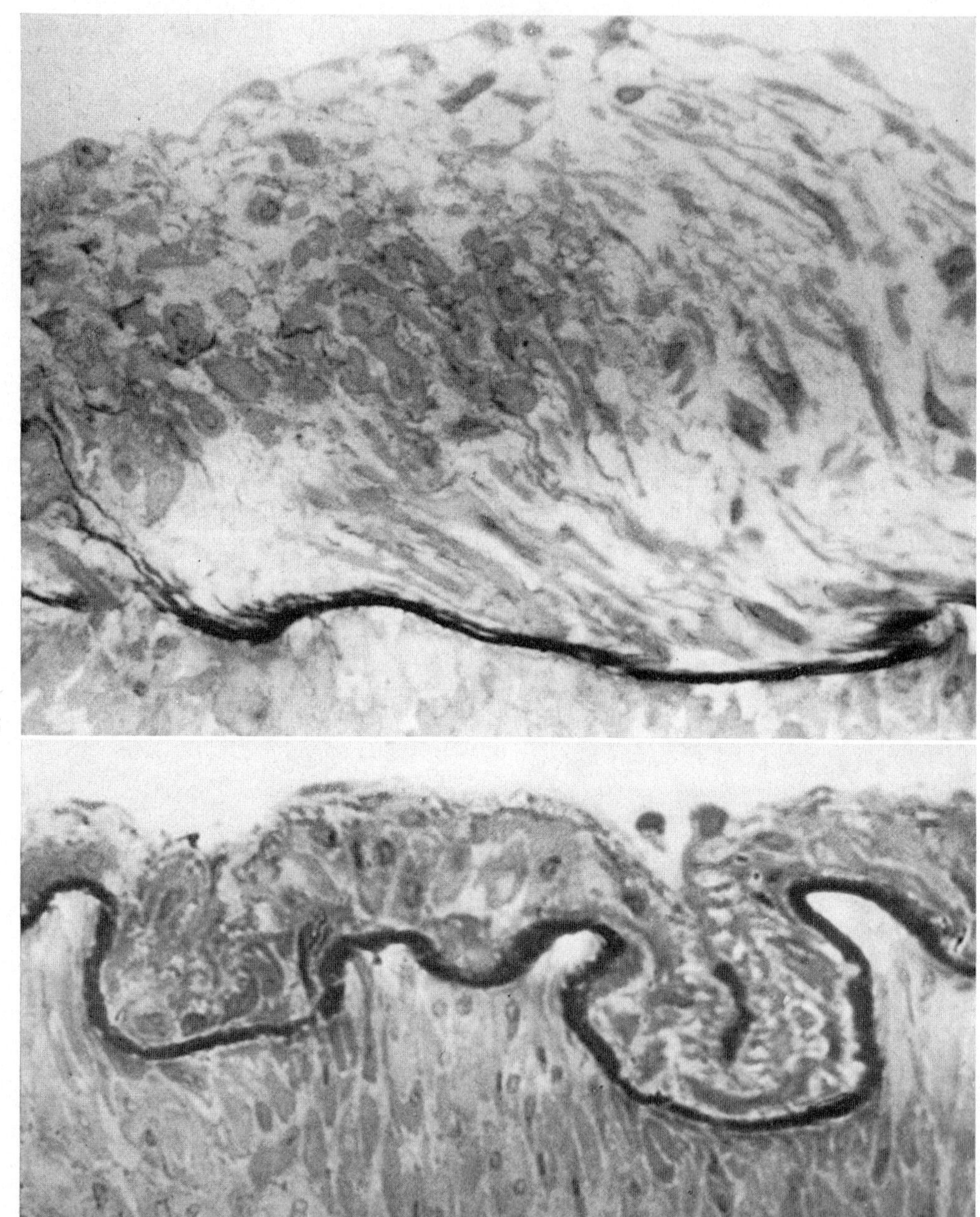

5

6

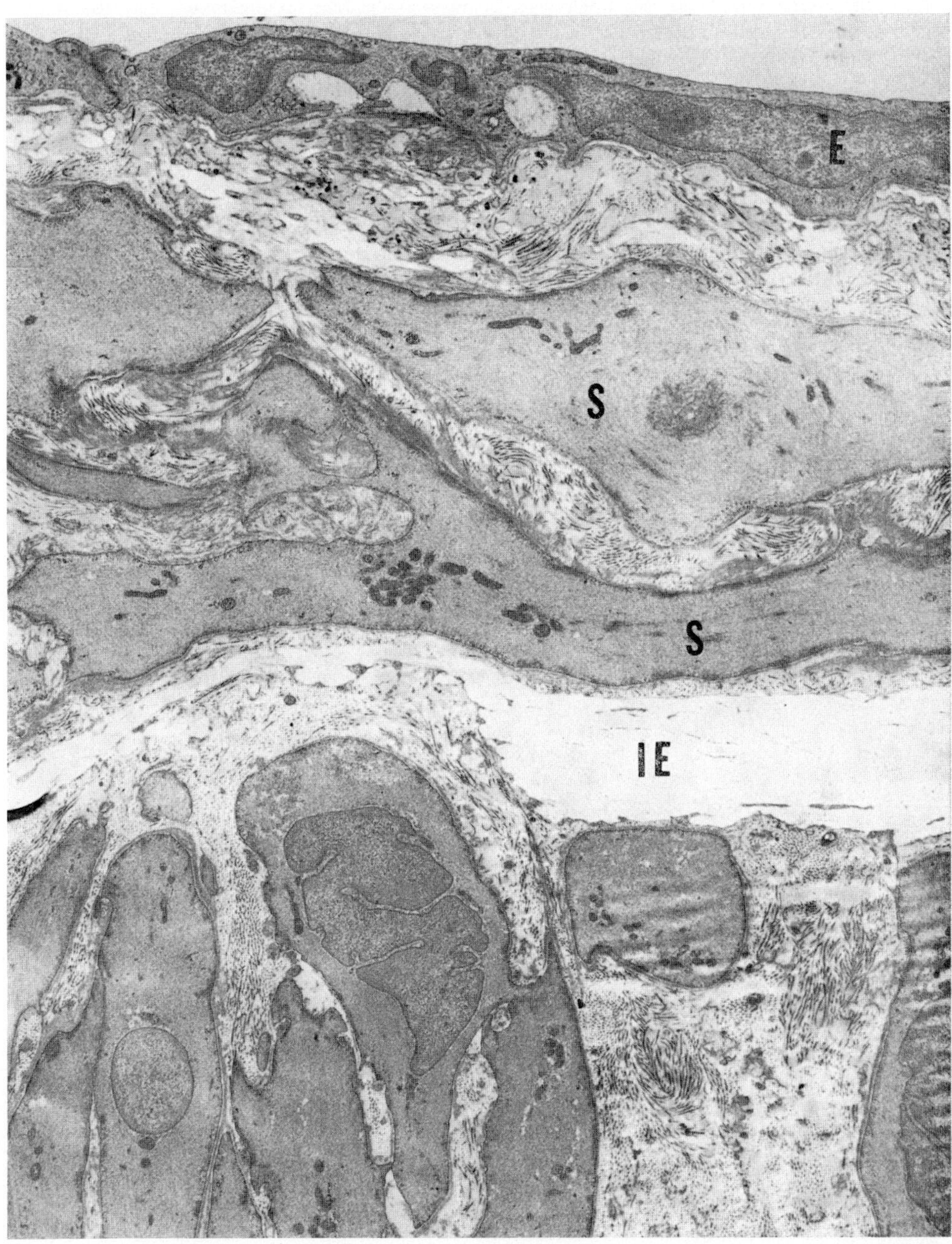
E
S
S
IE
7

Fig. 8. Diffuse intimal thickening adjacent to the thick intimal cushion at the apex of the bifurcation. There is a single layer of myofilament-rich smooth muscle cells (S). An endothelial cell on the surface of the intimal thickening contains numerous filaments (f) and several small lipid inclusions (L). Rhesus No. 26. × 16,900).

Fig. 9. The narrow peripheral part of a large intimal cushion near the bifurcation. One necrotic foam cell (F) is just beneath the endothelium (E). Portions of an intimal smooth muscle cell show degenerative change (d). The intercellular matrix is loose. Internal elastic lamina (IE). Rhesus No. 45. × 8,100.

Fig. 10. Continuation of the cushion shown in figure 9. This part of the intimal thickening is closer to the main body of the cushion than the part seen in figure 9. There are several strata of smooth muscle cells. One of two reticulum-rich smooth muscle cells has a lipid inclusion (L). Myofilament-rich smooth muscle cells with prominent basement membranes are at the base of the intimal thickening. Rhesus No. 45. × 8,100.

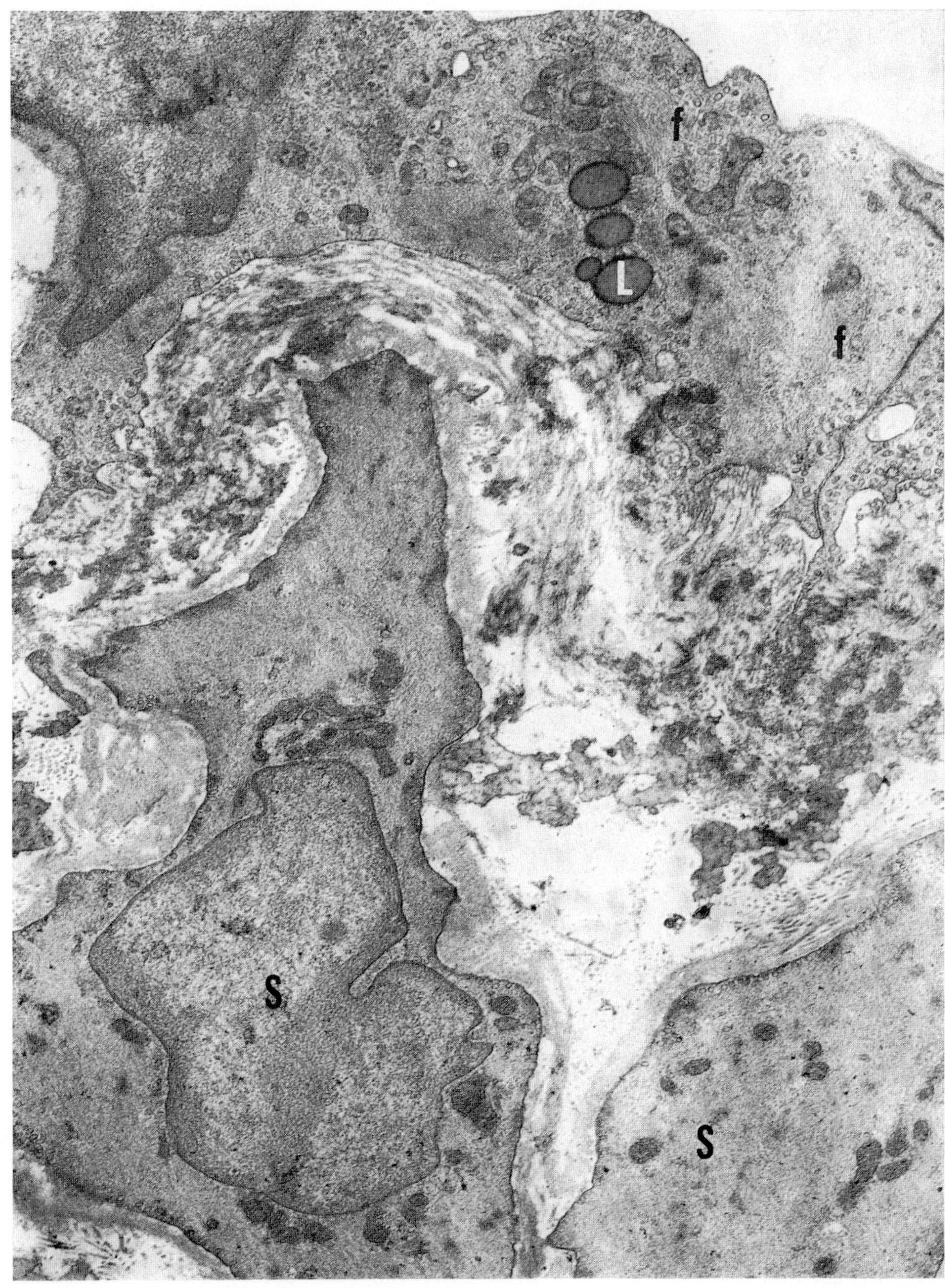
f
L
f
S
S
8

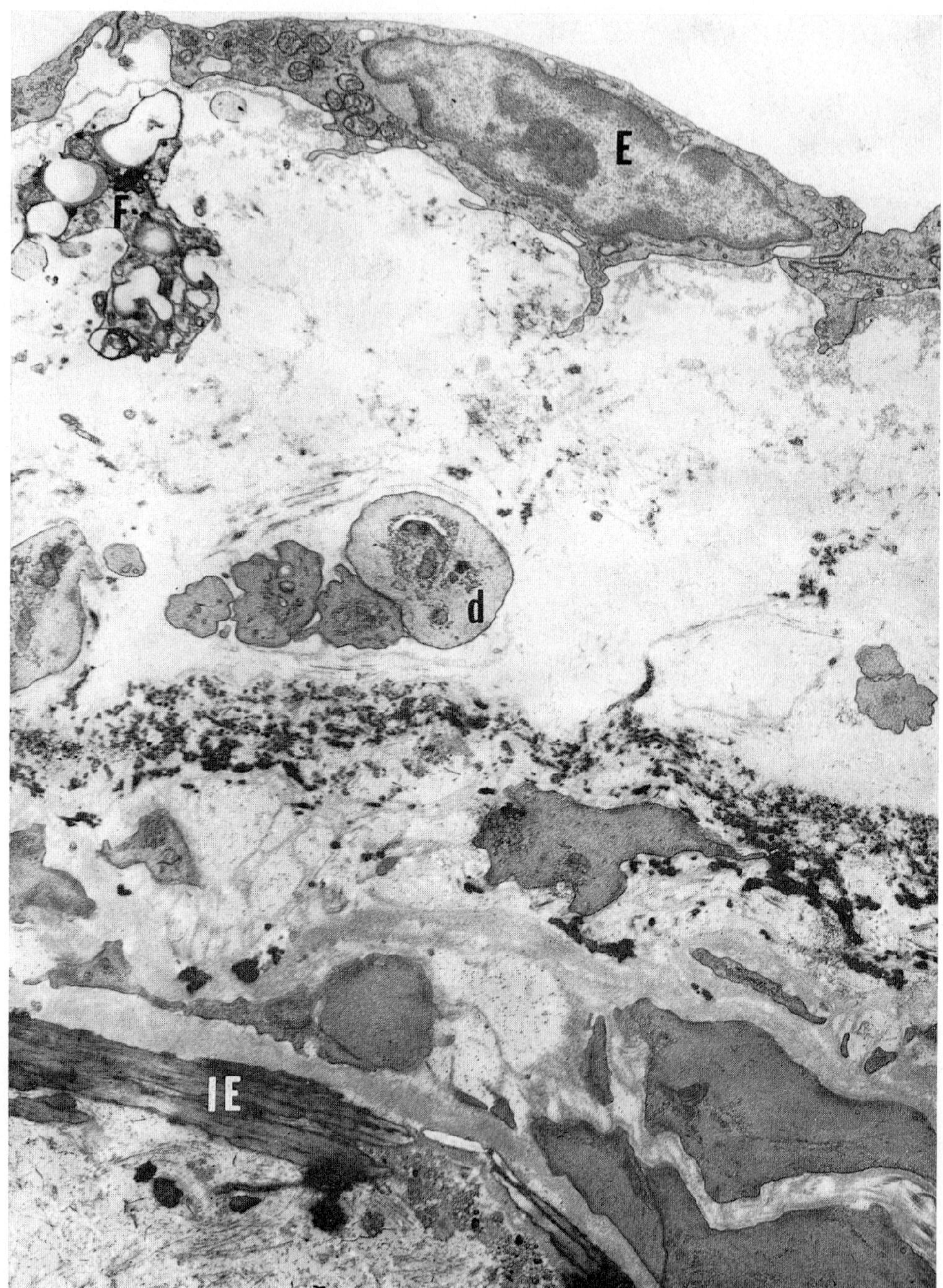
E
F
d
IE
9

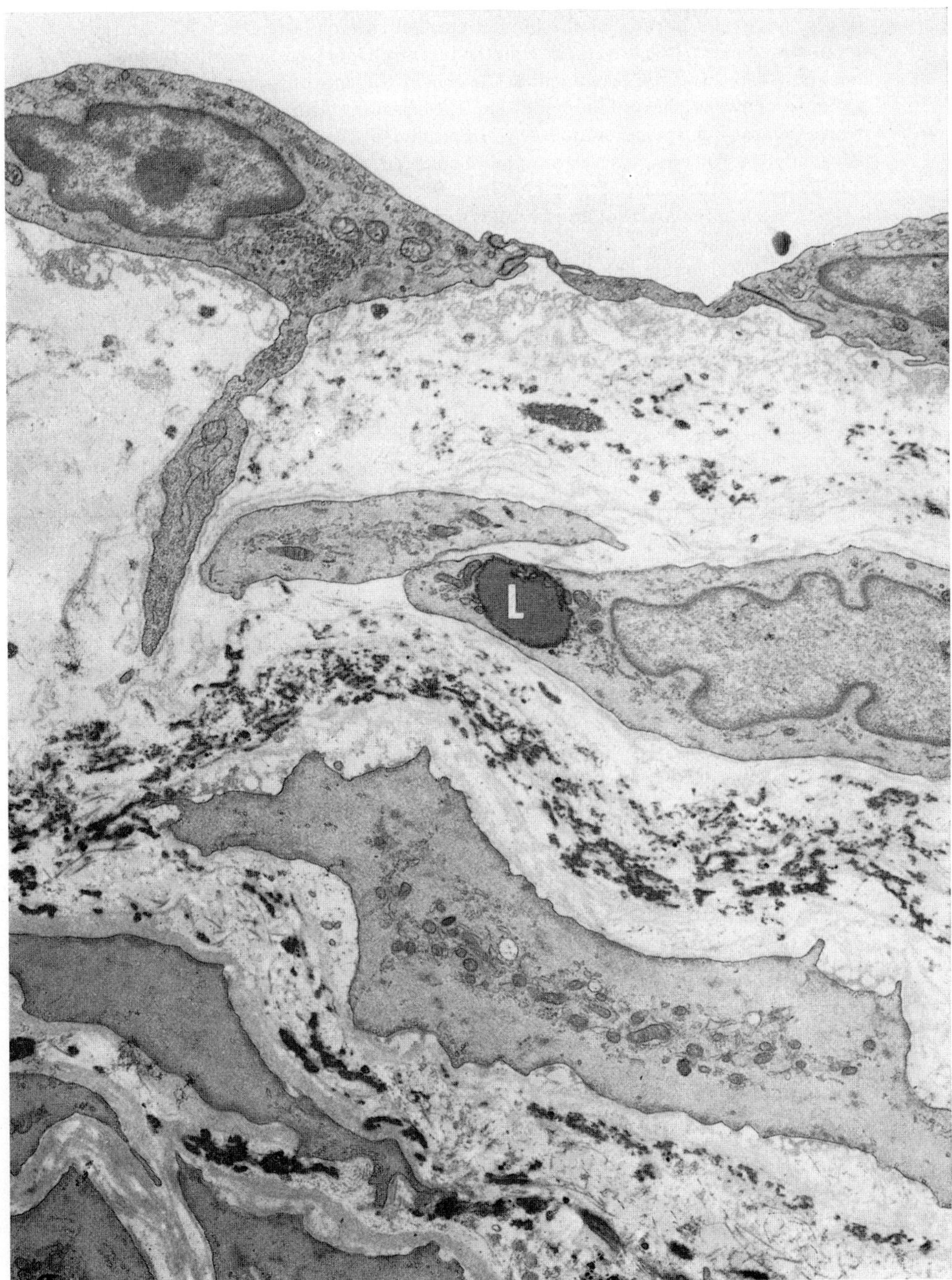

10

Fig. 11. Upper portion of the intimal cushion shown in figure 5. A reticulum-rich smooth muscle cell (RS) is near the endothelium (E). Rough-surfaced endoplasmic reticulum and Golgi stacks are prominent at the center, while myofilament bundles with dense bodies are at the periphery of the cytoplasm. The basement membrane is less prominent than in myofilament-rich smooth muscle cells. Intercellular matrix is loose. The endothelial cell has numerous filaments, vesicles of smooth-surfaced endoplasmic reticulum and a small lipid inclusion. Rhesus No. 26. × 17,800.

Fig. 12. Detail of a reticulum-rich smooth muscle cell from the intimal cushion shown in figure 11. Cisternae of rough-surfaced endoplasmic reticulum (r) are numerous, and Golgi stacks (G) are prominent. The intercellular matrix consists of finely reticulated ground substance (g). Rhesus No. 26. × 30,400.

Fig. 13. Superficial part of a large intimal cushion near the bifurcation. A reticulum-rich smooth muscle cell immediately below the endothelium (E) contains numerous lipid inclusions (L). Smooth-surfaced endoplasmic reticulum is in the form of small vesicles and multiple Golgi stacks. Cisternae of rough-surfaced endoplasmic reticulum (r) are numerous. Myofilaments at the periphery of the cell are indistinct. The extracellular space contains shreds of basement membrane (b), collagen (c), elastica (e), and finely reticulated ground substance (g). Rhesus No. 45. × 16,900.

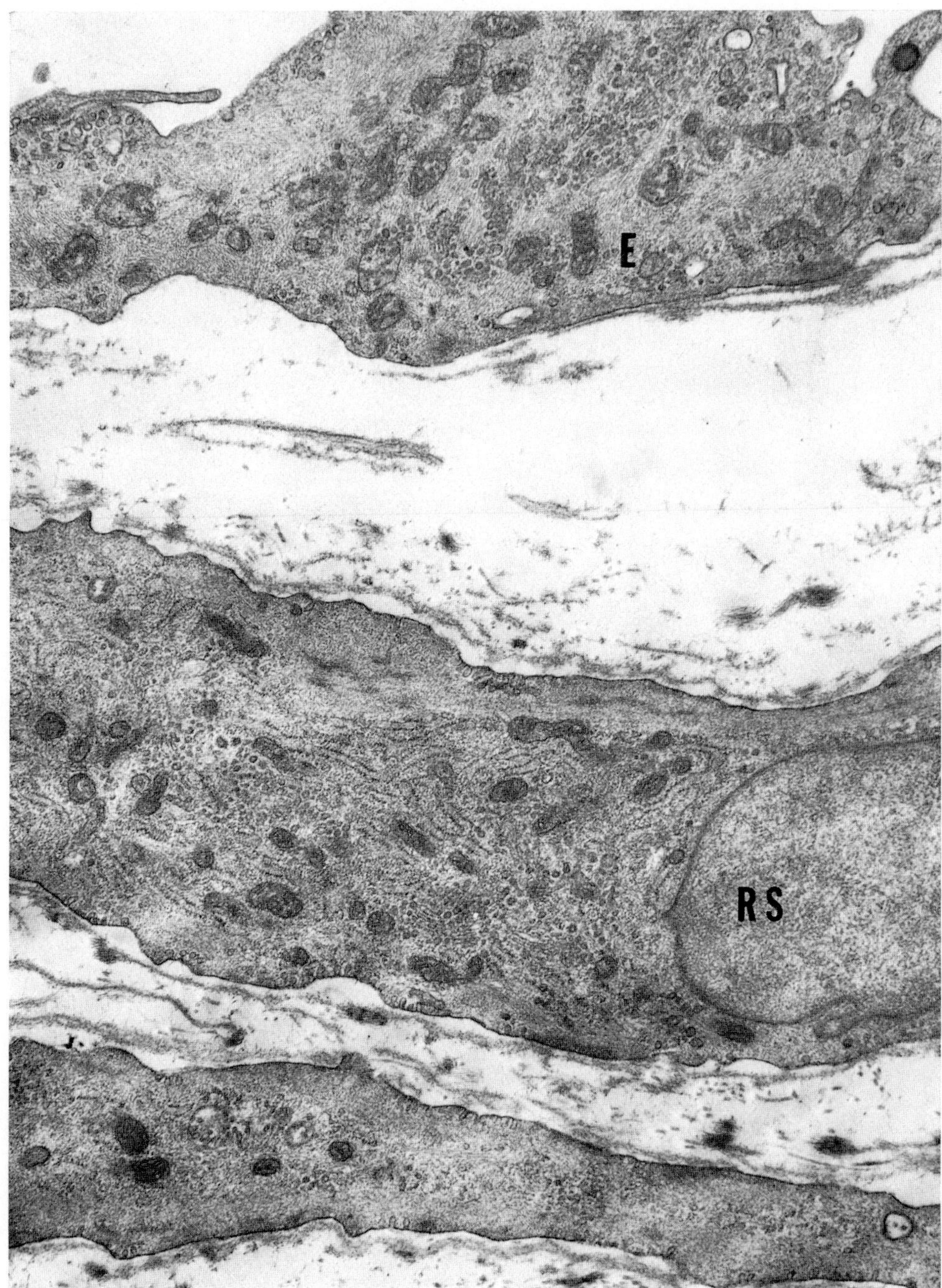
E
R S
11

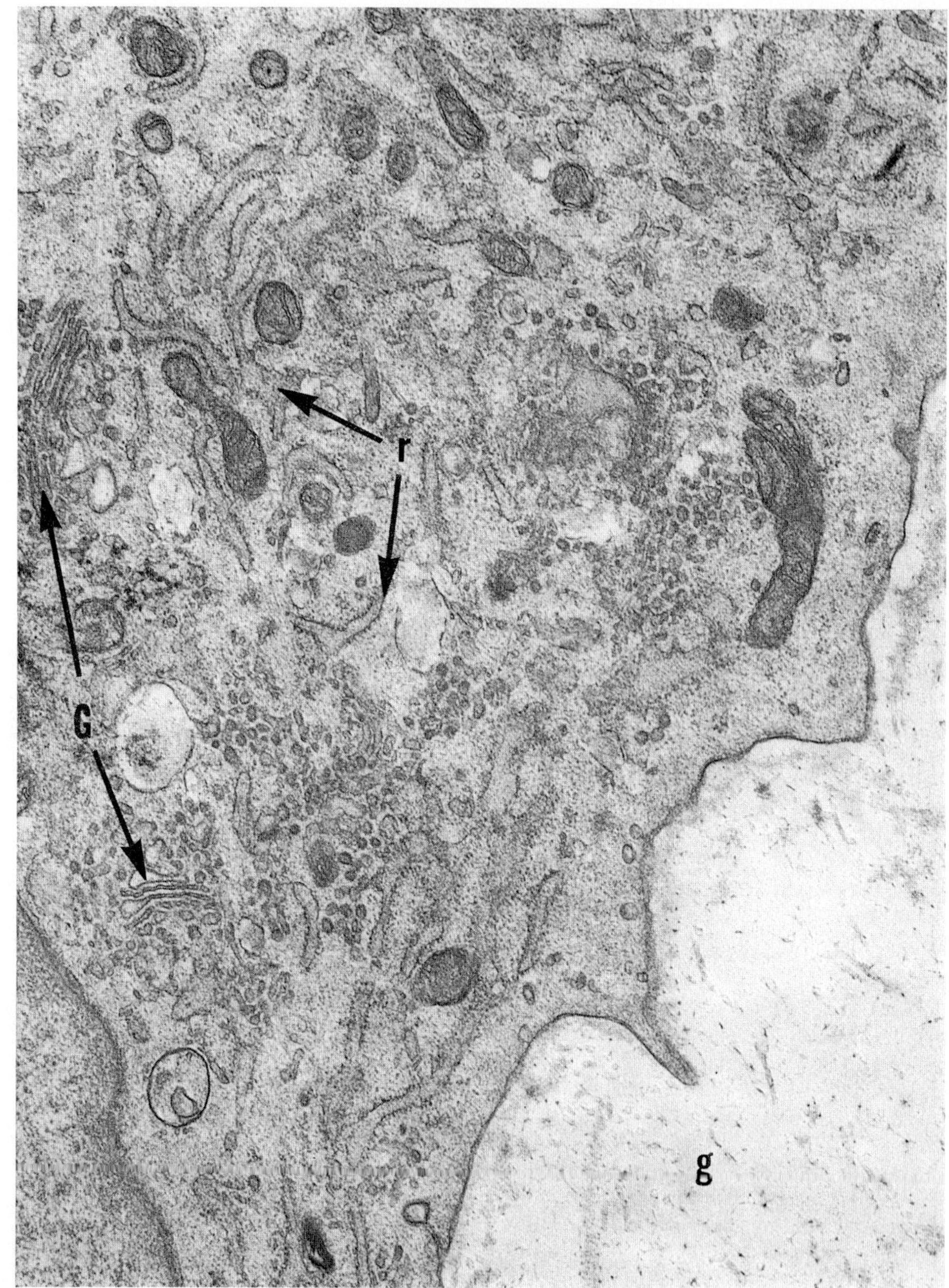
r
G
g
12

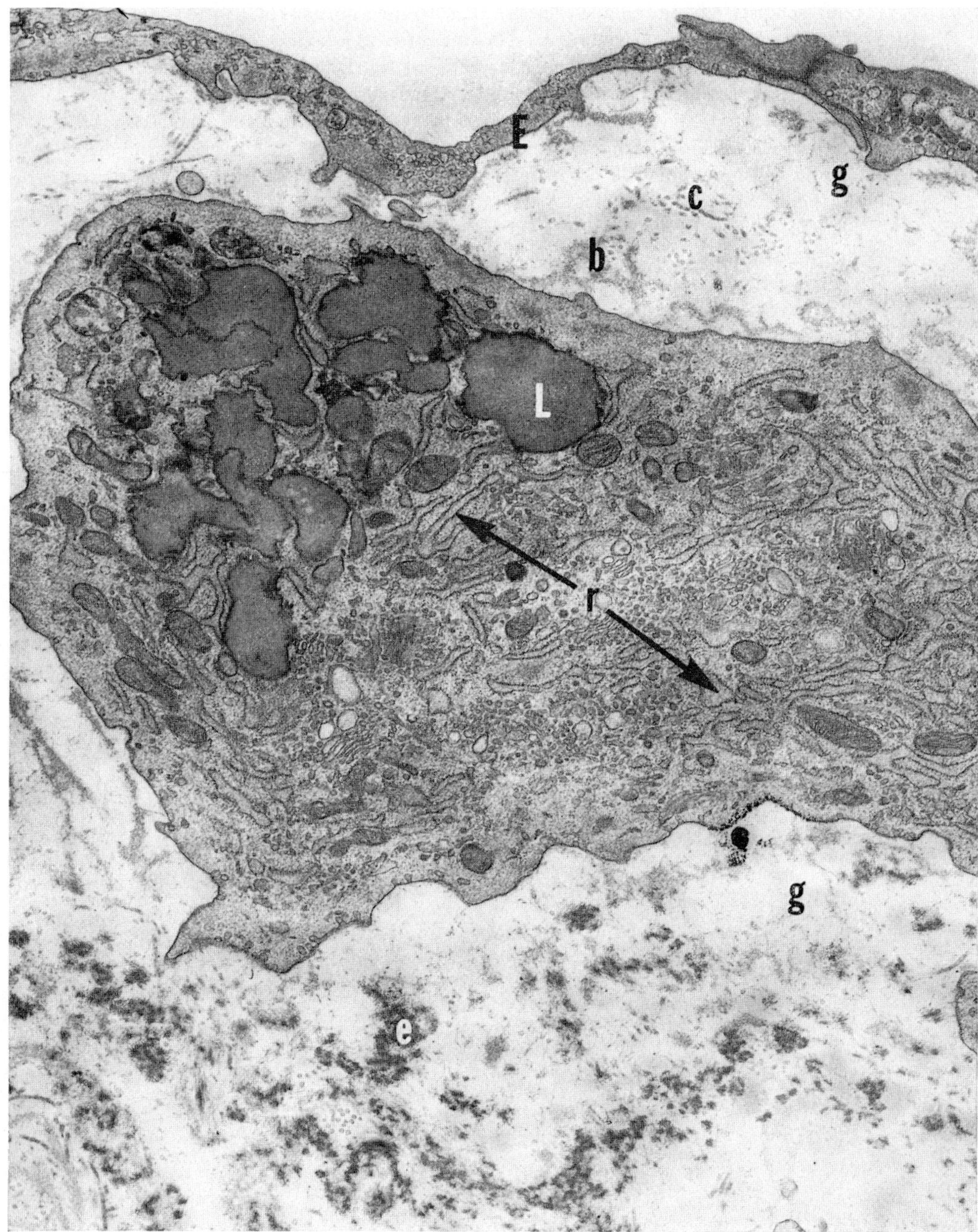
E
g
c
b
L
r
g
e
13

Fig. 14. Periphery of a small intimal cushion. The cushion consists of densely packed, myofilament-rich smooth muscle cells (S). Degenerating portions of cells (d) between the smooth muscle cells and the endothelium (E) are entirely without cytoplasmic constituents or contain only a few vesicles. A connection between degenerated cell portions and smooth muscle cells is sometimes present (fig. 16, 17). Rhesus No. 13. × 12,000.

Fig. 15. Diffuse intimal thickening. One myofilament-rich smooth muscle cell (S) is between the endothelium (E) and the internal elastic lamina (IE). In the underlying media numerous cell processes with degenerative change (d) are crowded against the internal elastic lamina. Rhesus No. 20. × 8,100.

Fig. 16. Intima and media adjacent to a small intimal cushion. The intima shows a portion of a single smooth muscle cell (IS). In the media several cell processes with degenerative change (d) are lodged against the internal elastic lamina (IE). One of these has a connection (arrow) with a medial smooth muscle cell (MS). Endothelial cell (E). Rhesus No. 13. × 12,000.

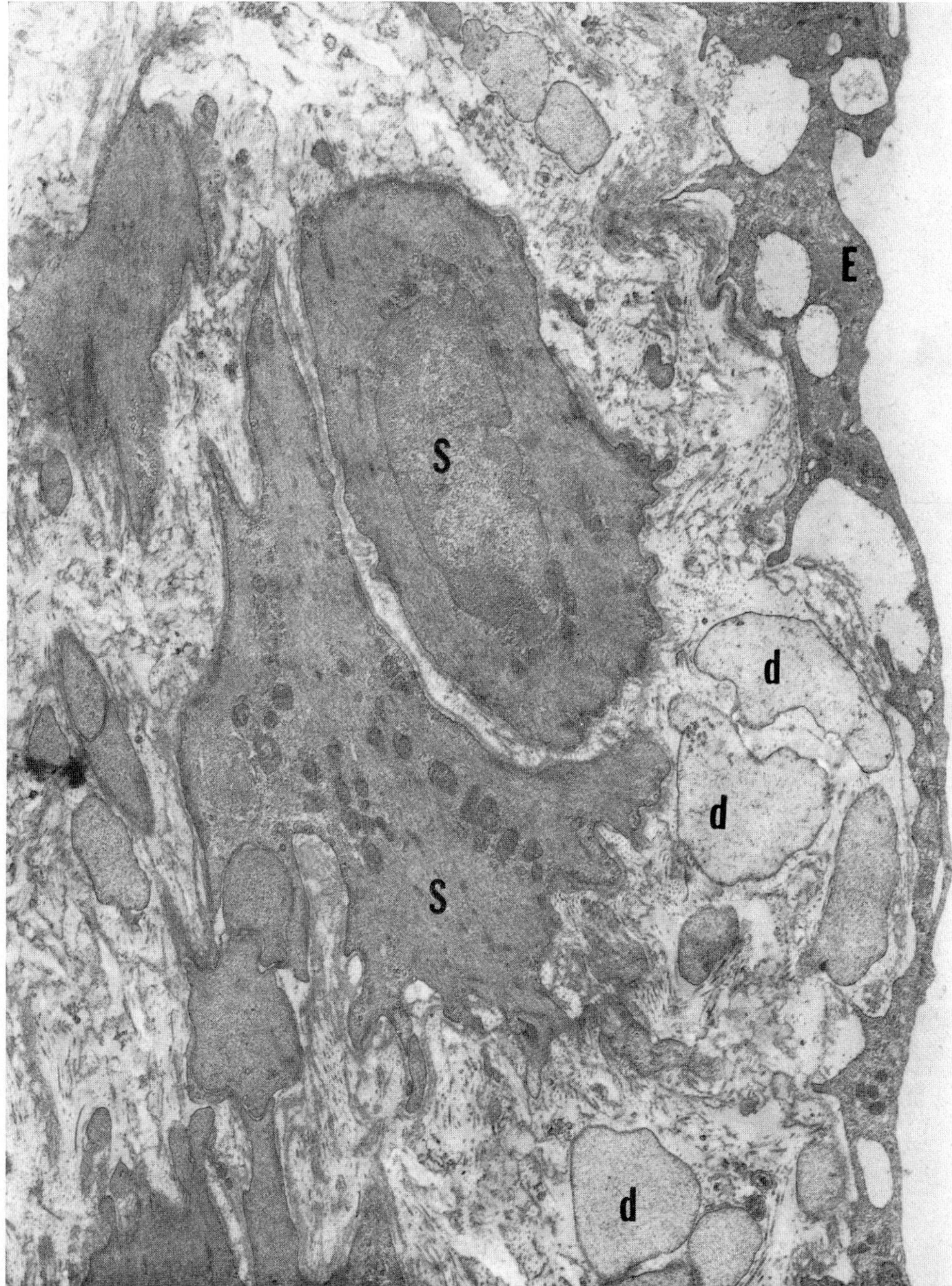

14

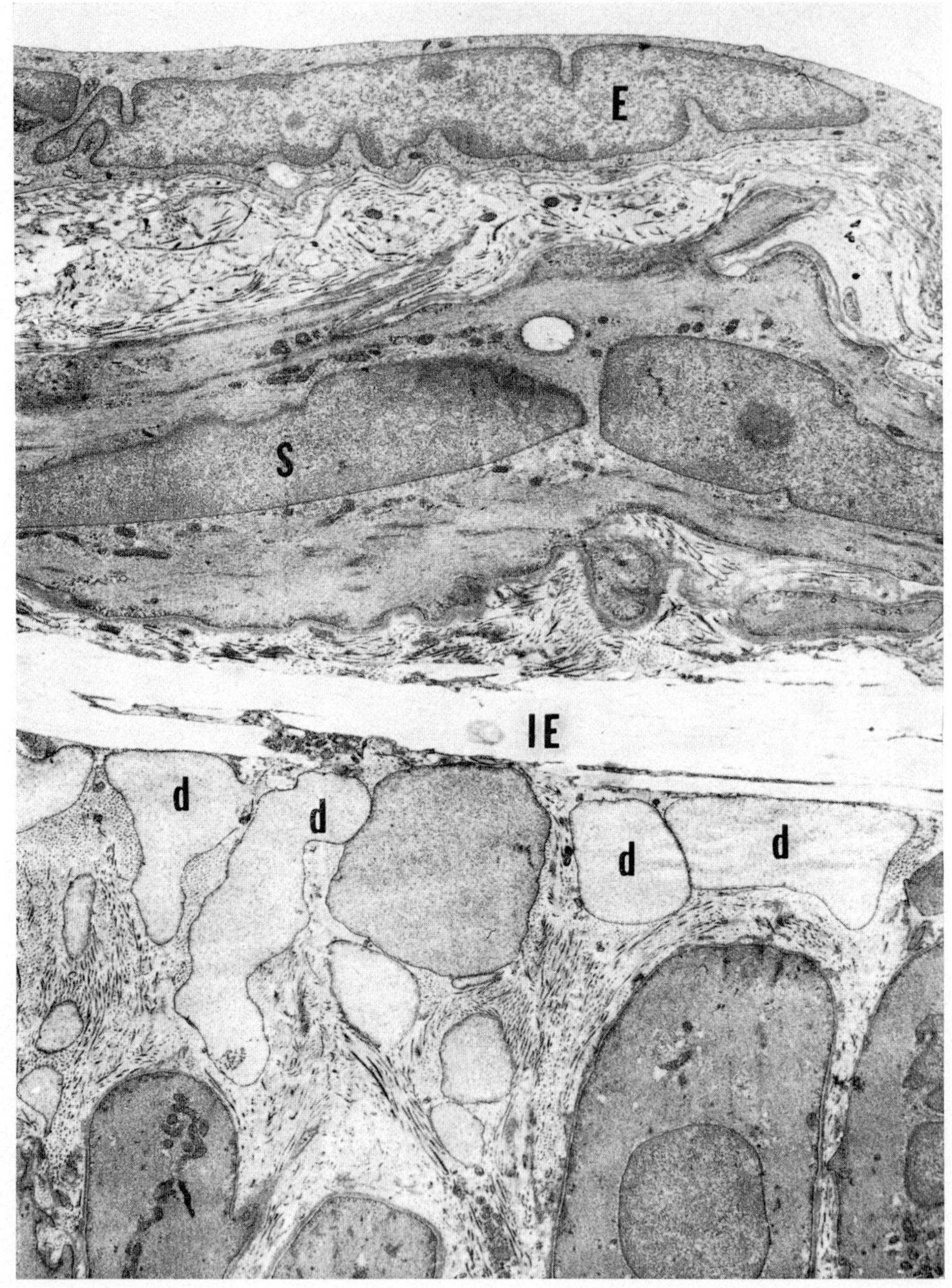
E
S
IE
d
d
d
d
15

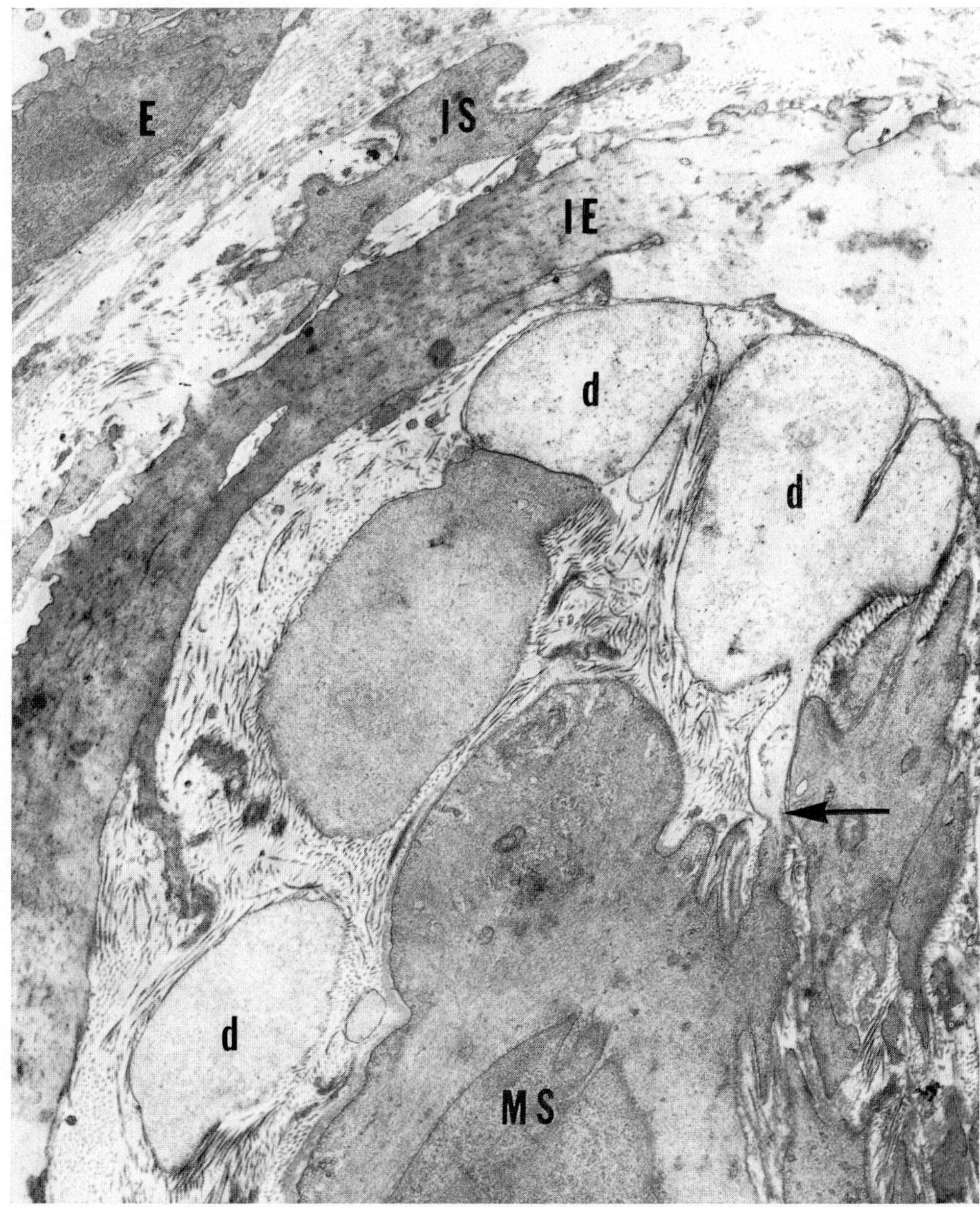
E
IS
IE
d
d
d
MS
16

Fig. 17. Detail of the medial smooth muscle cell shown in figure 16. The degenerated cell process (d) lacks myofilaments, pinocytotic vesicles, organelles, and basement membrane. The main body of the cell (MS) to which it is connected has all these components, but shows focal disorganization of myofilament bundles and the appearance of concentric membranes indicating lysosomal activity (ly). Rhesus No. 13. × 36,000.

Fig. 18. Media beneath a small intimal cushion. One degenerating medial smooth muscle cell shows transformation of cytoplasmic constituents into large residual bodies composed of complex multilaminated and vacuolated structures. The extreme peripheral cytoplasm is largely spared although there is some myofibrillar disorganization. There was no distinct internal elastic lamina between this part of the media and the intimal cushion. Rhesus No. 68. × 7,200.

Fig. 19. Detail from another section of the smooth muscle cell shown in figure 18. The contents of some of the multilaminated structures (la) are recognizable as a lipid droplet (L), granular material which appears to consist of disintegrated myofilaments (mf), and degenerating mitochondria. Some of the laminated structures have transformed into residual bodies. Rhesus No. 68. × 50, 600.

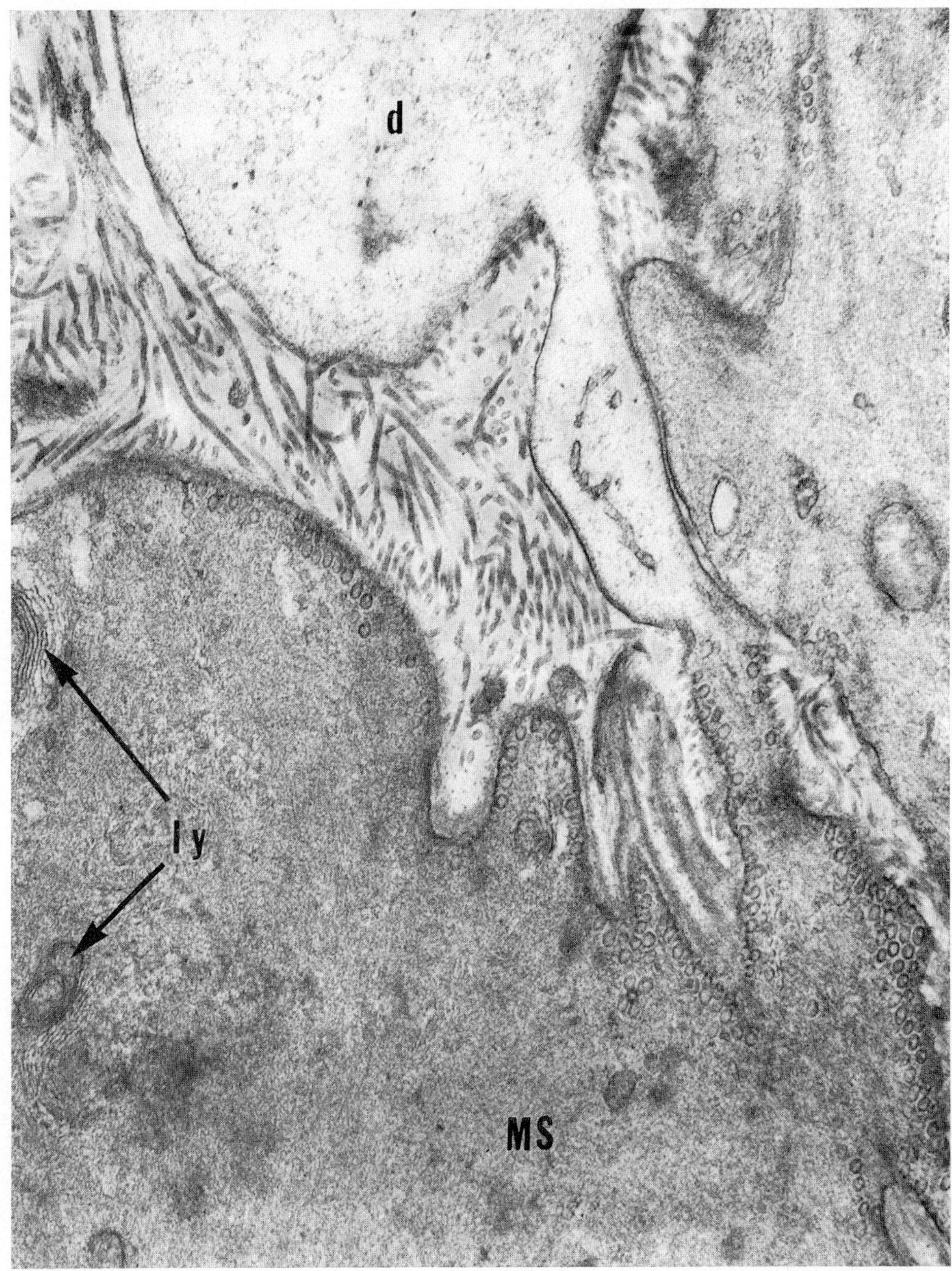
d
ly
MS
17

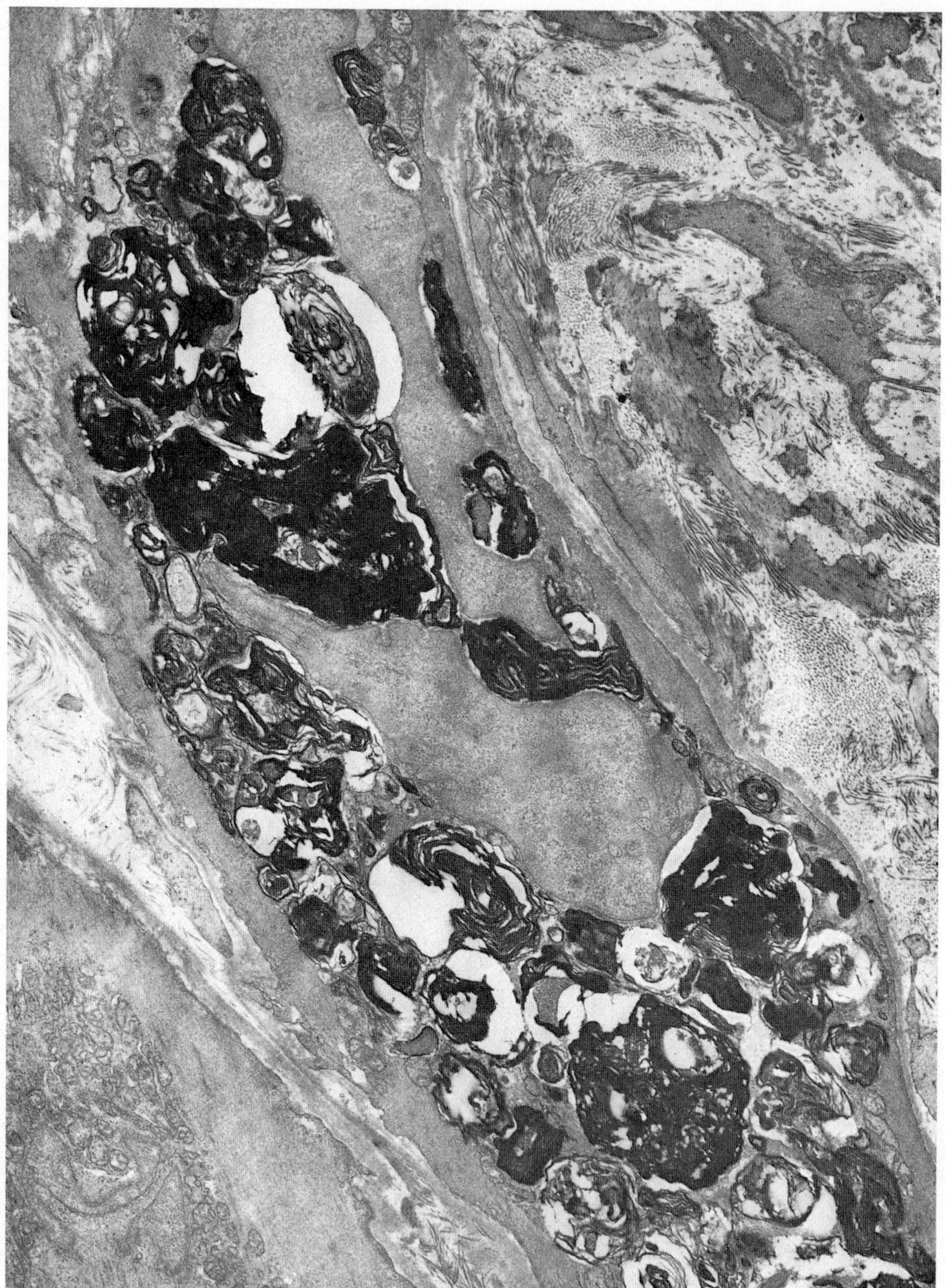

18

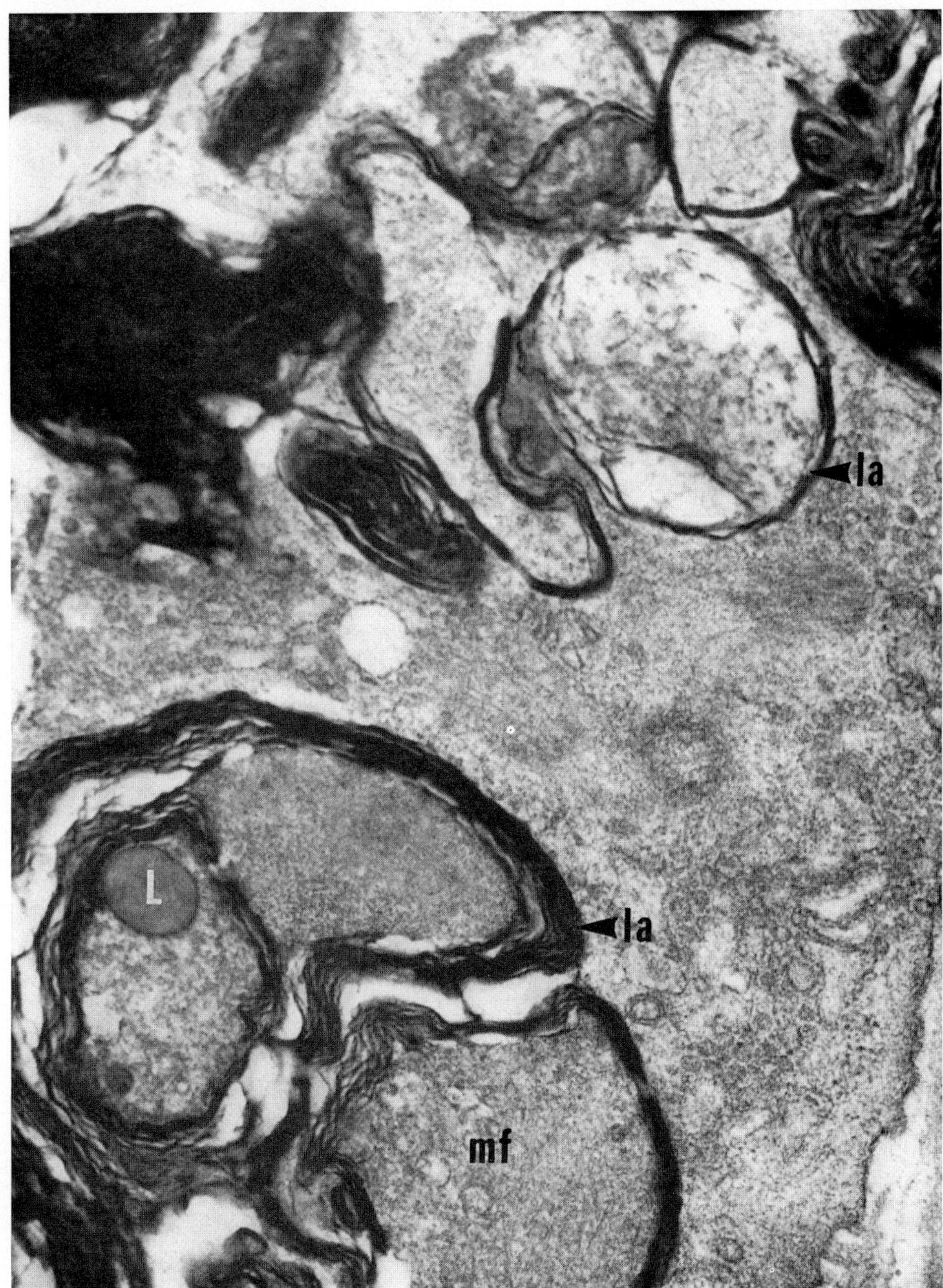
la
la
L
mf
19

References

1 ABELL, L. L.; LEVY, B. B.; BRODIE, B. B., and KENDALL, F. E.: A simplified method for the estimation of total cholesterol in serum and demonstration of its specificity. J. biol. Chem. *195:* 357–366 (1952).

2 ANDRUS, S. B. and PORTMAN, O. W.: Comparative studies of spontaneous and experimental atherosclerosis in primates; in FIENNES Some recent developments in comparative medicine, pp. 161–177 (Academic Press, London 1966).

3 BREMER, J. L.: On the variations of wall thickness in embryonic arteries. Anat. Rec. *27:* 1–13 (1924).

4 DOCK, W.: The predilection of atherosclerosis for the coronary arteries. J. Am. med. Ass. *131:* 875–878 (1946).

5 EDHOLM, G.: Über die arteria coronaria cordis des Menschen. Anat. Anz. *42:* 124–128 (1912).

6 EGGEN, D. A.: Cholesterol metabolism in the rhesus monkey, squirrel monkey, and baboon. J. Lipid Res. *15:* 139–145 (1974).

7 EGGEN, D. A.; NEWMAN, W. P., and STRONG, J. P.: Serum lipids and atherosclerotic lesions in baboon, rhesus and squirrel monkeys on a high fat – high cholesterol diet. Circulation *40:* suppl. III, p. 7 (1969).

8 EHRICH, W.; CHAPELLE, C. DE LA, and COHN, A. E.: Anatomical Ontogeny. B. Man. I. A study of the coronary arteries. Am. J. Anat. *49:* 241–282 (1931).

9 FANGMAN, R. J. and HELLWIG, C. A.: Histology of coronary arteries in newborn infants. Am. J. Path. *23:* 901–902 (1947).

10 GEER, J. C. and HAUST, M. D.: Smooth muscle cells in atherosclerosis. Monogr. Atheroscler., vol. 2 (Karger, Basel 1972).

11 GEER, J. C.; McGILL, H. C.; ROBERTSON, W. B., and STRONG, J. P.: Histologic characteristics of coronary artery fatty streaks. Lab. Invest. *18:* 565–570 (1968).

12 GROSS, L.; EPSTEIN, E. Z., and KUGEL, M. A.: Histology of the coronary arteries and their branches in the human heart. Am. J. Path. *10:* 253–273 (1934).

13 JAFFÉ, D.; HARTROFT, W. S.; MANNING, M., and ELETA, G.: Coronary arteries in newborn children. Acta paediat. scand. Suppl. *219:* 1–28 (1971).

14 LANGHANS, T.: Beiträge zur normalen und pathologischen Anatomie der Arterien. Virchows Arch. path. Anat. Physiol. *36:* 187–226 (1866).

15 MALINOW, M. R. and STORVICK, C. A.: Spontaneous coronary lesions in howler monkeys. J. Atheroscler. Res. *8:* 421–431 (1968).

16 McGILL, H. C.: The lesion; in SCHETTLER and WEIZEL Atherosclerosis III, pp. 27–38 (Springer, Berlin 1974).

17 McGILL, H. C.; STRONG, J. P.; HOLMAN, R. L., and WERTHESSEN, N. T.: Arterial lesions in the Kenya baboon. Circulation Res. *8:* 670–679 (1960).

18 McMILLAN, G. C.: The onset of plaque formation in arteriosclerosis. Acta cardiol. Suppl. XI, pp. 43–62 (1965).

19 MINKOWSKI, W. L.: The coronary arteries of infants. Am. J. med. Sci. *214:* 623–629 (1947).

20 MOON, H. D.: Coronary arteries in fetuses, infants, and juveniles. Circulation *16:* 263–267 (1957).

21 MOVAT, H. Z.; MORE, R. H., and HAUST, M. D.: The diffuse intimal thickening of the human aorta with aging. Am. J. Path. *34:* 1023–1031 (1958).

22 NEUFELD, H. N.; WAGENVOORT, C. A., and EDWARDS, J. E.: Coronary arteries in fetuses, infants, juveniles, and young adults. Lab. Invest. *11:* 837–844 (1962).

23 PFLIEGER, H. und GOERTTLER, K.: Konstruktionsprinzipien der Aortenwand im Ursprungsbereich der interkostalen, intestinalen und renalen Aortenäste. Arch. Kreislaufforsch. *62:* 223–248 (1970).

24 REALE E. und LUCIANO, L.: Elektronenmikroskopische Beobachtungen an Stellen der Astabgabe der Arterien. Angiologica *3:* 226–239 (1966).

25 ROBERTSON, J. H.: The significance of intimal thickening in the arteries of the newborn. Archs Dis. Childh. *35:* 588–590 (1960).

26 RYAN, G. B.; CLIFF, W. J.; GABBIANI, G.; IRLÉ, C.; MONTANDON, D.; STATKOV, P. R. and MAJNO, G.: Myofibroblasts in human granulation tissue. Hum. Path. *5:* 55–67, (1974).

27 RYAN, G. B.; CLIFF, W. J.; GABBIANI, G.; IRLÉ, C.; STATKOV, P. R., and MAJNO,G.: Myofibroblasts in an avascular fibrous tissue. Lab. Invest. *29:* 197–206 (1973).

28 SCHORNAGEL, H. E.: Intimal thickening in the coronary arteries in infants. Archs Path. *62:* 427–432 (1956).

29 SPURLOCK, B. O.; KATTINE, V. C., and FREEMAN, J. A.: Technical modifications in Maraglas embedding. J. Cell Biol. *17:* 203–207 (1963).

30 STARY, H. C.: Progression and regression of experimental atherosclerosis in rhesus monkeys; in GOLDSMITH and MOOR-JANKOWSKI Medical primatology 1972, part III, pp. 356–367 (Karger, Basel 1972).

31 STARY, H. C.: Proliferation of arterial cells in atherosclerosis; in WAGNER and CLARKSON Arterial mesenchyme and arteriosclerosis. Proc. Int. Workshop on Art. Mesenchyme and Arteriosclerosis, New Orleans 1973. Adv. exp. med. Biol., vol. 43, pp. 59–81 (Plenum Press, New York 1974).

32 STARY, H. C.: Cell proliferation and ultrastructural changes in regressing atherosclerotic lesions after reduction of serum cholesterol; in SCHETTLER and WEIZEL Atheroscelerosis III, pp. 187–190 (Springer, Berlin 1974).

33 STARY, H. C.: Coronary artery fine structure in rhesus monkeys: The early atherosclerotic lesion and its progression (this volume).

34 STARY, H. C. and McMILLAN, G. C.: Kinetics of cellular proliferation in experimental atherosclerosis. Archs Path. *89:* 173–183 (1970).

35 STEHBENS, W. E.: Focal intimal proliferation in the cerebral arteries. Am. J. Path. *36:* 289–301 (1960).

36 STEHBENS, W. E.: The renal artery in normal and cholesterol-fed rabbits. Am. J. Path. *43:* 969–985 (1963).

37 STEHBENS, W. E.: Intimal proliferation and spontaneous lipid deposition in the cerebral arteries of sheep and steers. J. Atheroscler. Res. *5:* 556–568 (1965)

38 STEHBENS, W. E. and LUDATSCHER, R. M.: Ultrastructure of the renal arterial bifurcation of rabbits. Expl molec. Path. *18:* 50–67 (1973).

39 THOMA, R.: Über die Abhängigkeit der Bindegewebsneubildung in der Arterienintima von den mechanischen Bedingungen des Blutumlaufes. Erste Mitteilung. Virchows Arch. path. Anat. Physiol. *93:* 443–505 (1883).

40 Thoma, R.: Über die Strömung des Blutes in der Gefässbahn und die Spannung der Gefässwand. I. Teil. Beitr. path. Anat. allg. Path. *66:* 92–158 (1920).

41 Vlodaver, Z.; Medalie, J., and Neufeld, H. N.: Coronary arteries in young monkeys. Israel J. med. Scis. *5:* 639–643 (1969).

42 Wilens, S. L.: The nature of diffuse intimal thickening of arteries. Am. J. Path. *27:* 825–839 (1951).

43 Wolkoff, K.: Über die histologische Struktur der Coronararterien des menschlichen Herzens. Virchows Arch. path. Anat. Physiol. *241:* 42–58 (1923).

44 Wolkoff, K.: Über die Altersveränderungen der Arterien bei Tieren. Virchows Arch. path. Anat. Physiol. *252:* 208–228 (1924).

H. C. Stary, M. D. and Jack P. Strong, M. D., Department of Pathology, Louisiana State University Medical Center, 1542 Tulane Avenue, *New Orleans, LA 70112* (USA)

Prim. Med., vol. 9, pp. 359–395 (Karger, Basel 1976)

Coronary Artery Fine Structure in Rhesus Monkeys: The Early Atherosclerotic Lesion and its Progression

H. C. STARY

Department of Pathology,
Louisiana State University Medical Center, New Orleans, La.

Contents

I. Introduction

In human coronary arteries, atherosclerotic lesions predominate at the bifurcation of the main stem of the left coronary artery and in the anterior descending branch just beyond the bifurcation [13, 35]. Lesions causing stenosis [2] or occlusion [16, 18, 20] of the coronary artery lumen occur most frequently in the same locations. In spite of the clinical importance of these coronary artery lesions, there has been no systematic study of the fine structure of atherosclerosis in this coronary artery segment. In general, there are few studies on the fine structure of coronary atherosclerosis in man [4, 8, 12] or on spontaneous or experimentally induced coronary atherosclerosis in animals [14, 15, 22].

The purpose of this study is twofold. Firstly, it describes the fine structural changes in naturally occurring nonatherosclerotic intimal thickening at and around the bifurcation of the left coronary artery when atherosclerosis is superimposed experimentally; secondly, the study characterizes the cellular elements of the resultant atherosclerotic lesions at various stages during development of the lesions. Some of the observations have been previously summarized [26].

II. Methods

A. Animals and Experimental Diet

All animals were young adult, male rhesus monkeys *(Macaca mulatta)*. Age was estimated by body weight and eruption of the third molar teeth. The body weight for each animal is given in table I. All animals were obtained from Shamrock Farms, Middleton, N.Y., where they had been conditioned for six to eight weeks after arrival from India. During this period they were dewormed, tested for tuberculosis, and given a low fat, low cholesterol diet. After arrival at the laboratory the animals received D & G research animal laboratory diet (Price Wilhoite Co., Frederick, Md.), which has a low content of fat and cholesterol, until the start of the experiment. The atherogenic diet consisted of D & G research animal laboratory diet (normal food) supplemented with butter and beef tallow (3:1), casein, cholesterol, and vitamin diet-fortification mixture. This diet contained 1 mg of cholesterol per calorie, or 0.37% cholesterol by weight.

B. Experimental Design

Fourteen monkeys were used. Two control monkeys received normal food throughout the experiment. Two animals received the atherogenic diet for four weeks, four for 12

Table I. Body weight at the beginning of the experiment and at the time of death, and mean value of all serum total cholesterol determinations made in each animal from the beginning of the experiment until the time of death

Weeks on atherogenic diet	Rhesus monkey No.	Weight at start of experiment, kg	Weight at end of experiment, kg	Mean serum total cholesterol, mg/100 ml
0 (controls)	45	4.7	4.5	130
	46	4.6	4.7	180
4	43	4.8	4.8	320
	44	5.0	4.9	560
12	27	4.0	4.7	500
	30	3.8	4.4	370
	33	3.9	4.7	760
	36	4.1	4.8	250
40	28	3.3	4.2	420
	29	3.8	4.6	760
	31	3.5	4.6	850
	32	4.6	4.4	680
	34	4.3	5.3	540
	35	4.1	4.4	200

weeks, and six for 40 weeks (table I). The animals were killed at the end of the diet periods. The experimental design has been previously reported in more detail [25].

C. Cholesterol Determination

Serum total cholesterol was determined on all animals by the method of ABELL *et al.* [1]. Serum samples were taken at one-week intervals initially, and less frequently later, from the time of the animals' arrival at the laboratory until their death. The mean value of all serum total cholesterol determinations for each animal was calculated before and after the start of the experimental diet (table I).

D. Electron Microscopy

Tissue taken for electron microscopy consisted of consecutive, complete cross-sections of the main stem of the left coronary artery, of the bifurcation of the main stem into anterior descending and circumflex branches, and of the proximal left anterior descending branch. The tissue was fixed in two changes of buffered osmium tetroxide, was dehydrated in graded alcohols, and embedded in Maraglas [24]. The complete cross-sections of the artery were thick-sectioned with a glass knife, mounted on slides, stained with Paragon multiple stain (Paragon C. & C. Co., Bronx, N.Y.), and examined with the light microscope. Areas of special interest were fine-sectioned with a diamond knife, stained with lead citrate and uranyl acetate, and examined and photographed in an RCA EMU 3F electron microscope.

III. Results

A. General Findings

The body weight and the mean value of all serum cholesterol determinations for each animal are shown in table I. The general results on growth and nutritional status of the monkeys and details of the serum cholesterol response of each animal to the atherogenic diet have been reported [25].

The heart of each animal was sectioned coronally at 1-cm intervals and examined grossly. There was no evidence of myocardial infraction in any of the animals.

B. Progression of Atherosclerotic Lesions

1. Animals on Normal Food

The nature of diffuse intimal thickening and intimal cushions in non-atherosclerotic control animals is characterized elsewhere in this volume [30].

2. Animals on the Atherogenic Diet for Four Weeks

Intimal thickening was not greater than in control animals. As in the controls, the thickest intimal segments were cushions at the bifurcation of the main stem of the left coronary artery. The number of intimal smooth muscle cells did not visibly increase and the ratio between the myofilament-rich and reticulum-rich types was similar to that of intimal thickening in animals on normal food. More smooth muscle cells had single or multiple lipid inclusions. Foam cells were infrequent and isolated and always limited to the zone adjacent to the endothelium and sometimes to the periphery of cushions (fig. 5). Macrophages, not seen in control animals, were sometimes near foam cells. Degenerative cell changes, cell necrosis, and the nature of the intercellular matrix were not perceptibly altered from normal.

3. Animals on the Atherogenic Diet for Twelve Weeks

Some increase in the thickness of the intima occurred because of the accumulation of foam cells, macrophages, and cell debris. In intimal cushions, accumulations were scattered and focal and limited to the superficial portion near the endothelium (fig. 1). In the slim periphery of cushions and in diffuse intimal thickening, single foam cells and a modest amount of cell debris often entirely filled the narrow space between the endothelium and the internal elastic lamina and obscured the normally finely reticulated ground substance (fig. 6). Foam cells did not form the continuous layers often found in later intimal lesions. Neutrophilic and eosinophilic granulocytes were seen in lesions for the first time, but they were rare. More smooth muscle cells had single or multiple lipid inclusions. Inclusions were more frequent in the loosely and irregularly arranged smooth muscle cells of the upper intima than in closely packed smooth muscle cells near the internal elastic lamina. The incidence of degenerative change and cell necrosis was higher than in control animals. Extracellular debris was derived primarily from necrotic foam cells, but some smooth muscle cells also were necrotic.

4. Animals on the Atherogenic Diet for Forty Weeks

At this time there were distinct and extensive atherosclerotic lesions in preexisting intimal cushions and in diffuse intimal thickening around the cushions. Coronary intima without preexisting intimal thickening did not have well-developed lesions, although there were, occasionally, solitary foam cells. The most severe lesions were in animals with the highest serum cholesterol levels. They occurred in the large saddle-shaped cushion at the apex of the bifurcation of the main stem into anterior descending and circumflex branches (fig. 2–4). The intima at this point was up to twelve cells thick, about twice the thickness of the intimal cushions of controls. A dense accumulation of cells replaced the usual loosely scattered arrangement of smooth muscle cells in the superficial half of intimal cushions and the finely reticulated ground substance normally found in between widely separated cells. The cellularity in the upper intima was composed of foam cells, smooth muscle cells, macrophages and infrequent granulocytes. At the surface, adjacent foam cells sometimes formed a complex meshwork by the loose interdigitation of numerous microvilli extending from the foam cell surfaces. Occasionally, a foam cell would be located between two endothelial cells in an interendothelial junction (fig. 15). Smooth muscle cells with and without lipid inclusions were scattered throughout the thickened intima and were the main cell type in deep layers. In the largest lesions foam cells also extended into the deepest intimal layer. Whereas cells at the surface were more crowded than normal, the usually densely arranged smooth muscle cells at the depth of the cushions were separated by coarsely grained cell debris and lipid derived from necrotic foam cells and necrotic smooth muscle cells (fig. 17). Such mixtures of cell debris and lipid were often pooled in the intima adjacent to the internal elastic lamina. At points where the internal elastic lamina had gaps, a few of the largest lesions showed extension of pooled cell debris into the adjacent media. Comparison between frozen sections stained with oil red 0 and electron micrographs indicated that a portion of these acellular pools consisted of lipid.

C. Cellular Elements of Atherosclerotic Lesions

The two types of intimal smooth muscle cells, reticulum-rich and myofilament-rich, found in nonatherosclerotic intimal thickening of rhesus monkeys [30] were also present in atherosclerotic lesions; however, the quantitative relationship between myofilament-rich and reticulum-rich smooth muscle cells was altered in favor of the reticulum-rich cells. In addition, cell types occurred which were not seen in the intima of normal animals.

1. Myofilament-Rich Smooth Muscle Cells

Myofilament-rich smooth muscle cells without lipid inclusions were identical to those in the intimal thickening of control animals [30] and similar to medial smooth muscle cells. Myofilament-rich cells contained intracytoplasmic lipid inclusions less often than other intimal cell types. Lipid inclusions were usually all of the homogeneous, moderately electron-dense type. Golgi stacks and vesicles of smooth-surfaced endoplasmic reticulum were more numerous in smooth muscle cells with lipid inclusions than in smooth muscle cells without. Indication of lysosomal activity, such as accumulation of small lamellated structures in some cells, was present. Coarsely laminated residual bodies, thought to represent the degenerated end products of various cell constituents were sometimes present; rarely did they fill the entire cytoplasm of a cell. Degenerative changes of another type occurred in the peripheral cell processes of some smooth muscle cells and consisted of loss of organelles and basement membrane, and swelling of cell projections (fig. 16). Identical changes occurred in normal animals and have been described in detail elsewhere in this volume [30]. Smooth muscle cells of lesions had degenerative changes in cell processes more frequently than smooth muscle cells in intimal thickening of normal animals. Some myofilament-rich smooth muscle cells contained large numbers of lipid inclusions. Such cells retained their usual elongated outline and a basement membrane and did not assume the ovoid or globular shape of foam cells. The term 'myogenic foam cell' has been applied to such cells [12] to distinguish them from the ovoid or globular form which are thought to be derived from macrophages. In control animals, myofilament-rich smooth muscle cells were the main cell type in intimal thickening and the only cell type in the deep part of the intima. As lesions developed other cell types accumulated, and the proportion of myofilament-rich smooth muscle cells in the intima decreased.

2. Reticulum-Rich Smooth Muscle Cells

Reticulum-rich smooth muscle cells differed from myofilament-rich cells by the presence of abundant rough-surfaced endoplasmic reticulum; by the scarcity, and in some cross-sections, the complete absence of myofilament bundles; and by a lesser prominence of basement membranes (fig. 7–9). Intimal lesions superimposed on cushions had more reticulum-rich smooth muscle cells than the intimal cushions of control animals. Reticulum-rich cells contained lipid inclusions of the same type as myofilament-rich cells, although more reticulum-rich than myofilament-rich cells contained lipid, and inclusions were more numerous. The preferential location of reticulum-rich cells in the superficial half of the intima near the endothelium might account for the greater amount of intracytoplasmic lipid. Like myofilament-rich cells, reticulum-rich cells with lipid inclusions had a more prominent Golgi complex and more vesicles of smooth-surfaced endoplasmic reticulum than smooth muscle cells without inclusions (fig. 9). Degenerative changes were identical to those in myofilament-rich cells. Cell necrosis was frequent in the more advanced intimal lesions and did not seem to depend entirely on a large number of lipid inclusions in the cells. Dissolution of cells with release of cell contents into the extracellular space was certain evidence of cell death. Like myofilament-rich cells, reticulum-rich cells filled with lipid inclusions still maintained their attenuated outline.

Cells ultrastructurally similar to reticulum-rich smooth muscle cells have been described in the aortic atherosclerotic lesions of various animal species and have been variously called modified smooth muscle cells [7, 9, 32], fibroblast-like smooth muscle cells [21], and fibroblast-like cells [5].

3. Foam Cells

Foam cells were large, ovoid or spherical cells (fig. 12–15). Their large size was due to numerous lipid inclusions within the cytoplasm. Foam cells lacked a basement membrane and they did not contain myofilament bundles with dense bodies characteristically associated with smooth muscle cells. Filaments of a different type were, however, frequently present. These were bundles of fine filaments arranged in linear fashion in the perinuclear region. They resembled the perinuclear filament bundles of macrophages. Lipid inclusions were of several morphologically distinct types (fig. 13). Although one type of inclusion might predominate in a cell, there usually was a mixture of several types. Early foam cells had lipid inclusions which were almost exclusively of a type that was easily extractable during tissue processing, leaving the cytoplasm full of globular, electron-lucent, usually at least partly membrane-bound vacuoles. A moderately electron-dense rim of lipid occasionally remained at the margin of vacuoles. In addition to vacuoles, some foam cells had lipid inclusions of moderate electron density identical to those seen in smooth muscle cells and stiletto-shaped spaces which presumably were the remnants of extracted cholesterol crystals. Coiled membranous structures, usually thought to play a role in lysosomal mechanisms, were closely associated with some lipid inclusions (fig. 14). Lipid droplets, cholesterol clefts, and residual bodies often appeared in between and most often at the center of the finely laminated membrane spirals. In foam cells, although sometimes overshadowed by lipid inclusions, lysosomal structures were larger and more frequent than in smooth muscle cells. Lysosomal structures were even more abundant in the foam cells of resolving atherosclerotic lesions in which the number of lipid inclusions had become stabilized [28]. The perinuclear area of foam cells contained a very prominent Golgi apparatus consisting of numerous stacks of flattened membranes and numerous vesicles of smooth-surfaced endoplasmic reticulum (fig. 13). Residual bodies frequently accompanied lipid inclusions. Residual bodies consisted of coarse granular material and crinkled, coarsely laminated structures (fig. 13). As in other cell types, residual bodies were particularly abundant in necrotic cells, and in the extracellular space surrounding disintegrated necrotic cells (fig. 17). Degenerative changes in foam cells and necrosis of foam cells with release of intracellular lipid into the extracellular space were frequent.

4. Macrophages

Intimal macrophages occurred near foam cells in the upper layers of intimal lesions (fig. 10, 11). Macrophages were variable in appearance. Some appeared immature and quiescent and had few organelles (fig. 10). In macrophages that appeared stimulated, the cytoplasm was richer in cell organelles and the plasma membrane was arranged as numerous microvilli (fig. 11). The Golgi apparatus was prominent and composed of several stacks of parallel smooth membranes, frequently with smooth membrane-bound vesicles among the lamellae; some of these vesicles appeared to bud from the lamellae. There were variable but usually small amounts of rough-surfaced endoplasmic reticulum and numerous free ribosomes. Bundles of filaments were often arranged in linear fashion in the perinuclear region (the 'Hof' of the nucleus [34]). They were similar to those described in peritoneal [3] and alveolar macrophages [17], and also resembled the perinuclear filaments of foam cells. Some macrophages had lipid inclusions and gave the impression of being transitional forms to foam cells. Cells electron microscopically similar to those classified as macrophages in the present study have been called monocytes [6] and undifferentiated cells [9] in human fatty streaks, and primitive cells in the experimental aortic lesions of rhesus monkeys [21].

5. Granular Leukocytes

Two types of granulocytes, eosinophil and neutrophil, occurred in the more advanced intimal lesions. Both types were rare and isolated in the superficial portion of the intima near foam cells and smooth muscle cells with lipid inclusions and near necrotic cell debris. Eosinophil granulocytes had bilobed nuclei and contained numerous specific granules. The characteristic granules were elongated oval bodies of similar size which varied in electron density at the core and at the periphery. The method of fixation used in this experiment produced a negative ('white') image of the disk-shaped granule core. Neutrophil granulocytes had a multilobed nucleus and contained numerous specific granules of uniform electron density, round or oval in shape.

6. Endothelial Cells

Endothelial cells on the surface of lesions were not uniform in appearance. Some endothelial cells had an increased cytoplasmic volume and more cell organelles. The Golgi apparatus was often more prominent than in endothelial cells on the surface of nonatherosclerotic thickening and consisted of several stacks of lamellae and numerous smooth-surfaced vesicles. Bundles of filaments of uniform periodicity were prominent and frequently distributed uniformly throughout the cell (fig. 8, 10). Pinocytotic vesicles were abundant at the luminal surface and at the base of cells. Some endothelial cells had one or more large lipid inclusions of uniform electron density, morphologically similar to the lipid inclusions of myofilament-rich smooth muscle cells. Rarely, there were grape-like clusters of small lipid droplets which resembled lipofuscin. Endothelial cells did not contain cholesterol crystals or residual bodies. Degenerative changes occurred in some endothelial cells on the surface of lesions and consisted of vacuole formation in the endoplasmic reticulum, dissolution of some organelles, and aggregation of chromatin at the periphery of nuclei.

IV. Discussion

A. Relationship of Atherosclerotic Lesions to Nonatherosclerotic Intimal Thickening

The largest atherosclerotic lesions were blended into the largest of the preexisting nonatherosclerotic intimal cushions normally found at the bifurcation of the main stem of the left coronary artery. Smaller lesions were found in the diffuse intimal thickening of the adjacent anterior descending branch, and in small intimal cushions at the ostia of distal branch vessels. In coronary segments without nonatherosclerotic thickening there was little intimal lipid or no lipid at all. The thickest lesions were in animals with the highest serum cholesterol level and the longest duration of hypercholesterolemia. The severity of early atherosclerotic lesions was influenced by three factors: the degree of preexisting intimal thickening, and the degree and the duration of the serum cholesterol elevation.

The largest experimental lesions produced in the coronary arteries of these rhesus monkeys fall into the category of thick fatty streaks and are comparable to large intimal fatty streaks in the coronary arteries of humans. Fibrous plaques did not occur, although the integration of advanced fatty streaks and preexisting intimal cushions created a superficial resemblance to fibrous plaques microscopically. Although both fibrous plaques and fatty streaks superimposed on nonatherosclerotic thickening are composed of similar structural elements, the distribution of the various cell types and the proportion of cell types within the two lesions is not the same.

The results confirm a close and preferential association between nonatherosclerotic intimal thickening and early atherosclerosis in the coronary arteries; however, the experiment has not determined whether this association will persist with the progress of atherosclerotic lesions beyond the stage of advanced fatty streaks, and whether fibrous plaques and complicated lesions will develop in the same locations. Nor has the cause of the relationship between nonatherosclerotic intimal thickening and atherosclerosis been determined. It was not the object of this study to answer the question whether nonatherosclerotic thickening is a predisposing factor for the deposition of lipid and atherosclerosis, or whether the configuration of the arterial tree and thus peculiarities of flow and intraluminal pressure predispose the intima to nonatherosclerotic thickening under conditions of normocholesterolemia, or to both nonatherosclerotic thickening and the added development of atherosclerotic lesions in hypercholesterolemia.

B. Cellular Composition of Atherosclerotic Lesions

Cellularity increased when atherosclerotic lesions developed in nonatherosclerotic intimal thickening. Apart from the emergence of foam cells, macrophages, and rare granulocytes, the number of reticulum-rich smooth muscle cells also increased.

Notwithstanding radioautographic evidence indicating *in situ* proliferation of intimal cells in experimental atherosclerosis [27, 29], the debate on the derivation of the increased number of cells found in atherosclerotic lesions has continued. Some authors have recently favored the concept of medial smooth muscle cell migration into the intima [11, 31]. Medial smooth muscle cells in gaps of the internal elastic lamina suggesting migration into the intima after vascular injury were observed by STEMERMANN and ROSS [31]. Medial smooth muscle cells arranged perpendicularly in gaps of the internal elastic lamina also occurred in the present study both at the base of nonatherosclerotic intimal thickening in control animals and at the base of lesions in hypercholesterolemic animals. Since the coronary artery segments examined in the present study had not been fixed in the distended state, there was substantial contraction of the vessel. It is difficult to determine, therefore, whether medial cells in gaps represent transmigration or whether they were forced into the gaps by postmortem contraction of the media.

Intimal macrophages occurred in atherosclerotic lesions although none had been seen in the nonatherosclerotic intimal thickening of control animals. The derivation of intimal macrophages is unknown. There are indications that in experimental atherosclerotic lesions macrophages can arise by proliferation of cells in the intima.

C. Degenerative Cell Change and Necrosis

In atherosclerotic lesions, degenerative changes in the peripheral cell processes of intimal smooth muscle cells resembled those of intimal smooth muscle cells in nonatherosclerotic intimal thickening of control animals [30], but they were more frequent. Medial smooth muscle cells near the internal elastic lamina showed identical changes, but degenerative changes in the media adjacent to lesions were no more frequent than degenerative changes in the media of control animals. It must be remembered that only a few of the most advanced intimal lesions marginally involved the media. Similar degenerative changes have been reported in the aortas of rhesus monkeys fed either stock or atherogenic diets [21, 33]. Observations in this experiment and in the companion study [30] indicate that degenerated cell processes separate from the main body of the cell and disintegrate. The degenerative change might be reversible under some circumstances, and shedding of the cell processes may not be inevitable. Separation of degenerated cell processes is not necessarily accompanied by necrosis of the main body of the cell. The cause of the degenerative changes is unknown.

In atherosclerotic lesions there was outright necrosis of numerous smooth muscle cells in addition to the focal cytoplasmic degenerations in some. Most necrotic cells in lesions were foam cells, however.

Necrotic cell debris accumulated progressively in the intimal lesions of animals on the atherogenic diet. It did not accumulate in nonatherosclerotic intimal thickening of control animals, although there was frequent dissolution of degenerated smooth muscle cell processes and occasionally necrosis of an isolated foam cell. The explanation for the difference may lie in the rate of formation of cell debris. In animals with normal serum cholesterol levels, formation of cell debris is probably balanced by its removal. Perhaps hypercholesterolemia unbalances the equilibrium by causing excessively rapid cell proliferation and cell necrosis, and normal removal mechanisms might fail to successfully contend with the excessive supply. With increasing age, a decrease in the metabolic performance of the arterial wall is possible and this might augment further retention of cell debris and extracellular lipid.

Necrotic cell debris has not generally been thought to play a major role in human fatty streaks. Instead, necrosis has been associated with the complicated lesions of advanced atherosclerosis [19]. The present observations indicate, however, that under experimental conditions, necrosis of cells is frequent in early lesions. Thus, there may be a difference between early human and early experimental lesions, the latter having more necrosis. The explanation for this discrepancy could lie in the exaggerated serum cholesterol level of the experimental condition.

D. Extracellular Lipid

GEER *et al.* [10] found abundant extracellular lipid in aortic fatty streaks experimentally induced in baboons. The extracellular lipid was not associated with necrotic smooth muscle cells, and necrotic foam cells were rare; the authors conclude that electron-dense extracellular material is derived from blood lipoprotein. The present experiment does not support these conclusions. Extracellular lipid was not visible with the electron microscope in the very earliest atherosclerotic lesions which consisted of lipid in smooth muscle cells and some foam cells. Since the lipoprotein concentration of the intima is proportional to that of plasma [23], elevated serum cholesterol levels must have been reflected in the extracellular fluid. Failure to demonstrate extracellular lipid in the earliest lesions with the electron microscope might be due to dissolution of lipid during tissue processing, or it might be that micellar lipoprotein is taken up by intimal cells instantly and completely.

Accumulation of extracellular lipid lagged behind the accumulation of intracellular lipid and was first seen by electron microscopy after the animals had been on the atherogenic diet for 12 weeks. After 40 weeks, extensive pools of extracellular lipid and cell debris had accumulated, particularly in deep parts of the intima. The extracellular material was associated with necrotic lipid-containing cells, from which it seems to have been derived. Micellar lipoprotein, after entering intimal cells, apparently undergoes fusion and complex metabolic change. Upon necrosis and disintegration of cells, the altered intracellular lipid is released into the extracellular space where it can now be seen as electron-dense particles of various sizes mingled with necrotic cell debris. The close morphologic resemblance between the cytoplasmic content of many of the degenerating cells and the extracellular particles is striking, and supports the view that the extracellular material which increasingly accumulates in progressing atherosclerotic lesions is derived from necrotic cells.

V. Conclusions

Rhesus monkeys receiving normal food had low serum cholesterol levels and nonatherosclerotic intimal thickening at and near the bifurcation of the main stem of the left coronary artery. Such intimal thickening contained an insignificant amount of intracellular and no extracellular lipid. The findings in animals on normal food are detailed and summarized elsewhere in this volume [30].

Rhesus monkeys receiving an atherogenic diet developed atherosclerotic lesions in nonatherosclerotic intimal thickening. The largest lesions were in and around the largest intimal cushion at the apex of the bifurcation of the main stem. Smaller lesions were in diffuse intimal thickening, in small distal cushions, and in intima without nonatherosclerotic thickening. The earliest changes after introduction of the atherogenic diet were lipid inclusions in some of the smooth muscle cells of intimal cushions and the emergence of isolated foam cells just beneath the endothelium. Factors that determined the progression of lesions were the size of the underlying nonatherosclerotic thickening, the degree of serum cholesterol elevation, and the duration of the hypercholesterolemia. The largest lesions contained layers of foam cells which had a preference for portions of intimal thickening adjacent to the lumen. Intimal smooth muscle cells were of two types, those rich in myofilament bundles, and others rich in rough-surfaced endoplasmic reticulum. Both types had lipid inclusions. Cells with lipid inclusions had a prominent Golgi apparatus and abundant smooth-surfaced endoplasmic reticulum. Other cells in lesions were macrophages and, less frequently, granulocytes. Degenerative change and necrosis occurred in both foam cells and smooth muscle cells. Necrosis was more frequent in foam cells. Necrotic cell remnants and extracellular lipid had a tendency to accumulate in pools adjacent to the internal elastic lamina. In the largest intimal lesions, extracellular debris and lipid extended into the adjacent media through small gaps in the internal elastic lamina.

Although the numerous smooth muscle cells of the associated nonatherosclerotic thickening, particularly those of the reticulum-rich type which resembled fibroblasts, created a resemblance to fibrous plaques, the largest experimental lesions in these coronary arteries were thick fatty streaks comparable to large fatty streaks of man.

Acknowledgements

The author is indebted to Mrs. KATHY COSENTINO and Mrs. CATHERINE VIAL for skillful technical assistance and to Dr. BARBARA LYNCH for criticism of the manuscript.

This work was supported by a grant-in-aid of research by the Louisiana Heart Association and by USPHS NIH, grant HL–08974.

Fig. 1. Intimal cushion proximal to the bifurcation of the main stem of the left coronary artery. An isolated foam cell (arrow) is immediately below the endothelium at the periphery of the cushion. The internal elastic lamina is compact and continuous. From a monkey on the atherogenic diet for 12 weeks with a mean serum cholesterol level of 500 mg/dl. Rhesus No. 27. Semithin Maraglas-embedded section. Paragon stain. × 800.

Fig. 2. Bifurcation of the left coronary artery. Two facing intimal cushions extend to suggest the lumina of the two branch vessels (A and B). Atherosclerotic lesions are superimposed on each of the cushions and extend into adjacent intima. The internal elastic lamina seems continuous although small gaps appear under higher magnification (fig. 3). A more proximal section through the same bifurcation reveals a more extensive gap in the internal elastic lamina (fig. 4). From a monkey on the atherogenic diet for 40 weeks with a mean serum cholesterol level of 850 mg/dl. Rhesus No. 31. Semithin Maraglas-embedded section. Paragon stain. × 200.

Fig. 3. Detail of one of the two cushions shown in figure 2. Foam cells are most frequent and lipid in intimal smooth muscle cells is most abundant in the part of the cushion near the lumen of the vessel. The internal elastic lamina consists of several delicate lamellae with small gaps. Semithin Maraglas-embedded section. Paragon stain. × 800.

Fig. 4. Detail of an intimal cushion with an atherosclerotic lesion. From the same bifurcation as in figure 2 but at a more proximal level. The internal elastic lamina has an extensive gap. Semithin Maraglas-embedded section. Paragon stain. × 800.

Fig. 5. Diffuse intimal thickening near the bifurcation of the main stem. The intima consists of one layer of smooth muscle cells (S) over a continuous internal elastic lamina (IE). Fragments of one necrotic cell, probably a foam cell (F), are immediately below the endothelium (E). The extracellular space of the intima contains elastica and collagen. From a monkey on the atherogenic diet for four weeks with a mean serum cholesterol level of 560 mg/dl. Rhesus No. 44. × 7,500.

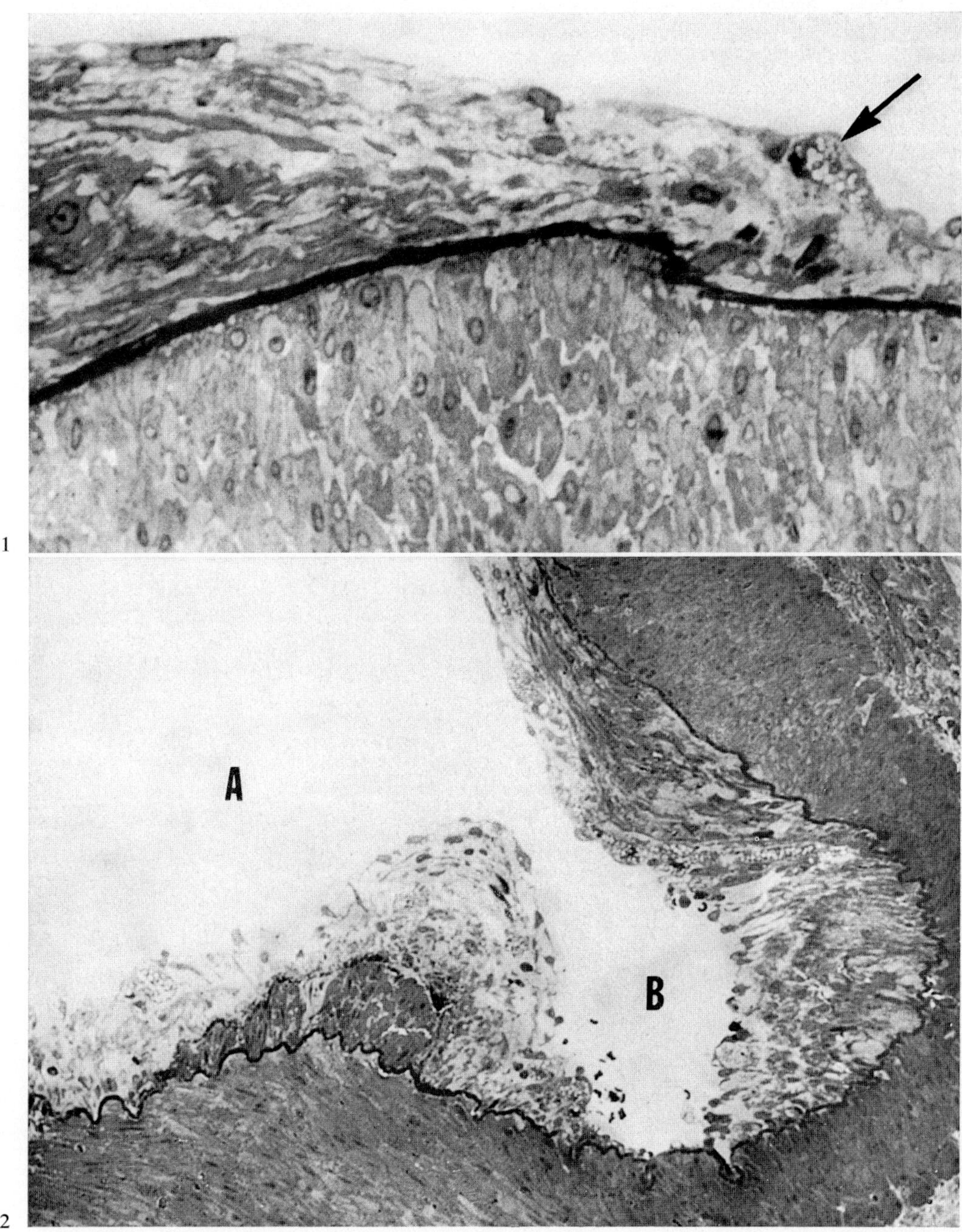
1
A
B
2

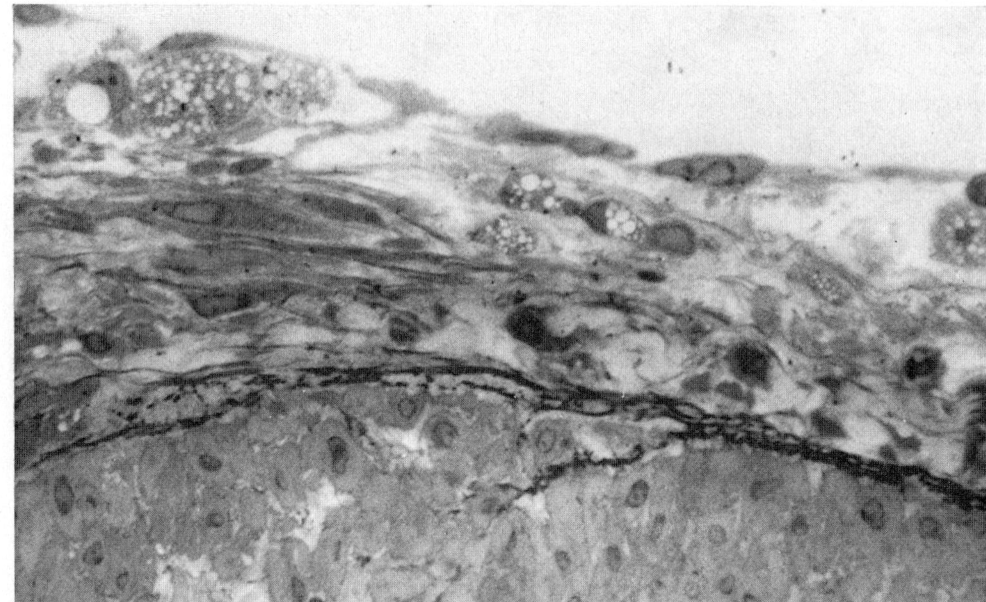

3

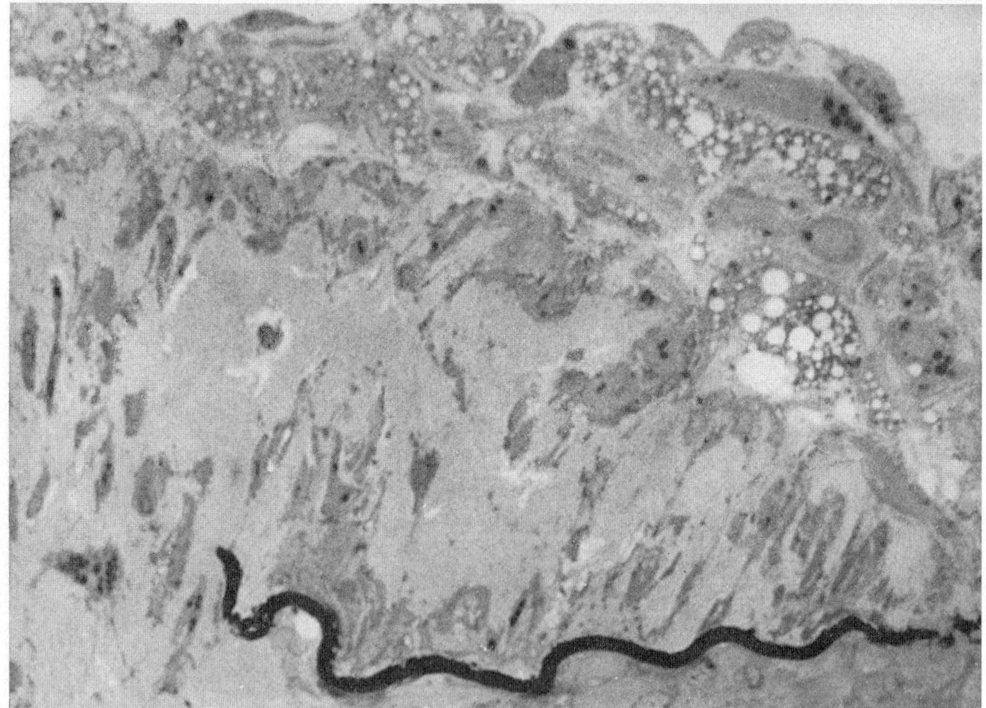

4

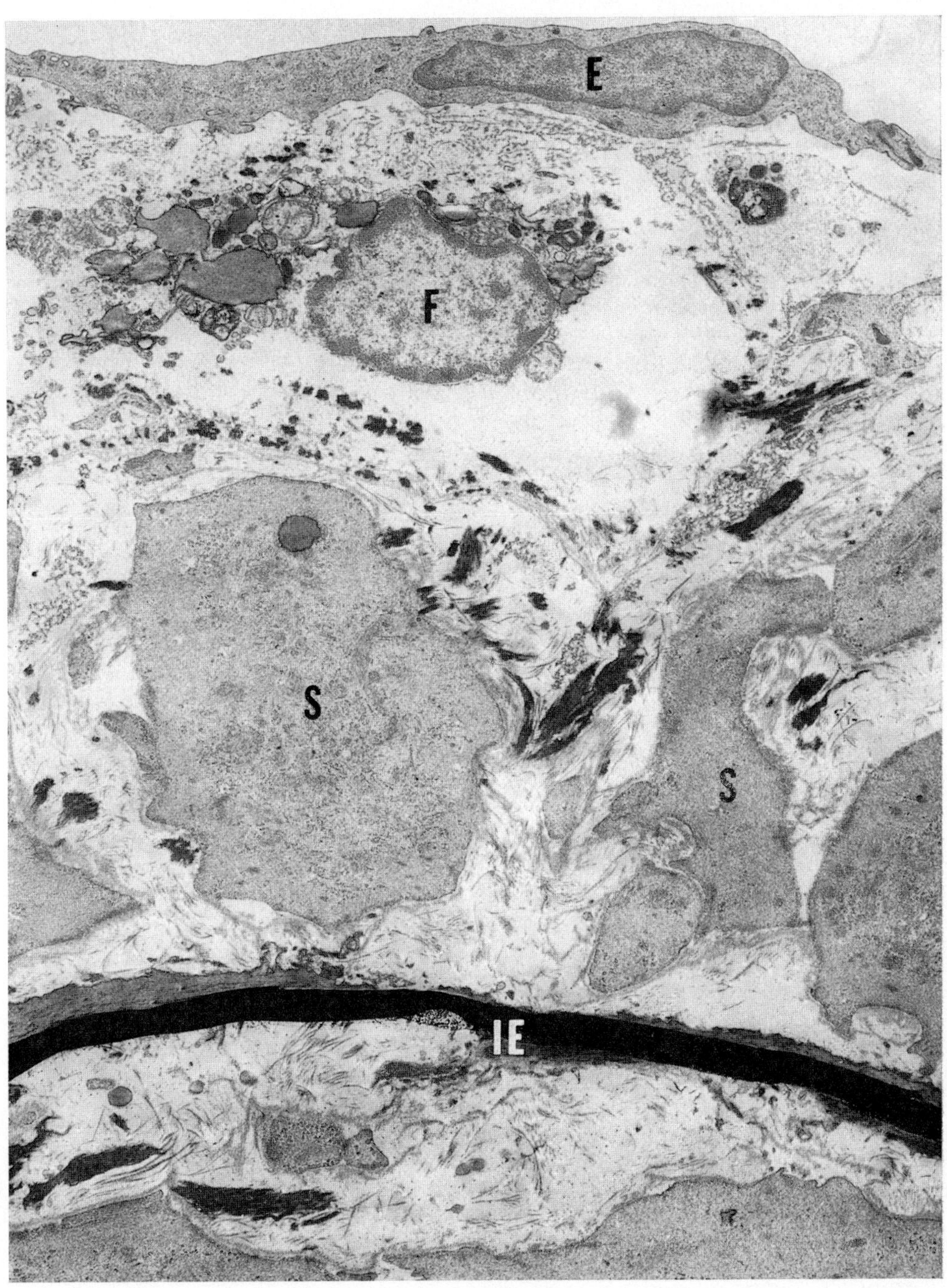
E
F
S
S
IE
5

Fig. 6. Peripheral portion of the small intimal cushion shown in figure 1. An isolated foam cell (F) is between the endothelium (E) and the internal elastic lamina (IE). The extracellular space of the intima contains an accumulation of lipid and cell debris (d). × 7,800.

Fig. 7. Large intimal cushion at the bifurcation shown in figure 2. Several smooth muscle cells in the midportion of the cushion indicate the variability in smooth muscle cell morphology when lesions are superimposed. A cell (1) shows extensive degenerative change; another (2) has a lipid inclusion, lysosome-like structures, rough-surfaced endoplasmic reticulum, and a prominent Golgi complex. Both cells lack myofilament bundles in this section but resemble smooth muscle cells in shape. Segments of other smooth muscle cells (3 and 4) have peripheral myofilament bundles. Extracellular lipid and cell debris (d) is pooled at the base of the intimal cushion. × 11,500.

Fig. 8. Detail of a large cushion with a superimposed lesion. From the bifurcation shown in figure 2. A reticulum-rich smooth muscle cell (RS) has numerous cisternae of rough-surfaced endoplasmic reticulum, pinocytotic vesicles and a basement membrane but lacks myofilaments in this section. Endothelial cells (E); foam cells (F). × 18,200.

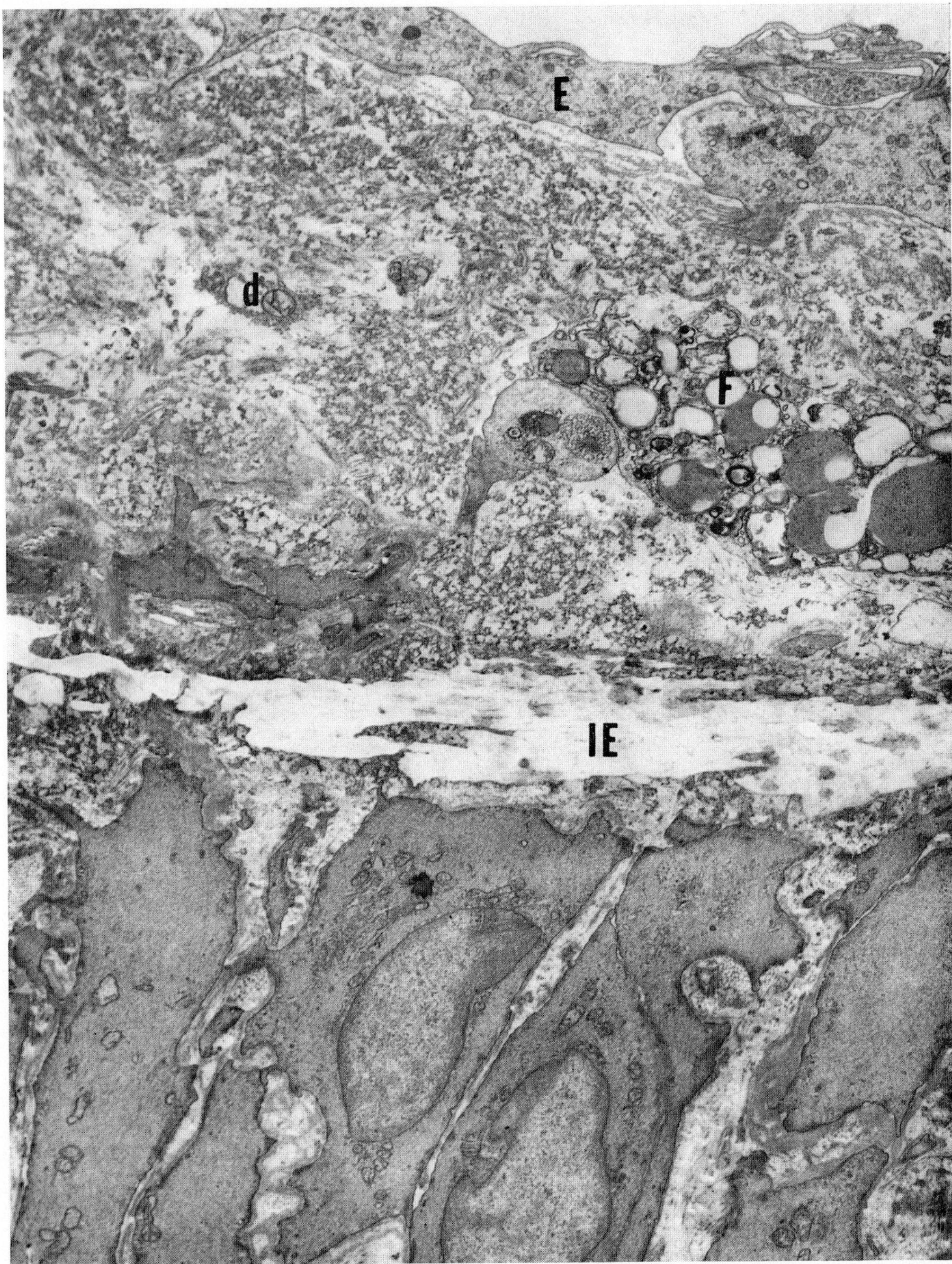

6

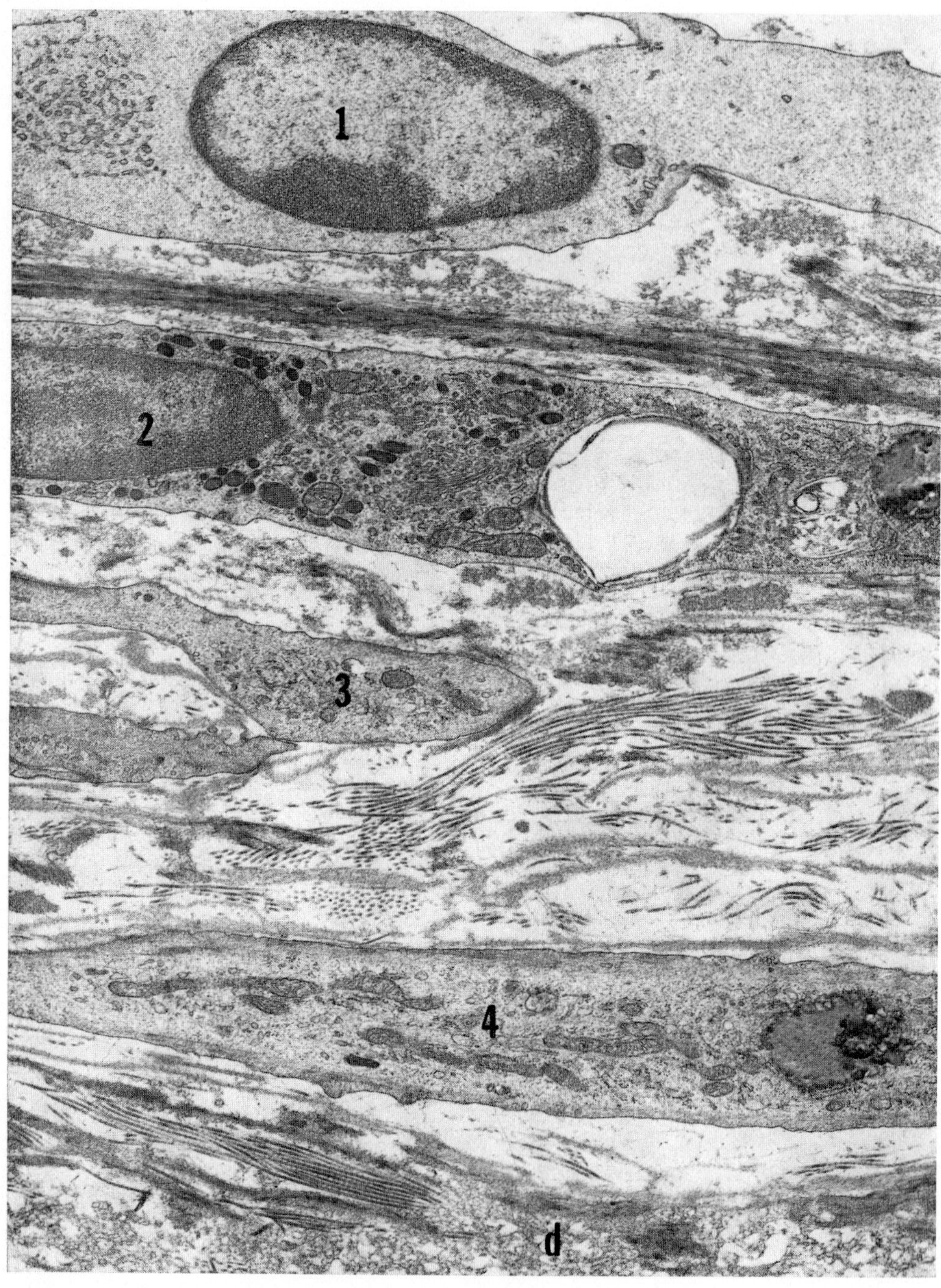

7

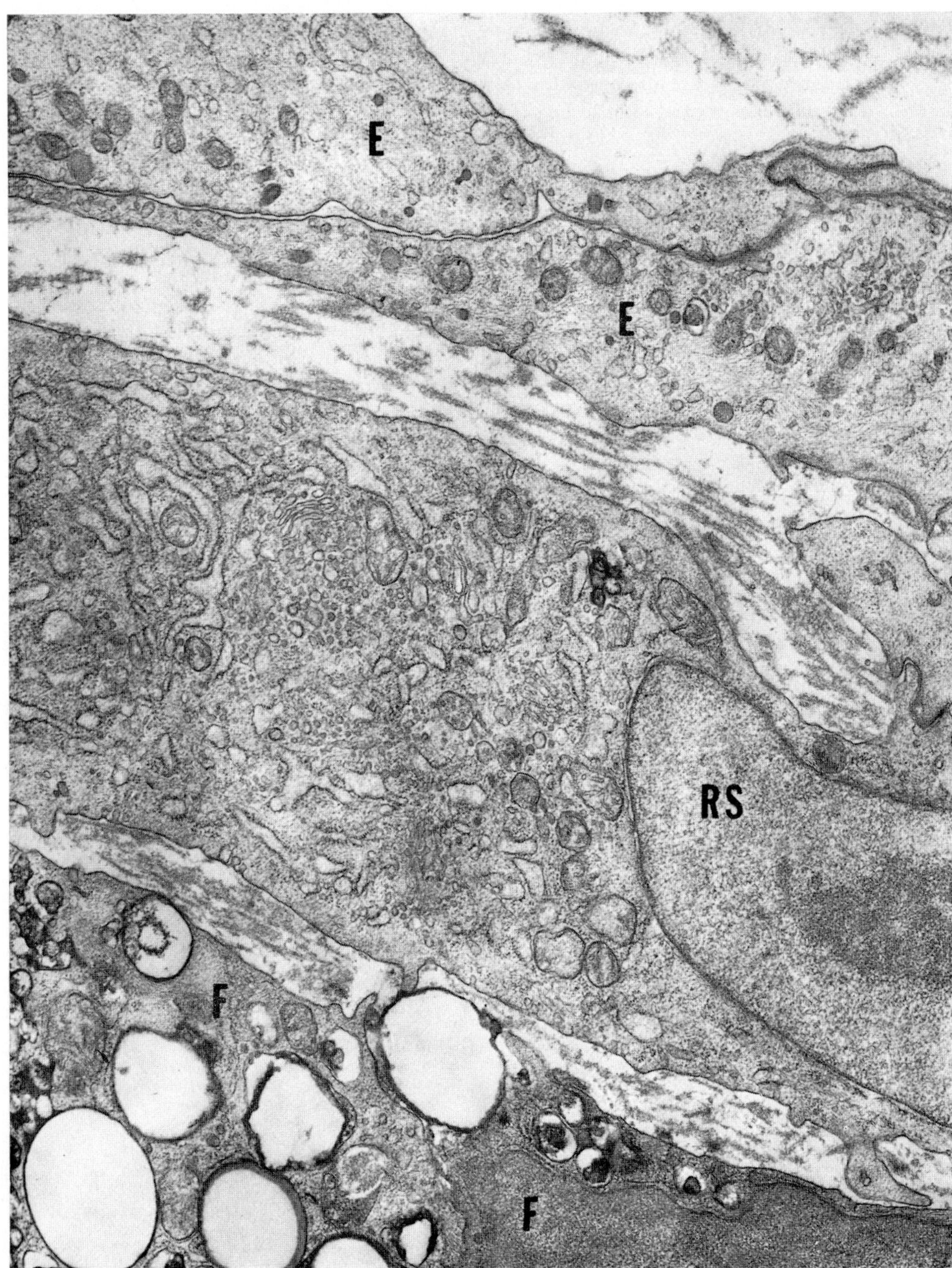
E
E
RS
F
F
8

Fig. 9. Detail of a reticulum-rich smooth muscle cell from one of the cushions shown in figure 2. Endoplasmic reticulum is mainly of the rough-surfaced type (r), but smooth-surfaced vesicles (s) are also prominent. Myofilament bundles with dense bodies are limited to the cell periphery. Lipid inclusions (L). × 25,600.

Fig. 10. Upper part of one of the intimal cushions shown in figure 2. A macrophage (M) is immediately adjacent to the endothelium (E). The macrophage is of the immature or quiescent type and has few organelles. Similar cells in aortic lesions have been described by other authors as monocytes, undifferentiated cells, and primitive cells (see text). The adjacent foam cell (F) has lipid inclusions and a prominent Golgi apparatus. × 15,400.

Fig. 11. Macrophage from a lesion superimposed on diffuse intimal thickening. The cytoplasm shows the features of a stimulated macrophage such as lysosome-like structures (ly), a prominent Golgi apparatus with several stacks of flattened membranes (G), numerous vesicles of smooth-surfaced endoplasmic reticulum, free ribosomes, and a plasma membrane arranged into numerous microvilli (v). Fragments of a necrotic foam cell (F) are adjacent to the macrophage. Endothelial cell (E). From a monkey on the atherogenic diet for 40 weeks with a mean serum cholesterol level of 680 mg/dl. Rhesus No. 32. × 24,000.

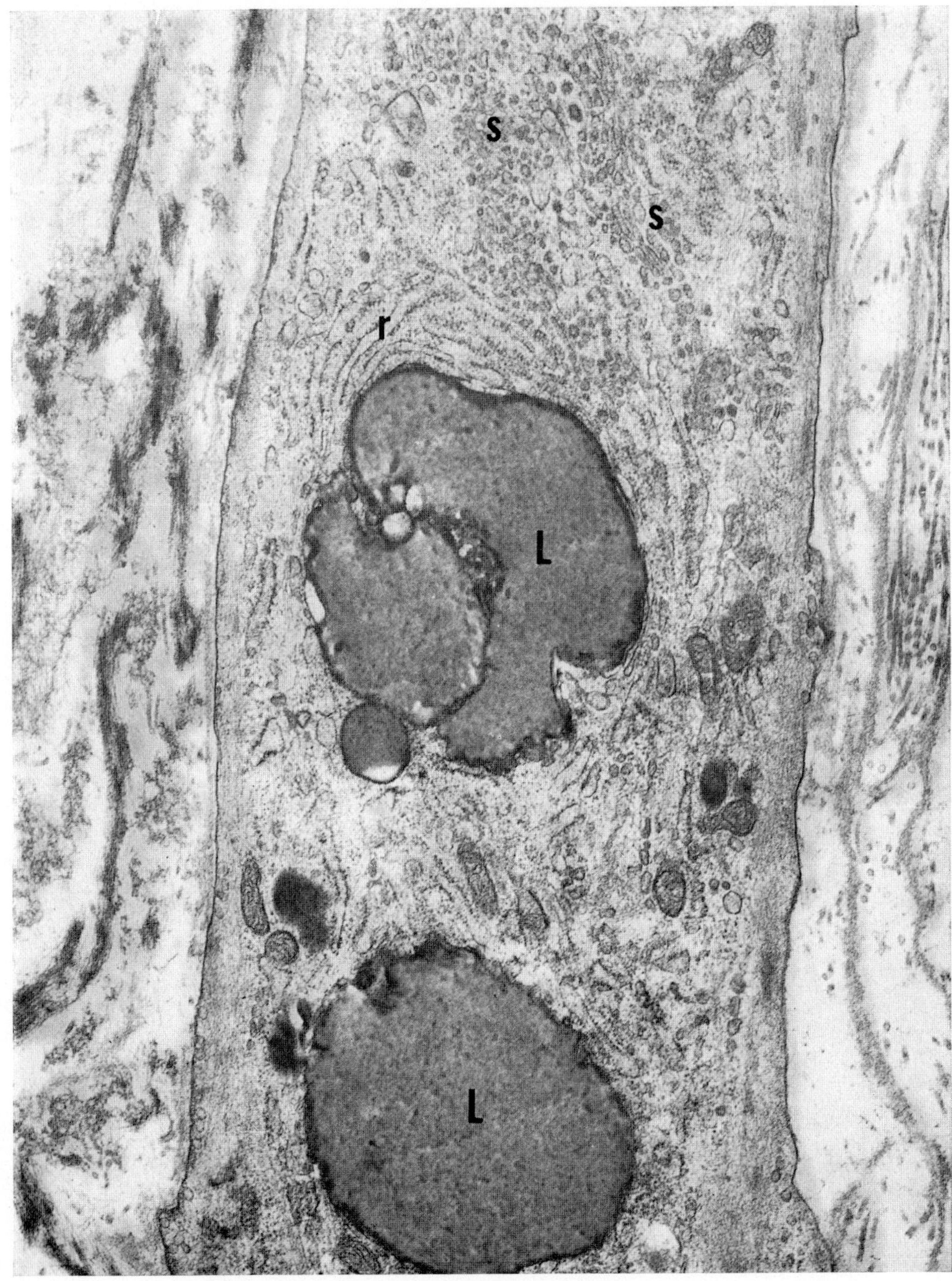
S
S
r
L
L
9

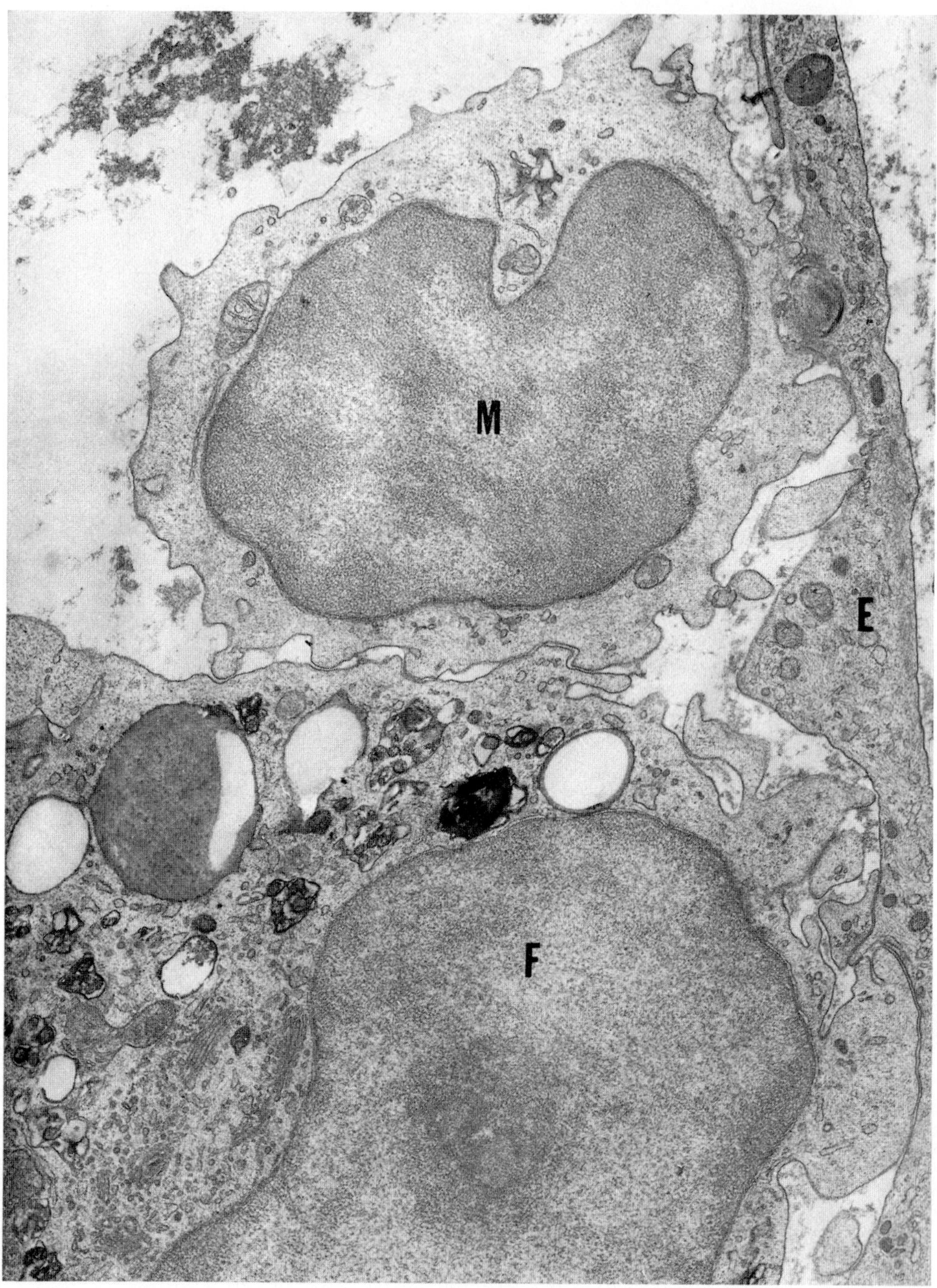

10

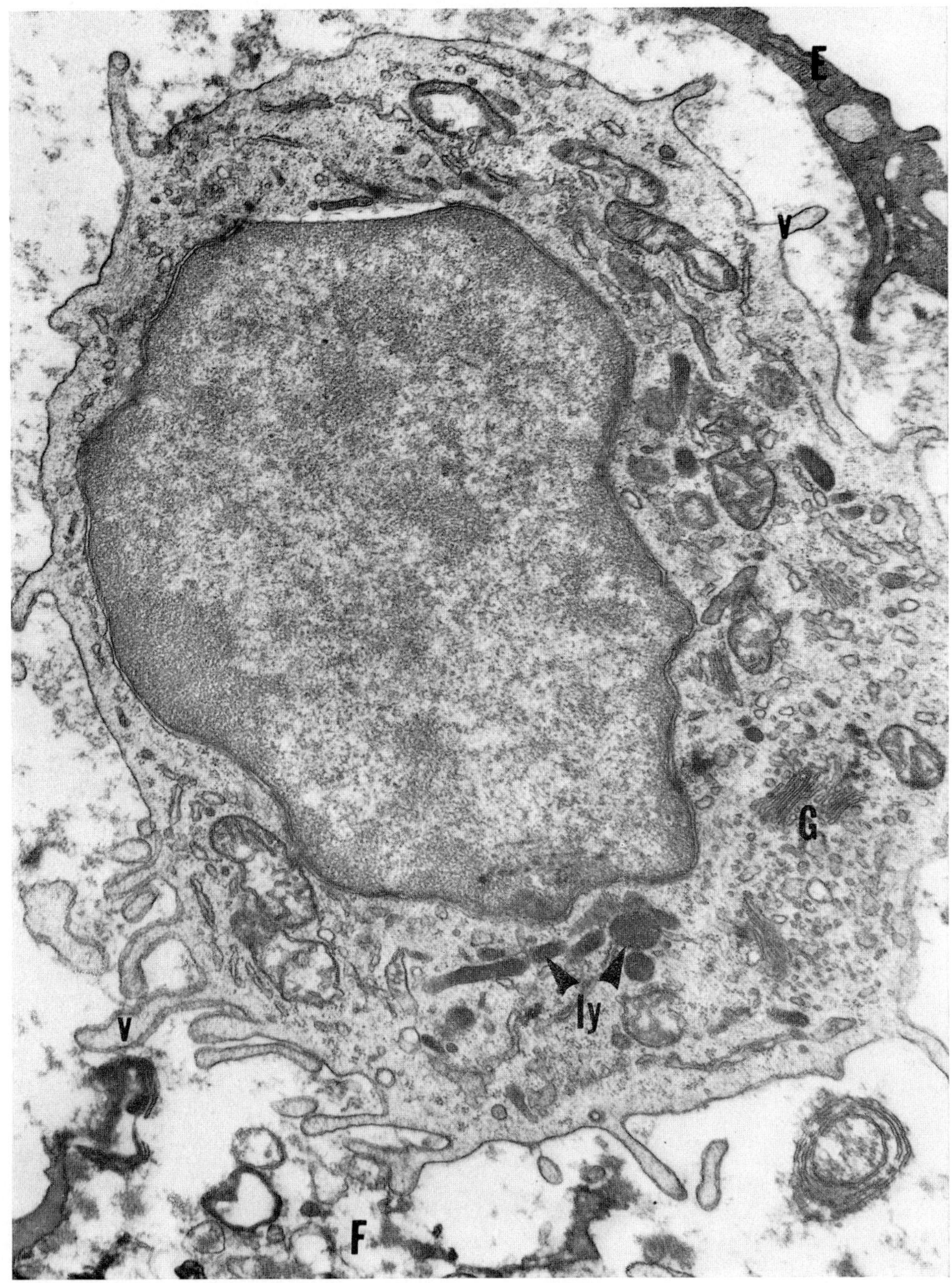
E
v
G
ly
v
F
11

Fig. 12. Small intimal cushion with a superimposed lesion. A foam cell obscures the loose intercellular substance near the lumen of the vessel. Lipid inclusions (L) in the form of vacuoles predominate. Golgi stacks (G) and vesicles of smooth-surfaced endoplasmic reticulum are more prominent in one region of the cell. Endothelial cells (E). From a monkey on the atherogenic diet for 40 weeks with a mean serum cholesterol level of 540 mg/dl. Rhesus No. 34. × 8,600.

Fig. 13. Detail of a foam cell from one of the cushions shown in figure 2. Lipid inclusions are of several morphologically distinct types. Partially membrane-bound lipid vacuoles (L) are usually more abundant than residual bodies (r). Stiletto-shaped spaces are remnants of extracted cholesterol crystals. The Golgi apparatus (G) consists of numerous stacks of flattened membranes and is, together with numerous vesicles of smooth surfaced endoplasmic reticulum, most abundant in the perinuclear area, the 'Hof' of the nucleus (N). × 22,300.

Fig. 14. Detail of a subendothelial foam cell with microvilli (v), numerous lipid vacuoles (L) and residual bodies (r). Smooth-surfaced membranes form a large, multilamellated structure of lysosomal nature with lipid in various configurations between and at the core of the lamellae. Nucleus (N); endothelium (E). Same lesion as in figure 11. × 17,600.

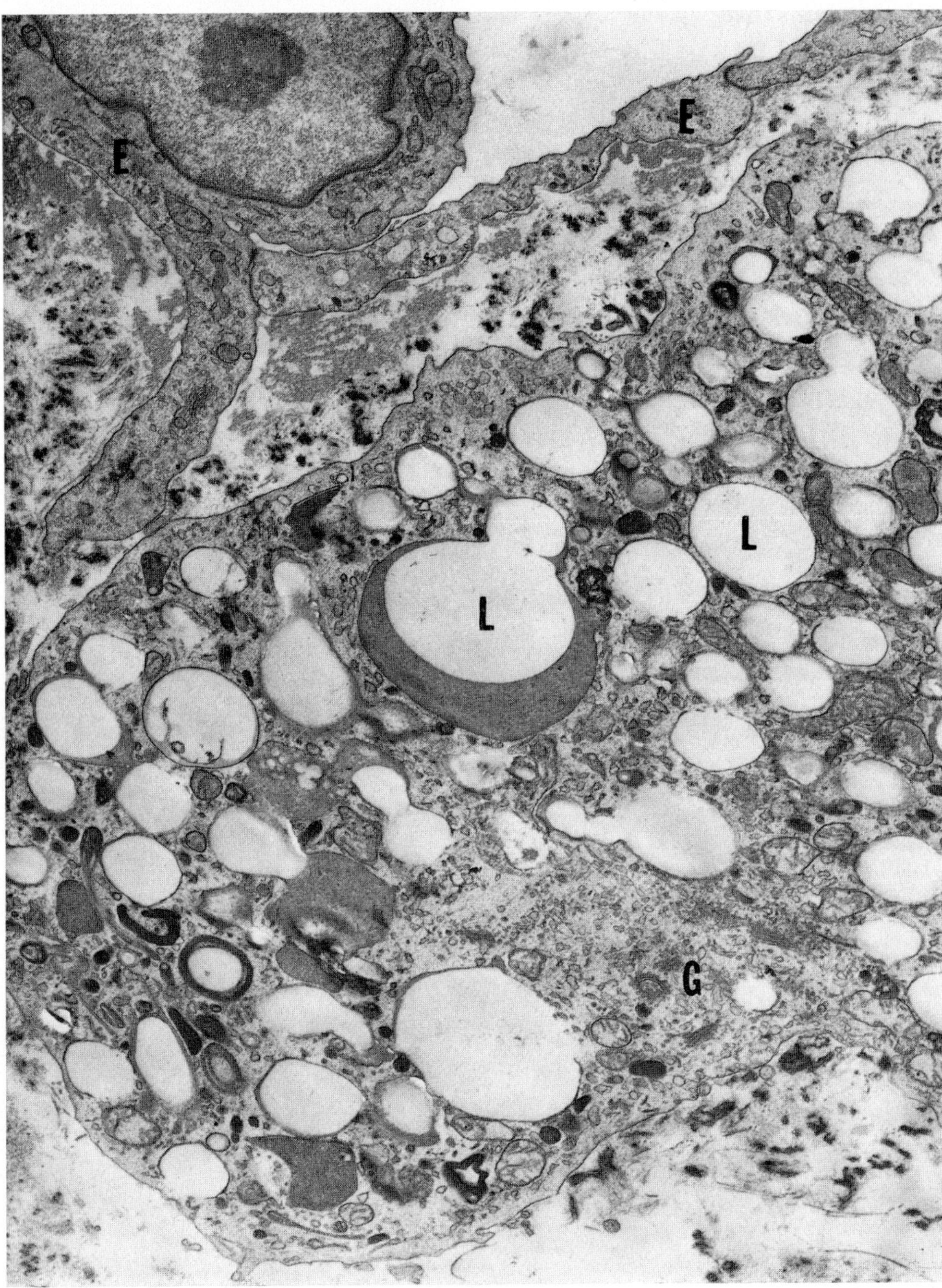
E
E
L
L
L
G
12

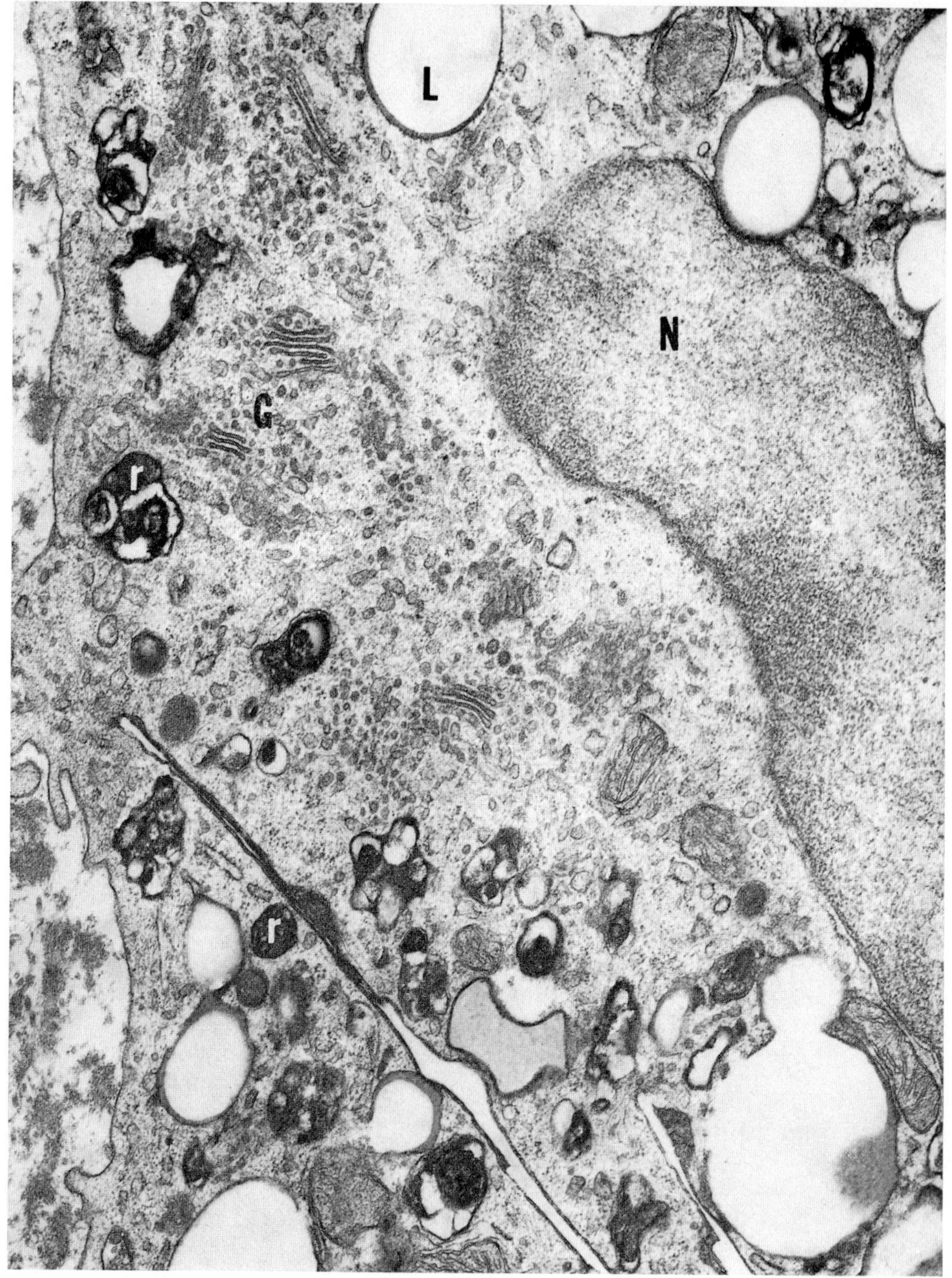
L
N
G
r
r
13

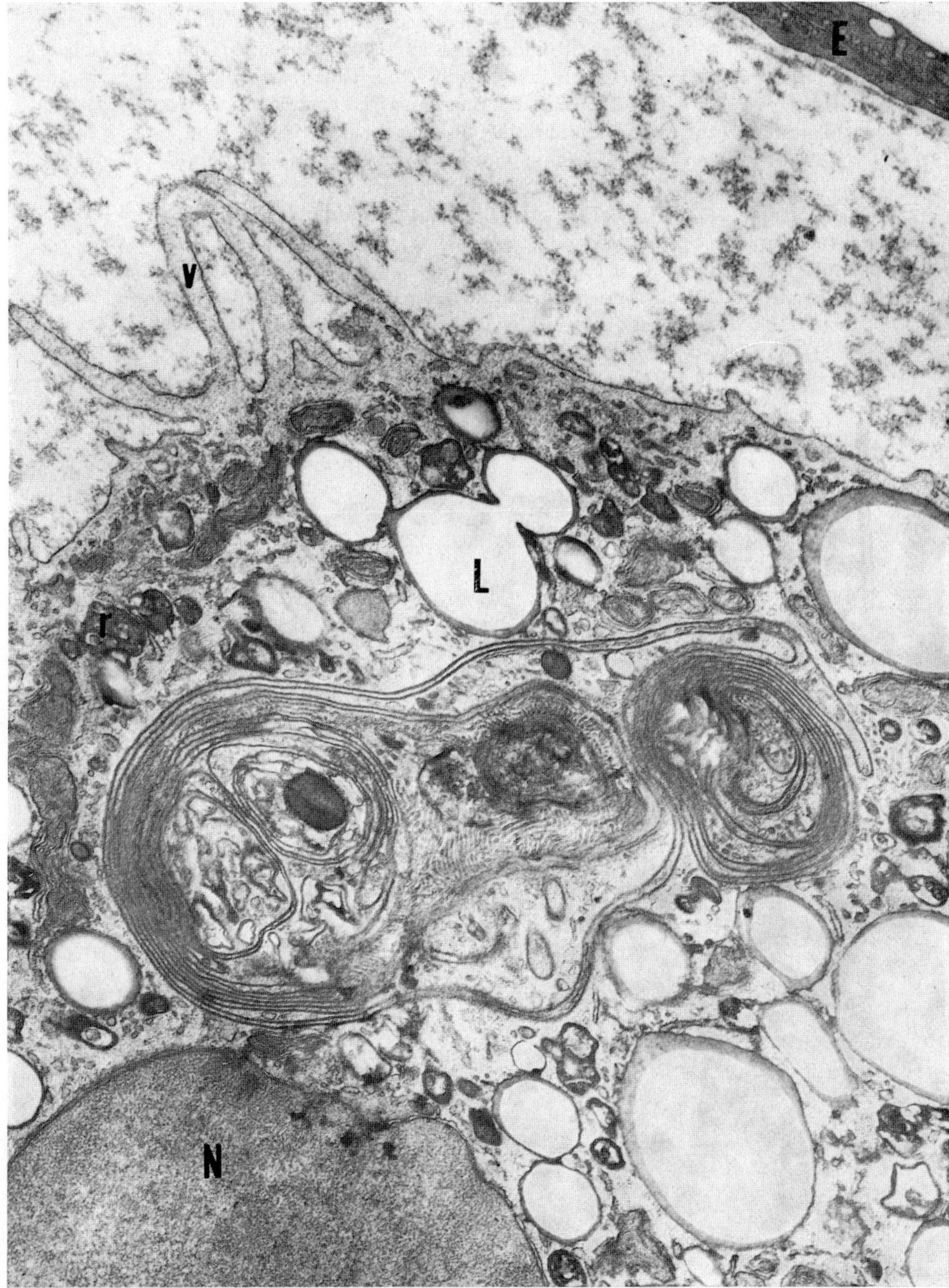
E
V
L
r
N
14

Fig. 15. Detail of a cushion with a superimposed lesion. From the bifurcation shown in figure 2. A foam cell in between two endothelial cells (E) gives the impression of transmigration. Lumen of the vessel (Lu). It is uncertain whether the foam cell is moving into the intima or out, or whether its position is due to arterial contraction that occurred when vascular attachments were released during dissection. $\times$ 9,000.

Fig. 16. Upper part of a cushion with a superimposed lesion. From the bifurcation shown in figure 2. The lesion contains numerous necrotic and degenerating (d) portions of cell cytoplasm. While many of these are free in the intercellular space, others are the prominent cytoplasmic projections of adjacent cells (arrow). Endothelial cell (E). $\times$ 9,000.

Fig. 17. Deep part of a cushion with a superimposed lesion. From the bifurcation shown in figure 2. The cytoplasm of a degenerating cell (D) is filled with numerous residual bodies. The cell is without distinguishing characteristics but the outline is that of a smooth muscle cell. The cell is nearly indistinguishable from extracellular material, which is morphologically identical to the residual bodies within the cell, and which probably represents debris derived from necrotic cells. Portions of two adjacent intimal smooth muscle cells (S) show neither lipid nor obvious degenerative cell change. $\times$ 9,000.

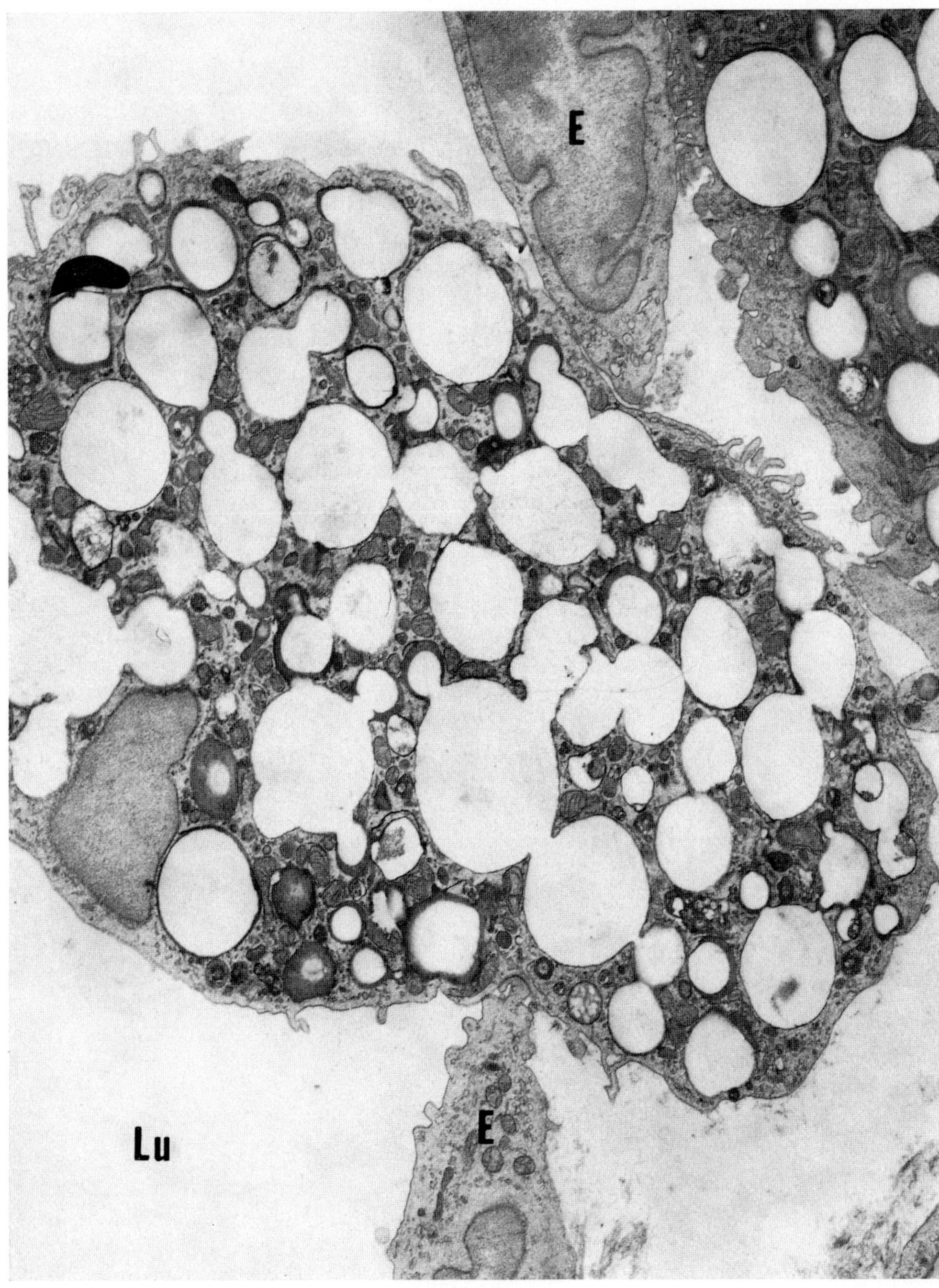
E
E
Lu
15

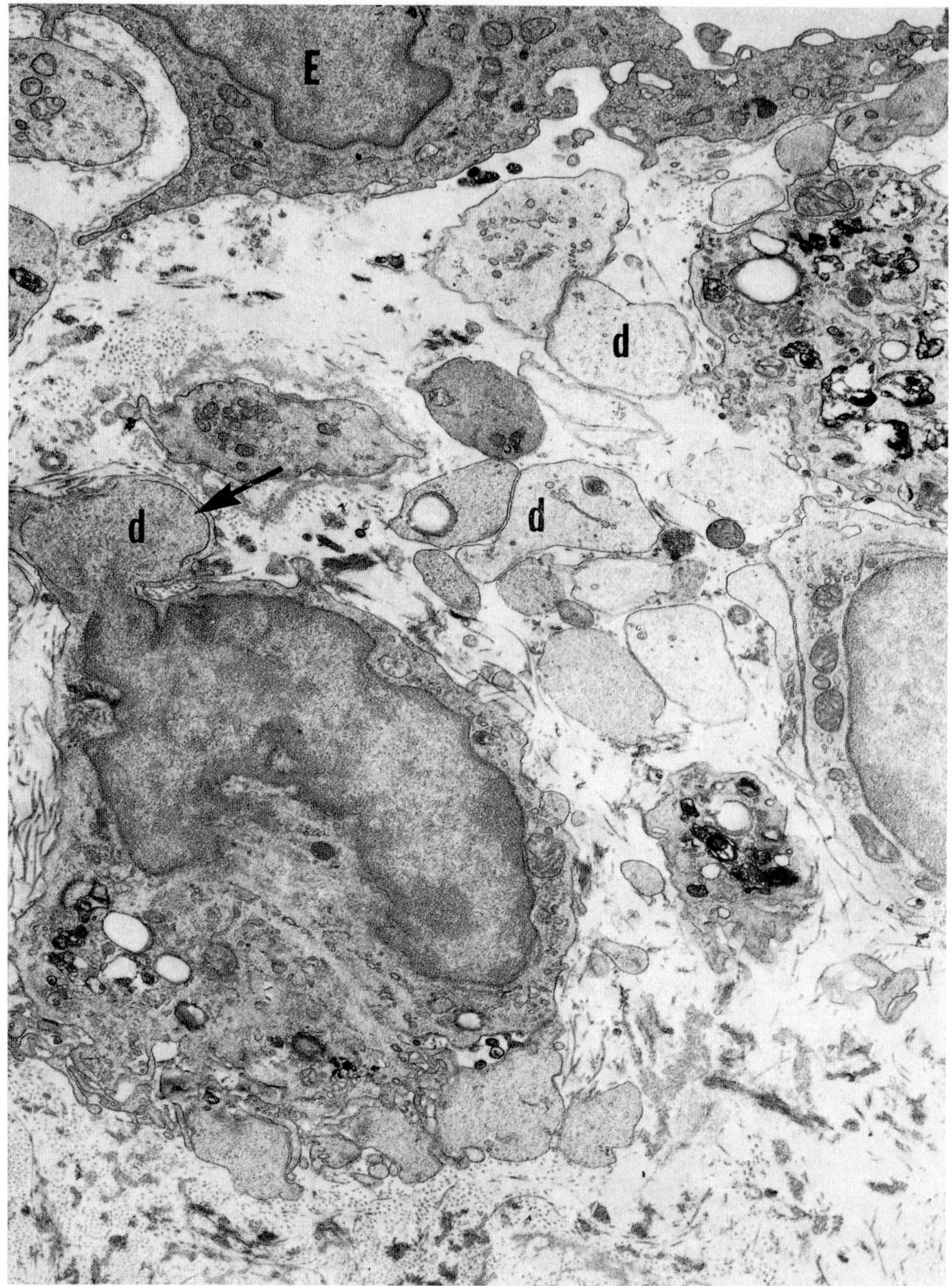

16

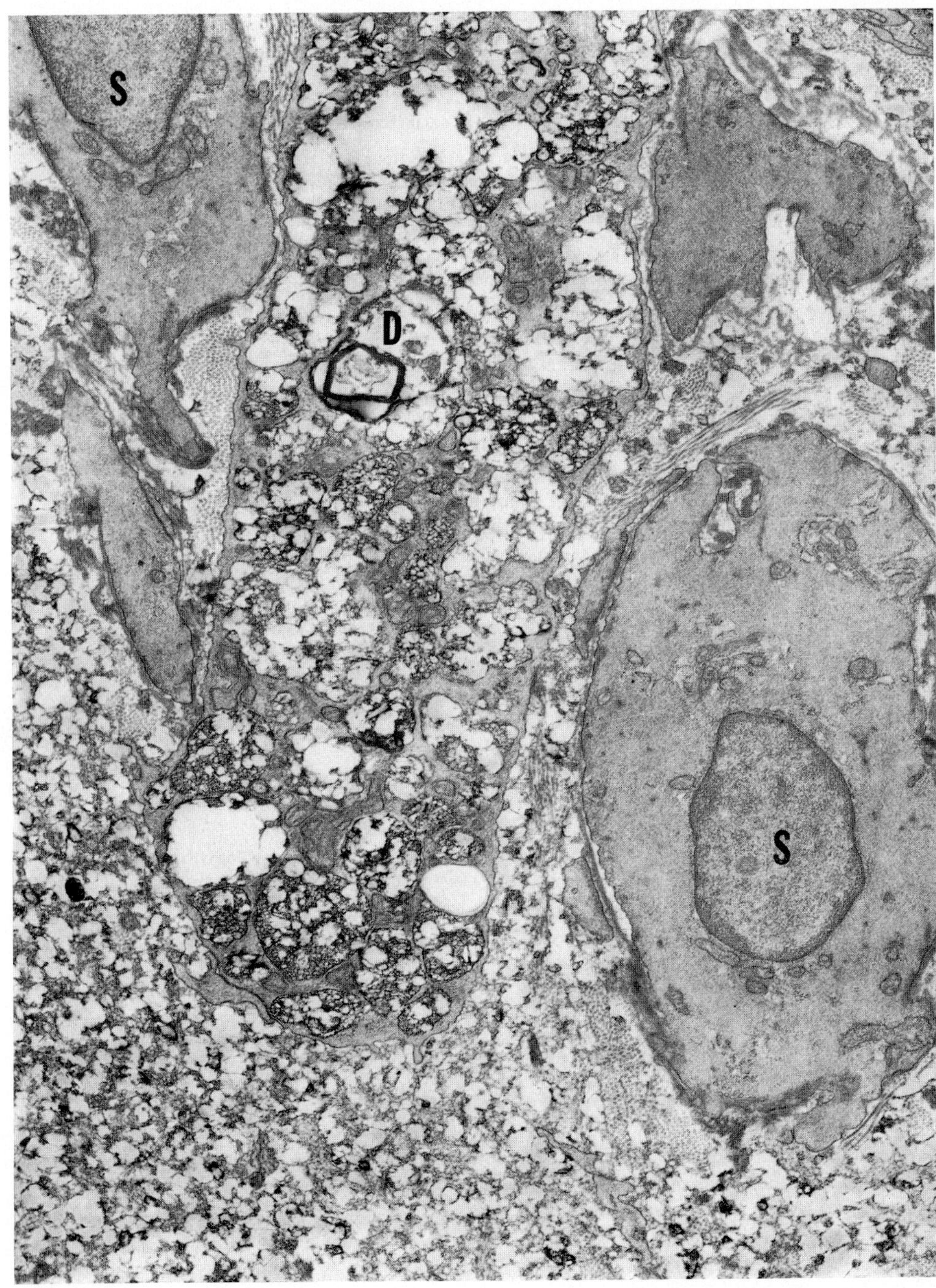

17

394 STARY

References

1 ABELL, L. L.; LEVY, B. B.; BRODIE, B. B., and KENDALL, F. E.: A simplified method for the estimation of total cholesterol in serum and demonstration of its specificity. J. biol. Chem. *195:* 357–366 (1952).

2 BERGER, R. L. and STARY, H. C.: Anatomic assessment of operability by the saphenous-vein bypass operation in coronary-artery disease. New Engl. J. Med. *285:* 248–252 (1971).

3 DAEMS, W. T. and BREDEROO, P.: Electron microscopical studies on the structure, phagocytic properties, and peroxidatic activity of resident and exudate peritoneal macrophages in the guinea pig. Z. Zellforsch. *144:* 247–297 (1973).

4 DAOUD, A.; JARMOLYCH, J.; ZUMBO, A.; FANI, K., and FLORENTIN, R.: 'Preatheroma' phase of coronary atherosclerosis in man. Expl. molec. Path. *3:* 475–484 (1964).

5 DAOUD, A. S.; JONES, R., and SCOTT, R. F.: Dietary induced atherosclerosis in miniature swine. II. Electron microscopy obervations. Characteristics of endothelial and smooth muscle cells in the proliferative lesions and elsewhere in the aorta. Expl. molec. Path. *8:* 263–301 (1968).

6 GEER, J. C.: Fine structure of human aortic intimal thickening and fatty streaks. Lab. Invest. *14:* 1764–1783 (1965).

7 GEER, J. C. and HAUST, M. D.: Smooth muscle cells in atherosclerosis. Monogr. atheroscler., vol. 2 (Karger, Basel 1972).

8 GEER, J. C. and MCGILL, H. C., jr.: The fine structure of coronary atheroma; in LIKOFF and MOJER Coronary heart disease, pp. 125–130 (Grune & Stratton, New York 1963).

9 GEER, J. C. and WEBSTER, W. S.: Morphology of mesenchymal elements of normal artery, fatty streaks, and plaques; in WAGNER and CLARKSON Arterial mesenchyme and arteriosclerosis. Adv. exp. med. Biol., vol. 43, pp. 9–33 (Plenum Publishing, New York 1974).

10 GEER, J. C.; CATSULIS, C.; MCGILL, H. C., jr., and STRONG, J. P.: Fine structure of the baboon aortic fatty streak. Am. J. Path. *52:* 265–286 (1968).

11 GETZ, G. S.; VESSELINOVITCH, D., and WISSLER, R. W.: A dynamic pathology of atherosclerosis. Am. J. Med. *46:* 657–673 (1969).

12 HAUST, M. D.: The morphogenesis and fate of potential and early atherosclerotic lesions in man. Hum. Path. *2:* 1–29 (1971).

13 MONTENEGRO, M. R. and EGGEN, D. A.: Topography of atherosclerosis in the coronary arteries. Lab. Invest. *18:* 586–593 (1968).

14 PARKER, F.: An electron microscopic study of experimental atherosclerosis. Am. J. Path. *36:* 19–53 (1960).

15 PARKER, F. and ODLAND, G. F.: A light microscopic, histochemical and electron microscopic study of experimental atherosclerosis in rabbit coronary artery and a comparison with rabbit aorta atherosclerosis. Am. J. Path. *48:* 451–481 (1966).

16 PITT, B.; ZOLL, P. M.; BLUMGART, H. L., and FREIMAN, D. G.: Location of coronary arterial occlusions and their relation to the arterial pattern. Circulation *28:* 35–41 (1963).

17 PRATT, S. A.; SMITH, M. H.; LADMAN, A. J., and FINLEY, T. N.: The ultrastructure of alveolar macrophages from human cigarette smokers and nonsmokers. Lab. Invest. *24:* 331–338 (1971).

18 RODRIGUEZ, F. L.; ROBBINS, S. L., and BANASIEWICZ, M.: Postmortem angiographic studies on the coronary arterial circulation. Incidence and topography of occlusive coronary lesions; relation to anatomic pattern of large coronary arteries. Am. Heart J. *68:* 490–499 (1964).

19 ROSS, R. and GLOMSET, J. A.: Atherosclerosis and the arterial smooth muscle cell. Science *180:* 1332–1339 (1973).

20 SCHLESINGER, M. J. and ZOLL, P. M.: Incidence and localization of coronary artery occlusions. Archs. Path. *32:* 178–188 (1941).

21 SCOTT, R. F.; JONES, R.; DAOUD, A. S.; ZUMBO, O.; COULSTON, F., and THOMAS, W. A.: Experimental atherosclerosis in rhesus monkeys. II. Cellular elements of proliferative lesions and possible role of cytoplasmic degeneration in pathogenesis as studied by electron microscopy. Expl. molec. Path. *7:* 34–57 (1967).

22 SIMPSON, C. F. and HARMS, R. H.: Electron microscopy of diethylstilbestrol induced coronary atherosclerosis of turkeys. Expl. molec. Path. *9:* 34–43 (1968).

23 SMITH, E. B. and SLATER, R. S.: Relationship between low-density lipoprotein in aortic intima and serum-lipid levels. Lancet *i:* 463–469 (1972).

24 SPURLOCK, B. O.; KATTINE, V. C., and FREEMAN, J. A.: Technical modifications in Maraglas embedding. J. Cell Biol. *17:* 203–207 (1963).

25 STARY, H. C.: Progression and regression of experimental atherosclerosis in rhesus monkeys; in GOLDSMITH and MOOR-JANKOWSKI Medical primatology 1972, part III, pp. 356–367 (Karger, Basel 1972).

26 STARY, H. C.: Electron microscopic observations on the progression of coronary atherosclerosis in rhesus monkeys. Am. J. Path. *70:* 39a (1973).

27 STARY, H. C.: Proliferation of arterial cells in atherosclerosis; in WAGNER and CLARKSON Arterial mesenchyme and arteriosclerosis. Adv. exp. med. Biol., vol. 43, pp. 59–81 (Plenum Press, New York 1974).

28 STARY, H. C.: Cell proliferation and ultrastructural changes in regressing atherosclerotic lesions after reduction of serum cholesterol; in SCHETTLER and WEIZEL Atherosclerosis III, pp. 187–190 (Springer, Berlin 1974).

29 STARY, H. C. and MCMILLAN, G. C.: Kinetics of cellular proliferation in experimental atherosclerosis. Archs Path. *89:* 173–183 (1970).

30 STARY, H. C. and STRONG, J. P.: Coronary artery fine structure in rhesus monkeys: Nonatherosclerotic intimal thickening (this volume).

31 STEMERMAN, M. B. and ROSS, R.: Experimental arteriosclerosis. I. Fibrous plaque formation in primates, an electron microscope study. J. exp. Med. *136:* 769–789 (1972).

32 THOMAS, W. A.; JONES, R.; SCOTT, R. F.; MORRISON, E.; GOODALE, F., and IMAI, H.: Production of early atherosclerotic lesions in rats characterized by proliferation of 'modified smooth muscle cells'. Expl. molec. Path. *2:* suppl. 1, pp. 40–61 (1963).

33 TUCKER, C. F.; CATSULIS, C.; STRONG, J. P., and EGGEN, D. A.: Regression of early cholesterol-induced aortic lesions in rhesus monkeys. Am. J. Path. *65:* 493–514 (1971).

34 VERNON-ROBERTS, B.: The macrophage; in HARRISON and MCMINN Biological structure and function, vol. 2 (Cambridge University Press, London 1972).

35 WOLKOFF, K.: Über die Atherosklerose der Coronararterien des Herzens. Beitr. path. Anat. allg. Path. *82:* 555–596 (1929).

H. C. STARY, M.D., Associate Professor of Pathology, Louisiana State University Medical Center, 1542 Tulane Avenue, *New Orleans, LA 70112* (USA)

Subject Index

To facilitate the use of the index and to economize space, the nomenclature has been simplified and standardized, i.e. 'diet' is used instead of 'dietary' or 'nutrition', etc.